# Simulations of Oscillatory Systems

with Award-Winning Software,
*Physics of Oscillations*

# Simulations of Oscillatory Systems

## with Award-Winning Software, *Physics of Oscillations*

## Eugene I. Butikov

*Department of Physics*
*Saint Petersburg State University*
*Saint Petersburg, Russia*

CRC Press
Taylor & Francis Group
Boca Raton   London   New York

CRC Press is an imprint of the
Taylor & Francis Group, an **informa** business

CRC Press
Taylor & Francis Group
6000 Broken Sound Parkway NW, Suite 300
Boca Raton, FL 33487-2742

First issued in paperback 2020

© 2015 by Taylor & Francis Group, LLC
CRC Press is an imprint of Taylor & Francis Group, an Informa business

No claim to original U.S. Government works

Version Date: 20150121

ISBN 13: 978-0-367-57591-5 (pbk)
ISBN 13: 978-1-4987-0768-8 (hbk)

**Visit the Taylor & Francis Web site at**
**http://www.taylorandfrancis.com**

**and the CRC Press Web site at**
**http://www.crcpress.com**

# Contents

## II  Nonlinear Oscillations                                      153

## 7  Free Oscillations of the Rigid Pendulum                      155

## 8  Rigid Planar Pendulum under Sinusoidal Forcing               189

# Preface

This textbook, *Simulations of Oscillatory Systems*, is accompanied by the award-winning educational software package, *Physics of Oscillations*. The textbook and the software are developed as exploration-oriented supplements to various courses in general physics and the theory of oscillations, and can be helpful to a wide range of students and researches. Successful usage of the software is possible without any knowledge of algorithmic languages or programming. The software, which is an essential part of the textbook, is designed to be a desktop laboratory that can serve as an electronic training course for individual, interactive work on a computer.

The software package, *Physics of Oscillations*, includes several highly interactive computer programs presenting an extensive set of computer-simulated experiments. Contemporary interactive media provides students with a powerful means to visualize and to explore the fundamental concepts of the physics of oscillations. With the textbook and the programs, students and their instructors get a powerful tool that enables them to investigate phenomena that are difficult to imagine and study in an abstract conventional manner. The simulation programs make visible the magnificence of mathematics in its application to the wonderful world of oscillations.

Students can work at a pace they can enjoy, varying parameters of the simulated systems and repeating the most interesting experiments several times on their own. The experience based on a student's own actions results in deeper understanding than the mere reception of someone else's knowledge. No doubt, for a great majority of human beings, a visual impression is much more intensive and permanent than a heard or read one. With some of the suggested programs, students have an opportunity to perform interesting mini-research projects in physics.

The structure of the textbook and programs allows students to study the subject at different levels of difficulty, depending on the time available and on the mathematical complexity of the course. The structuring of levels by degree of difficulty and mathematical complexity is especially convenient. At an introductory level, the underlying concepts and physical laws are discussed without much mathematics: The software suggests experiments that demonstrate typical examples of behavior of the simulated systems, and helps to develop physical intuition. At an intermediate level, a more detailed though rather elementary description is available. In Part I of the textbook each chapter is supplied with a section containing suggested activity, questions, exercises, and thought-out problems with a wide range of difficulty from straightforward to quite challenging. For those students who'd like an in-depth investigation, Part II of the textbook and the corresponding programs provide much more sophisticated, highly mathematical material, which delves into the serious theoretical background for the computer-aided study of oscillations.

Computer simulations provide very clear and impressively vivid illustrations of oscillations in various physical systems. The screen displays subtle details that usually escape us in direct observation. It is possible to change time scales and to widely vary parameters and other experimental conditions. We can investigate interesting situations that are inaccessible in a real experiment. The graphic representation of experimental results allows us to easily collect and understand large amounts of information.

The software package, *Physics of Oscillations*, allows users to observe the motion of various linear and nonlinear mechanical oscillatory systems directly on the computer screen and to obtain plots of the variables that describe the system along with phase diagrams and plots of energy transformations. The plots and phase diagrams appear on the screen simultaneously with the display of the motion. These simulations bring to life many abstract concepts of the physics of oscillations. The simulations aid greatly in developing our physical intuition and complement the analytical study of the subject in a manner that is mutually reinforcing.

The package, *Physics of Oscillations*, runs under the Windows operating system. To use the software, you should install it on your machine. To install the first part of the package called, "Simple Systems," download from the Web the archive file **butikov.faculty.ifmo.ru/MasterDiskOsc.zip** and unpack it to a temporary folder. Then, from Windows Explorer click on autorun.exe in the "Auto" subfolder, or on setup.exe in the "Software" subfolder, and follow instructions on the screen. The setup program will create an entry in the Start menu, and a shortcut (*Physics of Oscillations*) on your desktop. To start *Physics of Oscillations*, click the icon on the desktop, or open the *Physics of Oscillations* in the "Programs" menu, and choose the subprogram you need. To install the second part, *Nonlinear Systems*, download the archive file **butikov.faculty.ifmo.ru/Nonlinear.zip**.

Eugene I. Butikov
St. Petersburg State University, Russia

# Introduction

Welcome to the wonderful world of oscillations! Oscillations are everywhere. The Earth, like a giant bell, vibrates under our feet. Oscillating electric and magnetic fields carry light to our eyes. Vibrating air carries sound to our ears. Through electromagnetic waves and acoustic vibrations, we receive the major part of information about the world surrounding us.

Oscillations in various physical systems may differ in physical nature, but they also have much in common. A branch of science called the *theory of oscillations* deals with the analysis of laws common to all oscillations.

It is easier to understand common laws of oscillation processes if we analyze them in the most plain and obvious examples; e.g., in mechanical systems that are accessible to direct visual observation. For this purpose, the simulation experiments in the package *Physics of Oscillations* deal with commonly known mechanical systems such as the spring harmonic oscillator and the simple pendulum.

A *linear* (or harmonic) *oscillator* is a system in which a displacement from the equilibrium position causes a restoring force to appear that is proportional to the displacement. The time dependence of the state of the system is described by linear differential equations. Mechanical examples of such systems include a weighted spring (spring–mass system) and a torsion pendulum. The latter could be a flywheel that is attached to an elastic spiral spring. The spring twists when the flywheel is turned around its axis, much like the balance device of a mechanical watch. The vibrations of such a linear torsion pendulum are simulated in the suggested software.

A common example of a *nonlinear* mechanical system is an ordinary *pendulum* in the gravitational field. When the pendulum is moved from the equilibrium position, the restoring torque of gravity is proportional to the sine of the deflection angle. Therefore, for small oscillations, the pendulum may be considered a linear oscillator, but at large angular displacements, its behavior differs greatly from that of the linear oscillator.

In addition to natural (or free) oscillations, which are excited by some initial action on an isolated system that is then left to itself, the software simulates forced oscillations occurring under some periodic external action. It also treats parametric oscillations occurring during the periodic change of some parameter of the system to which the motion of the system is sensitive.

The simulation experiments allow the user to observe oscillations of various physical systems directly on the screen and at the same time to draw the time-dependent plots of the variables that characterize the system. The phase diagram as well as the plots of potential, kinetic, and total energy are also available for a detailed investigation of the system being analyzed. The experiments have been designed to be plain and obvious. They provide the capability of observing repeatedly and thoroughly the fine details of phenomena that usually escape notice during direct observation. The user can widely modify parameters of the physical system and the initial conditions. The graphic presentation of the results allows students to see and easily understand large amounts of information.

## Classification of Oscillations

1. According to the *physical nature* of the phenomena involved, oscillations in various systems are divided into mechanical oscillations and electromagnetic ones. *Mechanical oscillations* are characterized by alternating conversions of the kinetic energy into one (or several) kinds of potential energy and back. In *electromagnetic oscillations* alternating conversions occur between the electric field energy (which is analogous to the potential energy in mechanical systems) and the magnetic field energy (the analogue of the kinetic energy). Sometimes oscillations have a combined mechanical and electromagnetic nature, e.g., oscillations in plasma. Oscillations of different physical nature obey common laws. These laws common to all oscillations are studied by the theory of oscillations.

2. According to *kinematics*, i.e., the character of time dependence of some physical quantity $x(t)$ that characterizes the physical system, oscillations are classified as *periodic* and *non-periodic*. An important variety of periodic oscillations are *sinusoidal* or *harmonic oscillations*, when the time dependence of $x(t)$ is described by a sine (or cosine) function. The most important kinds of non-periodic oscillations are almost sinusoidal ones. Such oscillations can be treated approximately as sinusoidal with a slowly changing amplitude. Nearly sinusoidal oscillations with slowly varying amplitude— *modulated oscillations*—are widely used in radio communication: A bearing electromagnetic wave of a high frequency is modulated by oscillations of a low (e.g., acoustic) frequency.

3. According to the *means of excitation*, oscillations are divided into four main groups:

   - *Free* or *natural oscillations* occurring in a system that is left to itself after some initial excitation. Such oscillations take place about the position of stable equilibrium. In a conservative system the energy conversions during free oscillations are reversible, and (in a system with one degree of freedom) the oscillations are exactly periodic. In a linear system, in which a restoring force is proportional to the displacement from the equilibrium position, free oscillations without friction give an example of simple harmonic motion. Their period is independent of the amplitude. This property of a linear oscillator is called isochronism.

In real systems, conversions of the energy from one kind to the other and back are partially irreversible. Dissipation of the energy due to friction (or electric resistance) causes damping of free oscillations.

- *Forced oscillations* occurring in a system (that can execute free oscillations) under the influence of some periodic external action. Generally, the oscillations are called forced if the periodic external action on the oscillatory system can be expressed by a separate term in the differential equation describing the system. This term must be a given periodic function of time.

On the expiry of some time after the periodic external force begins to operate, forced oscillations become exactly periodic: They have a constant amplitude and acquire the period of the external force. The phase and the amplitude of these *steady-state forced oscillations* are independent of the initial conditions. The dependence of the amplitude of steady-state oscillations on the frequency $\omega$ of the external force has a resonant character—the amplitude grows appreciably as $\omega$ approaches the natural (resonant) frequency $\omega_0$ of the oscillator.

Initial conditions influence a *transient process*—the process of establishing steady-state oscillations. In a linear system the transient process can be represented as a superposition of steady-state oscillations whose frequency equals that of the external sinusoidal force and fading free oscillations that have the natural frequency. Usually a transient process lasts as long a time as is necessary for natural oscillations in the system to damp away.

- *Parametric oscillations* occurring in an oscillatory system excited by a periodic variation of some parameter of the system. The most familiar example of parametric oscillations is given by an ordinary child's swing excited by periodic changes of its length.

In the case of parametric oscillations the system is subjected to non-stationary forces that depend not only on time (explicitly) but also on coordinates. This case is more complicated for investigation. Let, for example, a restoring force $F = -kx$ arise when the system is displaced from the equilibrium position, but in contrast to the stationary case the parameter $k$ changes with time due to some periodic influence: $k = k(t)$.

In the differential equation of the system the coefficient of $x$ is not constant; it depends on time explicitly. Oscillations in such a system differ essentially both from natural free oscillations, described by a differential equation with constant coefficients, and from forced oscillations caused by an external force that depends only on time.

In the case of periodic changes of the parameter $k$, when $k(t + T) = k(t)$, oscillations in a system are called parametrically excited or simply parametric. When the process of oscillations caused by the periodic modulation of some parameter acquires an increasing character, the phenomenon is called *parametric resonance*. In the case of parametric resonance the state of equilibrium becomes unstable: The system leaves it executing oscillations with progressively increasing amplitude.

Parametric resonance differs considerably in some characteristics from ordinary resonance caused by a periodic external force exerted on the oscillatory system. Parametric resonance occurs when different relationships between the frequency of modulation of a parameter and the natural frequency of the system are fulfilled. To excite resonance, the amplitude of modulation must be large enough: The depth of modulation must exceed some *threshold* value in order to cause parametric resonance.

- *Self-excited* or self-sustained oscillations (or *auto-oscillations*) occurring without any periodic influence. They are possible in oscillatory systems that have a constant (non-periodic) source of energy and a positive feedback. Examples of self-sustained oscillations are given by mechanical watches or by electromagnetic oscillators. These nonlinear systems can regulate energy supply from a constant source to compensate losses caused by friction or resistance.

4. According to *complexity*, oscillatory systems are divided into *simple systems* (systems with only one degree of freedom) characterized by a single natural frequency; systems with lumped (concentrated) parameters with a finite number (or countable infinite number) of degrees of freedom, having as many natural frequencies (frequencies of normal oscillations) as the number of degrees of freedom; systems with distributed parameters having a continuous infinite number of degrees of freedom and consequently an infinite number of natural frequencies.

# Simulated Physical Systems

Part I of the textbook deals with the following types of oscillations in mechanical systems:

- Free oscillations of a linear torsion spring pendulum

- The torsion spring oscillator with dry and viscous friction

- Forced oscillations in a linear system under a sinusoidal force

- Non-sinusoidal (square-wave) external force in a linear system

- Parametric square-wave excitation of a linear torsion pendulum

- Parametric resonance at sinusoidal excitation a linear pendulum

Part II of the textbook covers more advanced and sophisticated topics of the physics of oscillations. Nonlinear systems that demonstrate chaotic behavior are considered. In particular, the package contains the following simulations:

- Free oscillations and rotations of a nonlinear rigid pendulum

- Forced swinging of a rigid planar pendulum and nonlinear resonance

- Parametric excitation of a rigid pendulum by periodic changes of its length

- Parametrically driven pendulum with vertical oscillations of the pivot

- Dynamic stabilization of inverted pendulum by oscillations of the pivot

- Torsion spring oscillator with dry and viscous friction under sinusoidal external force

# How to Use the Software

Each chapter of the textbook is related to one of the simulation programs of the package *Physics of Oscillations*. A chapter begins with a description of the simulated physical system and includes brief but important information on the theory of the phenomenon under consideration. In part I, much of this information may be familiar to you from lectures or traditional textbooks. If you feel sure you know the material well, you can skip it and begin with the part entitled "Questions, Problems, Suggestions." If you want to learn something new while using the software, this section of each chapter is very important. Try to solve the problems given therein beforehand. You will then be able to verify your solutions in a computation experiment. This will make your work on the computer much more interesting. If you find that the experimental results disagree with your predictions, try to find the reason for the differences. Much real research in physics amounts to the analysis of discrepancies between theoretical predictions and experimental results.

Below is a brief overview of the basic control functions carried out with the help of the command buttons and menu items.

- **Menu Bar** at the top of each window displays the commands used to operate the program. Also right-clicking the mouse anywhere in the program window invokes a **pop-up menu** with several commands.

- **Command Buttons** under the menu bar provide quick access to several commonly used commands. You click a command button once to carry out the action represented by that button.

- **Sliders** with the labels "Speed Up" and "Slow Down" allow you to vary the speed of animation for a convenient observation by changing the time scale in which the motion is simulated and displayed.

**Command Buttons** common to most of the programs execute the following actions:

- **Start, Pause, Go** – starts the simulation, makes a pause in the simulation, continues the interrupted simulation;

- **Restart** – restores the initial conditions and repeats the simulation from the beginning;

- **Erase** – clears the windows (erases all graphs and rescales), and continues the simulation;

- **Skip Over** – skips the simulation over several periods of the external action;

- **Erase** – removes the old plots (and rescales) before going on;

- **Change** – calls up the control panel to change properties;

- **Options** – calls up the control panel to choose options (see details below);

- **Exit** – returns to the starting page of the program;

- **Time Scale** – changes the number of periods displayed on the plots;
  **Animation** – speeds up or slows down the animation.

**Menu items** common to all the programs carry out the following functions:

- **File**:

  - **Exit** – closes all windows and terminates the program;
  - **Print** – opens the panel to make the printer settings;

- **Input** – offers to choose a parameter for input or opens the panel for entering several parameters of the simulated system (and the initial conditions for the simulation);

- **Options**:

  If you click the button **Options** or the menu item "Options," a control panel appears from which you can choose a mode of simulation and a mode of displaying information that best fits your requirements:

 – You can choose one of three option buttons in plotting functions: 1) the program is to pause in the simulation when all the space for the plots is used and is to wait for your next command, or 2) it should continue drawing new plots over the old ones, or 3) it should erase them before continuing plotting.

 – You can choose the thickness of lines for drawing the plots, and the background color (dark or white) by checking the switches;

 – You can make the image of the system more realistic (with the massive bobs on the rotor) by choosing "two-sided" from the option "needle image;"

 – If you wish to suppress the automatic choice of scale and to choose your own scale, select the option "Custom Scale" and indicate the maximal values of the angular displacement and angular velocity to be displayed on the plots.

 • **Examples** – offers a list of predefined examples to choose from or opens the panel with a set of examples provided;

 • **Zoom** – uses all the space available on the screen for drawing the plots and phase diagram.

Other controls specific for separate programs are described in the corresponding sections of the textbook. For more information on controls, search "How To ..." under the menu "Help" from within the programs.

Context-sensitive physical explanations can be called by pressing the F1 key or by clicking the menu item "Help on physics." Contents of the displayed help reference depend on the place where the program was interrupted. Each simulation program has a section entitled "Information about the Physical System." It might be a good idea to begin your work with a program by looking through this section. You will find a brief description of the physical system and some dynamic illustrations of its behavior, and gradually you will get used to its conventional image on the display screen. This image is intentionally made to appear very schematic to remind you that in the simulation experiment you are dealing not with a real physical system but with some idealized model of it.

It may prove helpful to look through the sections concerning an experiment in a survey mode (by choosing various examples from the menu item **Examples**) before you proceed to a detailed examination. In this mode, you need not enter data or choose options. The sets of available examples vary when you change from one screen configuration to another by clicking the "View" item. You can thereby get an idea of the contents and of the amount of work to be done. All quantities are then assigned suitable default values.

Clicking the menu item "View" allows you to choose one of the four screen configurations, which differ by the amount of information displayed:

1. Image of the physical system (only);

2. ...plus the plots of time dependence of the angular displacement and of the angular velocity;

3. ...plus the phase diagram;

4. ...plus the plots of energy transformations.

Command buttons in the Tutorial perform the following functions:

**Go, Pause** – begin the simulation, pause, resume;
<< **Back** – return to the preceding item of the Tutorial;
**More** >> – go to the next page of the Tutorial;
**Contents** – call up the contents of the Tutorial;
**Exit** – return to the starting page of the program;
**Animation** – speed up or slow down the animation.

To change properties of the physical system (such as the quality factor, the dead zone, the frequency and amplitude of the external action, etc.—depending on the system under consideration) or the initial conditions (values of the angular displacement and angular velocity at $t = 0$), click on the button **Change** or the item "Settings" (or "Input") in the menu. A control panel then appears in which you can type in the new values in corresponding dialog boxes, or you can change the values by dragging the slides on the scroll-bars. When all chosen values are specified, confirm your choice by clicking the **OK** button. Or click the **Cancel** button, if you wish to delete the changes and return to the previous values.

Instead of entering data, you can look for a suitable example of the case you are interested in by clicking the menu item "Examples." In each example all necessary properties and initial conditions are already specified. The set of available examples depends on the screen configuration you have chosen from the menu item "View."

The software allows you to make a hard copy of the plots with the help of your system printer. The plots will have the best resolution your printer can provide. All the graphs in this textbook are obtained in this way. To print the plots, perform the following steps:

- Input, if necessary, the values of parameters and initial conditions (the plot will correspond to the current values of the system parameters);

- Choose the menu item "Print." A printer control panel will appear;

- In the panel, choose the plot you wish to print (angular displacement or angular velocity time dependence, phase diagram, or energy transformations) by clicking the corresponding option button;

- Input the plot dimensions (width and height) in the units you are used to (inches, centimeters, pixels, or points). The program will round out the values in order to get the best result on your printer;

- Select the time interval for the plot: Indicate the number of the natural periods (or the external force periods, or the periods of modulation, depending on the program) to be reproduced (time scale), and the initial instant of the time interval;

- If you wish, you can add a caption to the plot (type its text into the corresponding box), and a header (for the whole page where you can print several plots);

- Choose whether or not to print the legend and the values of parameters (under the plot);

- Click the **Go** command button after all the necessary settings have been made.

- After the image of the plot is stored in the computer's memory (the label "Wait..." will change to the label "Ready"), you can choose several possibilities to continue:

  1. Get a hard copy of the plot by clicking the **Eject** command button. Only one plot will be printed out;

  2. If you click the **More** command button, then the next part of the process (one more interval ordered in the box "Time Scale") will be placed over the same plot (together with the previous part);

  3. You can store another plot image in order to print it below on the same sheet of paper. To do this, repeat the previous steps and make the settings for the second plot. The program will warn you if free space on the sheet is insufficient for the next plot;

- Click on the **Eject** command button to send all the stored images to the printer.

If you check the box "Write to ASCII file," all the data of the simulation experiment will be placed in a file (without any output to your printer). You can use this file afterwards, managing the data with a third-party software. The first row of the sheet identifies the values of the corresponding columns (time, angular displacement, angular velocity, energy, etc.). If you store plotting data in a file, you need not indicate the dimensions of the plot or choose other options (what to print or what captions or headers to make). The only significant parameter you need to specify is the number of periods of the simulation experiment that should be performed and stored. The stored data will apply to the current values of the properties of the physical system and the initial conditions.

On-line instructions concerning the operation are available from within the programs under the menu item "Help," "Help on Controls."

# Notes to the Instructor

This textbook, *Simulations of Oscillatory Systems*, with the software package, *Physics of Oscillations*, gives the instructor several choices on how to use it in physics courses. Certainly, it can be used in lecture demonstrations to illustrate various aspects of free, forced, and parametric oscillations in mechanical systems and in oscillatory circuits. The teacher can save much time and effort in avoiding long and dull explanations by using the dynamic simulations offered by these programs. But surely there is a much more efficient way to exploit the clear and plain character of the simulations.

A more important use of the textbook and the software is as a desktop laboratory equipped with a set of ready and well-adjusted experimental devices with which students may easily interact. When our aim is to learn physics, we should avoid wasting time on auxiliary technical details concerning the apparatus, in spite of their importance to the success of a real experiment. The same is true for the computation experiment in which we deal with idealized mathematical models rather than with real physical systems. If we are not interested in the inner workings of the simulation program itself, an evident advantage of ready and debugged software is that it allows us to concentrate entirely on the physical principles involved.

The simulation experiments in the *Physics of Oscillations* software involve mechanical systems primarily because the motion of such systems is easily represented on the computer screen. Such visualization makes the simulation experiments very convincing and easy to understand. When we begin with systems whose behavior is familiar (a simple torsion spring oscillator or a rigid pendulum) and see that the simulation shows us just what we would expect from a real system, the subsequent simulations of nonlinear and more complex systems (the driven pendulum or the parametrically excited oscillator) become more convincing. These are cases when our intuition based on common everyday experience often fails to work.

The most satisfactory pedagogical effect can be achieved if the learning process is organized so that students study the subject matter in the textbook beforehand and solve the problems and exercises assigned by the instructor. They are then more likely to know what to do when working with the program. It is also helpful for students to study the derivations of principal formulas and to attempt some of the derivations omitted in the textbook. However, especially important are the problems and the exercises.

Stars (asterisks) are added to problem numbers to grade them by difficulty. Simple problems have no stars. One star means that the problem requires thought. Two- and especially three-star problems are challenging in some way. The best learning will take place while a student is struggling with the tough problems.

Most of the exercises ask students to calculate some value they should input in order to get a desired result. Students thus have an opportunity to verify their answers (and thus the understanding of the subject as a whole) not by peeping in the back of the textbook, but rather by a real experiment with the mathematical model of the simulated physical system. The students' work on the computer is thus made much more interesting and fruitful.

The textbook and the software are designed for use at a wide range of levels and can be adapted to meet the needs of many physics courses. The structure of the textbook and the software makes them accessible to some extent even to introductory physics students who are unfamiliar with the calculus and are unable to follow all the derivations given in the textbook. Nevertheless, such students can acquire a qualitative understanding of the phenomena and can improve their physical intuition through experimental work on the computer. Animation and the simultaneous drawing of plots of different variables make it easy to understand a large amount of information.

Advanced undergraduate and graduate students can find in the textbook and the software a lot of both interesting and useful optional material. With it they may perform small research projects and thereby come to appreciate the beauty of oscillatory phenomena.

The choice of subjects to be studied depends mainly on the content of the physics course in question and on the classroom hours available. In any case, certain material ought to be considered obligatory: The program "Free Oscillations in a Linear System" (including damping caused by viscous friction); the program "Forced Oscillations of Linear Torsion Spring Pendulum" (which may be restricted to steady-state oscillations, with only a notion of transients—especially at the resonance).

Material that can be optional includes the program "Torsion Spring Oscillator with Dry Friction" (interesting because it illustrates (1) the physical cause for random errors arising in measuring instruments using a needle, and (2) the method of joining the solutions of linear differential equations describing the motion during successive time intervals); the program "Non-Sinusoidal External Force in a Linear System"—an impressive example of the spectral decomposition of a periodic process, and of the concept of how the spectrum of a signal is transformed in a linear system (an explanation is given of how a square-wave input voltage is distorted by an oscillatory circuit and transformed into the output voltage of a different shape); the programs "Parametric Excitation of Linear Oscillator" and "Sinusoidal Modulation of the Parameter."

The physical systems that are simulated in these latter two programs may seem exotic and even somewhat ridiculous. However, they give a very clear example of the parametric excitation of a linear mechanical system. All peculiarities of parametric resonance can be exhaustively investigated in this case, and its physical properties are completely explained. The possibility of manual control of the moment of inertia aids a great deal in understanding the phenomenon. This simulation provides a good background for the study of more complicated nonlinear parametric systems like a pendulum whose length is periodically changed (model of a swing), or a pendulum with the suspension point driven periodically in the vertical direction. (These systems are included in Part II of the textbook).

In Part II of the textbook several nonlinear physical systems are investigated with the help of simulation programs dealing with more advanced and sophisticated problems of the physics of oscillations. In particular, the following physical systems are considered:

- Free oscillations and rotations of a planar rigid pendulum. Dependence of the period on the amplitude. Phase portrait. Limiting motion along the separatrix. Oscillations with extremely large amplitudes and revolutions.

- Forced swinging and revolutions of the rigid planar pendulum by the periodic external torque. Nonlinear resonance, phase locking, hysteresis at slow scanning of the external force frequency, chaotic behavior of a simple dynamical system, strange attractor.

- Parametric excitation of the rigid planar pendulum by square-wave modulation of its length (a simplified model of the playground swing). Intervals of parametric instability.

- Constrained vertical oscillations of the suspension point of a rigid pendulum. Dynamic stabilization of the parametrically driven inverted pendulum at vertical oscillations (Kapitza's pendulum).

- Forced oscillations in a mechanical system with dry (Coulomb) and viscous friction.

# Part I

# Oscillations in Simple Systems

# Chapter 1

# Free Oscillations of a Linear Oscillator

**Annotation.** In Chapter 1 free (unforced) oscillations in a simple mechanical system and its electromagnetic analogue—series $LCR$-circuit—are investigated both analytically and with the help of computer simulations. Basic general concepts of the theory of oscillations are introduced and discussed. Chapter 1 includes a description of the simulated physical system and a summary of the relevant theoretical material for students as a prerequisite for the virtual lab "Free Oscillations of Linear Torsion Pendulum" from the software package *Physics of Oscillations* (Part I, Simple Systems). A set of theoretical and experimental problems to be solved by students on their own is included, as well as various assignments that the instructor can offer students for possible individual work.

## 1.1    Summary of the Theory

### 1.1.1    General Concepts

Equilibrium of a physical system is called *stable* if under any disturbance a *restoring force* (or a restoring torque) arises that tends to return the system to the equilibrium position. If the system is disturbed from this state by some external initial action and then left to itself, the system subsequently oscillates about its equilibrium position. Such unforced oscillations are called *free* or *natural,* because they are not driven by an external force. Any system that is able to execute free oscillations in the vicinity of its position of stable equilibrium—a repetitive motion back and forth around the equilibrium position—is called an *oscillator*.

In the absence of friction, energy transmitted to the system at the initial excitation is conserved, that is, remains constant during subsequent oscillations. In such an idealized *conservative* system with one degree of freedom, the motion is strictly *periodic* and continues indefinitely without damping. If there is friction in

3

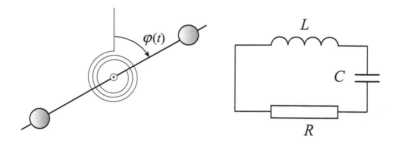

Figure 1.1: Schematic image of the torsion spring oscillator simulated in the program and its electromagnetic analog—$LCR$-circuit.

the system, free oscillations are *damped*: they gradually fade away because of the dissipation of energy, and the system eventually comes to rest in the equilibrium position.

If the restoring force that tends to return a disturbed conservative system to its equilibrium position is proportional to the displacement from this position, oscillations of the system are *harmonic*. Simple harmonic motion occurs when there is a linear restoring force. The simplest example is a mass on a spring (a *spring–mass system*). If there is no friction or other dissipation, or when the damping force is proportional to the velocity and oppositely directed (viscous friction), the differential equation describing the motion of the system is *linear* because the displacement and its time derivatives are to the first power. Such a physical system is called a linear oscillator.

## 1.1.2    Differential Equation of a Linear Torsion Oscillator

The linear oscillator simulated in the suggested computer program is a balanced flywheel (rotor) whose center of mass lies on the axis of rotation. Such a flywheel may consist, for example, of a rigid rod with two equal masses, as shown in the left-hand panel of Figure 1.1.

The rod of the flywheel can rotate about an axis that passes through its center. A spiral spring with one end fixed and the other attached to the flywheel flexes when the flywheel is turned, and creates a restoring torque $N$ which tends to return the flywheel to the equilibrium position. In our model we assume that the spring obeys Hooke's law: The torque $N(\varphi)$ is proportional to the angular displacement $\varphi$ from the equilibrium position:

$$N = -D\varphi. \tag{1.1}$$

Here $D$ is a constant of proportionality called the torsion *spring constant*. Measured in units of torque per radian, its value depends on the strength of the spring. The angular displacement $\varphi(t)$ of the flywheel from its equilibrium position $\varphi = 0$

(vertical in Figure 1.1) is measured by a needle attached to the flywheel, and a fixed dial.

The right-hand panel of Figure 1.1 shows an $LCR$-oscillatory circuit that can be regarded as an electromagnetic analogy of the mechanical device. Both systems are described by identical differential equations and thus are dynamically isomorphic. However, the mechanical system has a definite didactic advantage for exploration of oscillations because it allows us to observe a direct visualization of motion.

The law of rotation of a flywheel whose moment of inertia about its axis of symmetry is $J$, and which is acted upon by a torque $N$ in the absence of friction, gives the following differential equation:

$$J\ddot{\varphi} = -D\varphi \qquad \text{or} \qquad \ddot{\varphi} + \omega_0^2\varphi = 0, \tag{1.2}$$

where we have introduced the notation $\omega_0^2 = D/J$ for the coefficient of $\varphi$. The general solution of Eq. (1.2) can be written as a superposition of two harmonic oscillations with the same frequency $\omega_0$, one cos-like and the other sin-like, with arbitrary amplitudes $C$ and $S$, respectively:

$$\varphi(t) = C\cos\omega_0 t + S\sin\omega_0 t. \tag{1.3}$$

For a particular solution, the values of $C$ and $S$ depend on the initial conditions. Another equivalent form of the general solution (with two other arbitrary constants $A_0$ and $\delta_0$) represents a simple harmonic oscillation:

$$\varphi(t) = A_0\cos(\omega_0 t + \delta_0). \tag{1.4}$$

The two arbitrary constants $A_0$ and $\delta_0$ in the general solution (1.4) have the physical sense of the amplitude and the initial phase, respectively. Their values, as well as the values of $C$ and $S$ in Eq. (1.3), depend on the *initial conditions* (that is, on the angular displacement $\varphi(0)$ and the angular velocity $\dot{\varphi}(0)$ at the initial moment $t = 0$). In other words, these characteristics of the motion depend on the way in which such free oscillations are excited.

The subsequent natural oscillations occur with the frequency $\omega_0$, the squared value of which is proportional to the spring constant $D$ and inversely proportional to the moment of inertia $J$ of the flywheel. The frequency $\omega_0$ and the corresponding period $T_0 = 2\pi/\omega_0$, unlike the amplitude and initial phase, do not depend on the initial conditions – they are entirely determined by the properties of the system, i.e., by the values of the physical parameters $D$ and $J$. Free oscillations of the system always occur with the same *natural frequency $\omega_0$* at arbitrary initial conditions, that is, independently of the mode of excitation.

When the flywheel is also acted upon by a force of viscous friction, which is proportional to and oppositely directed to the angular velocity $\dot{\varphi}$, the differential equation of motion has the form:

$$\ddot{\varphi} + 2\gamma\varphi + \omega_0^2\psi = 0, \tag{1.5}$$

where the *decay constant* (or *damping constant*) $\gamma$ characterizes the strength of viscous friction in the system.

Equations (1.2) and (1.5) are *linear* differential equations because the dependent variable, $\varphi(t)$, and its time derivatives occur only to the first power. We shall therefore refer to this system as a *linear oscillator*. These equations are also *homogeneous*, because $\varphi(t)$, or its derivatives, appear to the same power in every term of the equation. The homogeneity of Eqs. (1.2) and (1.5) implies that an external driving force, independent of $\varphi(t)$ or its time derivatives, is not present. We call the unforced oscillations described by Eqs. (1.2) and (1.5) *free* or *natural* oscillations. They occur when there is no external driving force.

As for any homogeneous equation, Eqs. (1.2) and (1.5) have the trivial solution, $\varphi(t) = 0$ and $\dot{\varphi}(t) = 0$. This solution describes a system that is always at rest in its equilibrium position. Since the solutions to this second-order differential equation are completely determined by the values of $\varphi$ and $\dot{\varphi}$ at some particular moment, it is clear that if the initial values of $\varphi$ and $\dot{\varphi}$ are zero, they must remain zero forever. Hence, if non-zero values $\varphi$ and $\dot{\varphi}$ are to be found, $\varphi(0)$ and $\dot{\varphi}(0)$ cannot both be zero. Oscillations of the system are produced only when there is some initial excitation of the system.

When friction is sufficiently weak, so that $\gamma < \omega_0$, the general solution of Eq. (1.5) can be written in the form:

$$\varphi(t) = A_0 \exp(-\gamma t) \cos(\omega_1 t + \delta_0). \tag{1.6}$$

This solution describes *damped oscillations* whose amplitude $A_0 \exp(-\gamma t)$ decreases exponentially with time. The amplitude constant $A_0$ (the initial amplitude) and the initial phase $\delta_0$ depend on the initial conditions. The frequency $\omega_1$ appearing in the cosine term in Eq. (1.6) is given by

$$\omega_1 = \sqrt{\omega_0^2 - \gamma^2} = \omega_0 \sqrt{1 - (\gamma/\omega_0)^2}. \tag{1.7}$$

In the case of relatively weak damping, when constant $\gamma$ is small compared to the natural frequency $\omega_0$ ($\gamma/\omega_0 \ll 1$), frequency $\omega_1$ is very close to $\omega_0$:

$$\omega_1 \approx \omega_0 - \gamma^2/(2\omega_0). \tag{1.8}$$

The fractional difference $(\omega_0 - \omega_1)/\omega_0$ of these frequencies is proportional to the square of the small parameter $\gamma/\omega_0$.

Graphs of the deflection angle and of the angular velocity for oscillations damped by the force of viscous friction are shown in Figure 1.2.

### 1.1.3    The Time of Damping and the Quality Factor $Q$

In the cases of weak and moderate damping, in Eq. (1.6) the time-dependent factor $A_0 \exp(-\gamma t)$ can be treated as a slowly decreasing amplitude of diminishing oscillations. After an interval $\tau = 1/\gamma$, the amplitude is $e \approx 2.72$ times smaller than its initial value. The time $\tau$ is called the *decay time* or the *time of damping*.

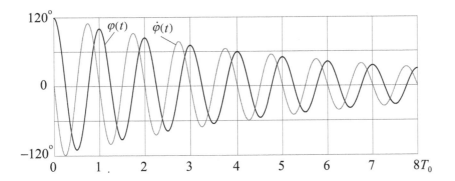

Figure 1.2: Plots of the deflection angle $\varphi(t)$ and of the angular velocity $\dot{\varphi}(t)$ for damped natural oscillations.

When $\gamma \ll \omega_0$, or $\tau \gg T_0 = 2\pi/\omega_0$ (condition of *weak damping*), the oscillator executes a large number $N$ of oscillations during the decay time $\tau$: $N = \tau/T_0 \gg 1$. Consecutive maximal deflections from the equilibrium position diminish in a geometric progression. Letting $\varphi_n$ be the maximal angular displacement of the $n$-th oscillation, we have

$$\varphi_{n+1}/\varphi_n \approx \exp(-\gamma T_0) \approx 1 - \gamma T_0. \tag{1.9}$$

That is, the ratio of successive terms in this infinite geometric progression is less than unity by the small value $\gamma T_0 = T_0/\tau \ll 1$.

The strength of viscous friction in the system is usually characterized either by the damping constant $\gamma$, which, as can be seen from Eq. (1.5), has the dimension of frequency, or by a dimensionless quantity $Q$, called the *quality factor*. The quality factor is defined by:

$$Q = \frac{\omega_0}{2\gamma} = \pi \frac{\tau}{T_0}. \tag{1.10}$$

The number of cycles during which the amplitude of oscillations decreases by a factor $e \approx 2.72$ is given by $Q/\pi$, and the number of cycles $N_{1/2}$ during which the amplitude is halved is given by:

$$N_{1/2} = (\ln 2/\pi)Q = 0.22\,Q = Q/4.53. \tag{1.11}$$

It follows from Eq. (1.11) that $N_{1/2} = 4$ if $Q = 18.13$, that is, for this value of the quality factor the amplitude of natural oscillations halves under the viscous friction after each 4 cycles. The graphs in Figure 1.2 obtained in computer simulation of natural oscillations at $Q = 18.13$ show clearly that indeed the amplitude halves after the first four cycles, and halves again after the next four cycles.

When $\gamma \geq \omega_0$ (condition of *strong damping*), a disturbed oscillator returns to the equilibrium position without oscillating. In this motion, the oscillator either approaches the equilibrium position asymptotically from one side or overshoots

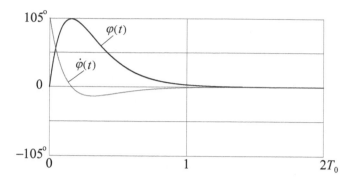

Figure 1.3: Plots of the deflection angle $\varphi(t)$ and of the angular velocity $\dot{\varphi}(t)$ for the case of critical damping ($\gamma = \omega_0$, $\varphi(0) = 0$ and $\dot{\varphi}(0) = 5\omega_0$).

the equilibrium position only once and then asymptotically reapproaches it from the other side. This latter case occurs only when the initial angular velocity of the oscillator is directed toward the equilibrium position, and its magnitude is large enough.

When $\gamma = \omega_0$, the system is said to be *critically damped*. The general solution to the differential equation of motion, Eq. (1.5), for the critically damped system takes the form:

$$\varphi(t) = (C_1 t + C_2) \exp(-\gamma t), \tag{1.12}$$

where $C_1$ and $C_2$ are constants defined by the initial conditions. For example, if the system is given an initial velocity $\Omega_0$ at the equilibrium position, that is, if $\varphi(0) = 0$ and $\dot{\varphi}(0) = \Omega_0$, then $C_1 = \Omega_0$, $C_2 = 0$, and the motion of the system is described by the function:

$$\varphi(t) = \Omega_0 t \exp(-\gamma t). \tag{1.13}$$

The graphs of the deflection angle $\varphi(t)$ and of the angular velocity $\dot{\varphi}(t)$ for this case of excitation at critical damping are shown in Figure 1.3.

An interesting feature of the critically damped system is that, after an initial disturbance, it returns to rest in the equilibrium position usually sooner than it does in any other case (i.e., than it does for any other value of the damping constant $\gamma$ for a given value of $\omega_0$). It is seen from Eq. (1.10) that the value of the quality factor that corresponds to critical damping ($\gamma = \omega_0$) is $Q = 0.5$.

Non-oscillatory motion at strong friction, when $\gamma > \omega_0$, can be represented as a superposition of two exponential functions, which have different time constants $\tau_1 = -1/\alpha_1$ and $\tau_2 = -1/\alpha_2$:

$$\varphi(t) = C_1 e^{\alpha_1 t} + C_2 e^{\alpha_2 t}, \qquad \text{where } \alpha_{1,2} = -\gamma \pm \sqrt{\gamma^2 - \omega_0^2}. \tag{1.14}$$

Values of $C_1$ and $C_2$ are determined by the initial conditions.

In measuring instruments such as moving-coil galvanometers, damping is introduced intentionally in order to overcome the difficulty of taking a reading from an oscillating needle. A measuring instrument is said to be *critically damped* if the needle just fails to oscillate and comes to rest in the shortest possible time.

If the instrument is *underdamped* ($Q > 0.5$), the needle oscillates repeatedly before coming to rest. If the instrument is *overdamped* ($Q < 0.5$), the needle does not oscillate, though generally it takes longer to come to rest than it does when the instrument is critically damped.

## 1.1.4 The Phase Diagram of a Linear Oscillator

The *mechanical state* of a torsion pendulum at any instant is determined by the two quantities: the *angular displacement* $\varphi$ and the *angular velocity* $\dot{\varphi}$ (or instead of $\dot{\varphi}$ by the *angular momentum* $J\dot{\varphi}$). The evolution of the mechanical state of the system during its entire motion can be graphically demonstrated very clearly in a *phase diagram*. This is a graph that plots the angular velocity $\dot{\varphi}$ (or the angular momentum $J\dot{\varphi}$) versus the angular displacement $\varphi$.

The mechanical state of the system at any instant is represented by a point, called the *representative point*, in the phase plane. If the motion of the physical system is *periodic*, the representative point, moving clockwise, generates a *closed path* in the phase plane. The phase trajectory of periodic motion is closed, because the system returns to the same mechanical state after a full cycle.

The phase diagram for harmonic oscillations (e.g., for oscillations in a linear system without friction) is an ellipse (or a circle at the appropriate choice of the scales). The points of intersection of the phase curve with the $\varphi$-axis (the *extreme, or turning points*) correspond to maximal deflections of the oscillator from the equilibrium position. At these points, the sign of the angular velocity $\dot{\varphi}$ changes, and the tangent to the phase curve is perpendicular to the abscissa axis.

As noted above, the period of harmonic oscillations is determined entirely by the parameters of the physical system, specifically by the values of the spring constant $D$ and the moment of inertia $J$. Unlike the amplitude and the initial phase, the period does not depend on the initial conditions, that is, on the way oscillations are excited. This property of the harmonic oscillator is called *isochronism*. Because of this property, the representative point generates ellipses of different sizes (which correspond to various amplitudes of oscillations in the same system) during the same time $T_0$.

In the presence of relatively weak viscous friction ($\gamma < \omega_0$) the extreme displacements, as well as the extreme values of the angular velocity, gradually diminish with each subsequent cycle of oscillation. Consequently the phase trajectory for free oscillations is transformed from a closed curve (an ellipse or a circle) into a shrinking spiral that winds around a focal point located at the origin of the phase plane. The lower panel of Figure 1.4 shows the phase diagram of damped oscillations. In the upper panel the parabolic potential well $E_{\text{pot}}(\varphi)$ of the linear oscillator is shown together with the graphs $E_{\text{tot}}(\varphi)$ of the total energy and $E_{\text{kin}}(\varphi)$ of the kinetic energy. During the oscillations, the point representing the total energy

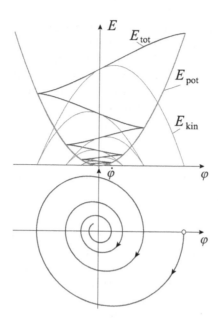

Figure 1.4: The parabolic potential well (upper panel) and the phase diagram (lower panel) of damped oscillations.

$E_{tot}$ travels in this well from one slope to the other descending gradually to the bottom of the well.

The family of phase trajectories, which correspond to the same values of the parameters of the system but to different initial conditions, forms a *phase portrait* of the system. This phase portrait gives a clear graphic representation of all possible motions of the system. The phase portrait of a conservative linear oscillator is formed by a set of similar ellipses with a common *center* at the origin of the phase plane. The center represents a state of rest in the equilibrium position.

When friction is relatively weak ($\gamma < \omega_0$), this center becomes an *attractor* of the phase trajectories called the *focal point*. That is, all phase trajectories of damped oscillations spiral in toward the origin, forming an infinite number of gradually shrinking loops, as in Figure 1.5a.

When friction is relatively strong ($\gamma > \omega_0$), the attractor of the phase trajectories becomes a *node*: all phase trajectories of non-oscillatory motion approach this node directly, without spiraling.

The phase portrait of an overdamped system is shown in Figure 1.5b. The phase curves asymptotically approach the origin, where they have a common tangent $\dot{\varphi} = \alpha_1\varphi$, where $\alpha_1 = -\gamma + \sqrt{\gamma^2 - \omega_0^2}$. At specific initial conditions, when $\dot{\varphi}(0) = \alpha_1\varphi(0)$, the representative point moves towards the node directly along this tangent. The other rectilinear phase trajectory $\dot{\varphi} = \alpha_2\varphi$ ($\alpha_2 = -\gamma - \sqrt{\gamma^2 - \omega_0^2}$) occurs at initial conditions of the type $\dot{\varphi}(0) = \alpha_2\varphi(0)$.

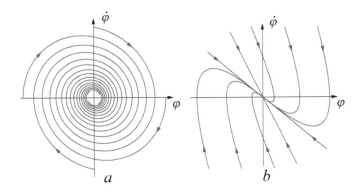

Figure 1.5: Phase portrait of (*a*) an underdamped ($\gamma = 0.1\omega_0$) and of (*b*) an over-damped ($\gamma = 1.1\omega_0$) linear oscillator with viscous friction.

In general, only one phase trajectory passes through a given point of the phase plane. Indeed, if we consider this arbitrary point as an initial state of the system, the further motion of the system is defined uniquely. This motion is represented by a single phase trajectory, passing through the point. However, there may be exceptions in that either no phase trajectory passes through a phase point, or there are several trajectories at once. Phase points of this kind are called *singular*.

For a linear oscillator, there is only one singular point: the origin of the phase plane. It corresponds to the state of rest in the equilibrium position, where both $\varphi$ and $\dot{\varphi}$ are zero. When $\gamma = 0$, this point is a center, and no phase trajectory passes through it. When $0 < \gamma < \omega_0$, or when $\gamma > \omega_0$, it is respectively a focus or a node, to which all phase trajectories are attracted. For nonlinear systems such as a planar rigid pendulum, there exists another kind of singular points, namely the saddle points.

## 1.1.5 Energy Transformations

The total energy $E$ of the torsion spring pendulum is the sum of the elastic *potential energy* $E_{\text{pot}}$ of the strained spring and the *kinetic energy* $E_{\text{kin}}$ of the rotating flywheel:

$$E = E_{\text{pot}} + E_{\text{kin}} = \frac{1}{2}D\varphi^2 + \frac{1}{2}J\dot{\varphi}^2. \tag{1.15}$$

Free oscillations in the absence of friction are characterized by the *exchange of energy* between its potential and kinetic forms. At the points of maximum displacement from the equilibrium position, the kinetic energy is zero, and the total energy of the oscillator is the potential energy of the strained spring. A quarter of a period later, the oscillator passes through the equilibrium point, where the potential energy is zero and the total energy of the oscillator is the kinetic energy of

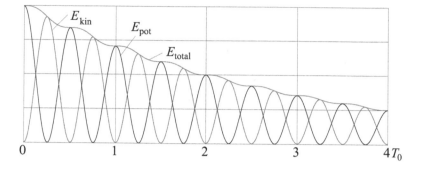

Figure 1.6: Energy transformations during damped natural oscillations.

the flywheel. During the next quarter period, the reverse exchange of energy occurs: the kinetic energy is transformed into potential energy. Such transformations happen twice during one period. That is, oscillations of the two kinds of energy, 180° out of phase with one another, are executed between zero and a maximal value $\frac{1}{2}DA_0^2$ with the frequency $2\omega_0$, i.e., with double the natural frequency $\omega_0$ of the system.

The exchanges between potential and kinetic energy described above are characteristic of a conservative system, in which such transformations are reversible. The sum of the kinetic and potential energy, i.e., the total mechanical energy $E$ of the oscillator, is the same at every instant and is equal to the maximum values of both kinetic energy and potential energy:

$$E = \frac{1}{2}DA_0^2 = \frac{1}{2}J\omega_0^2A_0^2. \tag{1.16}$$

The total energy of the system is proportional to the square of the amplitude $A_0$. The values of the two forms of energy, averaged over a period, are each equal to one half of the total energy:

$$\langle E_{\text{pot}} \rangle = \langle E_{\text{kin}} \rangle = \frac{1}{2}E = \frac{1}{4}DA_0^2 = \frac{1}{4}J\omega_0^2A_0^2. \tag{1.17}$$

In the presence of friction, the exchanges between kinetic and potential energy are partially irreversible because of the dissipation of mechanical energy. This dissipation occurs nonuniformly during a complete cycle: Its instantaneous rate, $-dE/dt$, is zero when the flywheel, in a given cycle, is at the extremes of its motion and its angular velocity, $\dot{\varphi}$, is zero. This is clearly seen in Figure 1.6, which shows the transformations of energy occurring in damped oscillations. The rate of dissipation is greatest when the flywheel moves in the vicinity of the equilibrium position, where its angular velocity is maximal. Indeed, the rate of energy dissipation $-dE/dt = -N_{\text{fr}}\dot{\varphi}$ caused by viscous friction is proportional to the square of the angular velocity and hence to the momentary value of the kinetic energy

$J\dot{\varphi}^2/2$ of the oscillator. The statement, frequently encountered in textbooks, that the energy of a damped oscillator decays exponentially, applies only to values of the total energy averaged over a period.

The graphs in Figure 1.6 correspond to the same value of the quality factor ($Q = 18.13$) as the graphs of $\varphi(t)$ and $\dot{\varphi}(t)$ in Figure 1.2. We see clearly that for this quality factor the amplitude of oscillations halves after each four cycles of natural oscillations (Figure 1.2), while the total energy halves after each two cycles (Figure 1.6).

### 1.1.6   The Computer Simulation of a Linear Oscillator

The computer simulation of the torsion spring pendulum in the suggested simulation program of the package PHYSICS OF OSCILLATIONS is based on numerical integration of the differential equation, Eq. (1.5). Although this equation can be integrated analytically, the analytic solution is not used in those parts of the computer program that demonstrate the evolution of the system with time. The analytic solution is used only to determine the maximal values of the angular displacement and of the angular velocity in order to establish suitable scales for the corresponding plots. The agreement between the results of the numerical integration in the computer simulation and the analytic predictions can serve as a confirmation of the quality of the algorithm used (the fourth order Runge – Kutta method). This verification gives us confidence in the reliability of the the computer simulations of nonlinear systems in subsequent programs of this software package since the simulations are based on the same numerical method.

In this numerical simulation of the linear oscillator we adopt a unit of time that is appropriate for the system under consideration, namely the period $T_0 = 2\pi/\omega_0$ of free oscillations in the absence of friction. Thus the simulated oscillator may be characterized by only one physical parameter: either by the dimensionless ratio of the damping constant to the natural frequency $\gamma/\omega_0$, or by the equivalent dimensionless quantity – the quality factor $Q = \omega_0/(2\gamma)$, inversely proportional to $\gamma/\omega_0$.

The angular displacement, $\varphi$, is expressed in radians in the program, though for convenience of observation, the dial on the screen and the plots involving the angle of deflection are graduated in degrees. The angular velocity, $\dot{\varphi}$, is measured in units of the natural frequency, $\omega_0$. When the initial conditions are set in a simulation computer experiment, the initial angular velocity also must be expressed in units of $\omega_0$.

## 1.2   Review of the Principal Formulas

The differential equation of a free (unforced) linear torsion oscillator:

$$\ddot{\varphi} + 2\gamma\dot{\varphi} + \omega_0^2\varphi = 0. \tag{1.18}$$

The frequency and the period of free oscillations without friction (at $\gamma \ll \omega_0$):

$$\omega_0 = \sqrt{\frac{D}{J}}, \qquad T_0 = \frac{2\pi}{\omega_0}. \tag{1.19}$$

An oscillatory solution (valid at $\gamma < \omega_0$):

$$\varphi(t) = A_0 e^{-\gamma t} \cos(\omega_1 t + \delta_0), \tag{1.20}$$

where the constants $A_0$ and $\delta_0$ are determined by the initial conditions $\varphi(0)$, $\dot{\varphi}(0)$. The frequency $\omega_1$ of damped oscillations

$$\omega_1 = \sqrt{\omega_0^2 - \gamma^2}. \tag{1.21}$$

An equivalent form of the general solution:

$$\varphi(t) = e^{-\gamma t}(C \cos \omega_1 t + S \sin \omega_1 t), \tag{1.22}$$

where the constants $C$ and $S$ are determined by the initial conditions. They are related to $A_0$ and $\delta_0$:

$$A_0 = \sqrt{C^2 + S^2}, \qquad \tan \delta_0 = -S/C. \tag{1.23}$$

In the case of weak damping ($\gamma \ll \omega_0$)

$$\omega_1 \approx \omega_0 - \gamma^2/(2\omega_0). \tag{1.24}$$

The decay time (during which the amplitude is reduced by the factor $e \approx 2.72$):

$$\tau = 1/\gamma. \tag{1.25}$$

A non-oscillatory motion at $\gamma = \omega_0$:

$$\varphi(t) = (C_1 t + C_2)e^{-\gamma t}. \tag{1.26}$$

The quality factor $Q$ of an oscillator:

$$Q = \pi \frac{\tau}{T_0} = \frac{\omega_0}{2\gamma}. \tag{1.27}$$

The number of oscillations, during which the amplitude is halved:

$$N_{1/2} = \frac{\ln 2}{\pi} Q = 0.22\, Q = \frac{Q}{4.53}. \tag{1.28}$$

The total mechanical energy of the oscillator consists of elastic potential energy of the strained spring and kinetic energy of the flywheel:

$$E = E_{\text{pot}} + E_{\text{kin}} = \frac{1}{2}D\varphi^2 + \frac{1}{2}J\dot{\varphi}^2. \tag{1.29}$$

The values of the potential energy and kinetic energy of the oscillator, averaged over a cycle, equal one another, each of them constituting one half the total energy:

$$\langle E_{\text{pot}} \rangle = \langle E_{\text{kin}} \rangle = \frac{1}{2}E = \frac{1}{4}DA_0^2 = \frac{1}{4}J\omega_0^2 A_0^2. \tag{1.30}$$

## 1.3 Questions, Problems, Suggestions

### 1.3.1 Free Undamped Oscillations

**1.3.1.1 The Initial Conditions and the Shape of the Plots.** In the absence of friction a linear oscillator executes simple harmonic motion, which is characterized by purely sinusoidal time dependence of the angular displacement and of the angular velocity.

(a) What initial conditions give rise to oscillations of cosine time dependence, and of sine time dependence? Suppose that you want to get oscillations with the angular amplitude of 90°. What initial angular displacement $\varphi(0) = \varphi_0$ at zero initial angular velocity $\dot{\varphi}(0) = 0$ ensures the desired amplitude?

(b) What initial angular velocity $\dot{\varphi}(0) = \Omega$ ought you to impart to the oscillator, at rest in the equilibrium position, in order to obtain the same amplitude of 90°? Remember that the initial angular velocity $\Omega$ must be expressed for input in units of the natural frequency $\omega_0$. Verify your answer with a computer experiment, using the appropriate initial conditions.

**1.3.1.2 Maximal Deflection and Conservation of Energy.** Imagine exciting an oscillator initially at rest in the equilibrium position by a push that produces an initial angular velocity $\Omega = 2\omega_0$.

(a) Calculate the angle $\varphi_{max}$ of maximal deflection using the law of the conservation of energy.

(b) Verify your result experimentally. Note that the simulation program performs the numerical integration of the differential equation independently of conservation laws, such as the conservation of energy. That is, these laws are not used in the program.

**1.3.1.3 The Phase Trajectory and the Initial Conditions.** Compare the motion of the representative point along the phase trajectory of a conservative oscillator with the time-dependent plots of the angle of deflection and of the angular velocity.

(a) How is the phase trajectory changed if you change the initial conditions?

(b) Does the direction of the motion of the representative point along the phase trajectory depend on the initial conditions?

(c) Is it possible that phase trajectories for different initial conditions coincide? If so, formulate the requirements for the coincidence.

**1.3.1.4 Elliptical and Circular Shape of the Phase Trajectory.**

(a) Prove analytically that the phase trajectory of a conservative linear oscillator is an ellipse with its center at the origin of the phase plane. Use the general solution of Eq. (1.2), expressed by Eq. (1.4). What are the semiaxes of the ellipse?

(b) Show that the elliptical shape of the phase diagram of a conservative linear oscillator follows immediately from the law of the conservation of the energy.

(c) What scale on the axis of the ordinate (the angular velocity axis) of the phase plane produces a circular phase trajectory?

(d) Does the time interval during which the representative point passes along one loop of the phase trajectory depend on the initial conditions?

**1.3.1.5  The Phase Diagram and Energy Transformations.** Compare the phase trajectory with the plot of potential energy versus the angle of deflection. The positioning of plots on the display screen (if you open the window "Phase diagram") is convenient for such comparison. Pay special attention to the positions of the extreme points (turnarounds) on the phase trajectory and in the parabolic potential well. For the initial conditions $\varphi(0) = \varphi_0$, $\dot{\varphi}(0) = \Omega$, what are the values of the potential energy and the kinetic energy at the extreme points and at the equilibrium position?

What are the extreme deflection $\varphi_{\max}$ and the maximal angular velocity $\omega_{\max}$ of the flywheel?

**1.3.1.6  The Shape and the Frequency of Energy Oscillations.** Consider the plots of the time dependence of kinetic energy and potential energy.

(a) What can you say about their maximal and average values? Compare these plots with the plots of the angular displacement and the angular velocity.

(b) At what frequency do the oscillations of each kind of energy occur? What are the limits (the extreme values) and the mean (averaged over a period) values of each kind of energy in these oscillations?

**1.3.1.7  The Phase Trajectories with the Same Energy.** Consider the oscillations of a conservative oscillator at different initial conditions but with the same total energy. What differences do you observe in the plots and the phase trajectories in these cases?

## 1.3.2   Damped Free Oscillations

**1.3.2.1  The Sequence of Maximal Deflections.** Under the action of a weak force of viscous friction, the sequence of maximal deflections of a free, damped linear oscillator forms a decreasing geometric progression: Each consecutive maximal deflection is smaller than the preceding one by the same factor, $\exp(-\gamma T_0) \approx 1 - \gamma T_0$ [see Eq. (1.10)].

(a) Calculate the value of the quality factor $Q$ at which the amplitude halves during every two complete oscillations.

(b) Input this value in a computer experiment and verify the theoretically predicted constant ratio of successive maximal deflections. Note that this ratio does not depend on the initial conditions.

(c) Evaluate the increment of the period of oscillations at this value of the quality factor with respect to the period $T_0$ in the absence of friction (in percent). Can you detect the increment in the simulation experiment? The marks on the time axis correspond to integer numbers of periods $T_0 = 2\pi/\omega_0$ without friction.

**1.3.2.2\*  Maximal Deflection after an Initial Push.** Imagine that we excite oscillations with an initial push which imparts an initial angular velocity of $2\omega_0$ to the flywheel in its equilibrium position.

(a) Calculate the first maximal deflection of the flywheel after the excitation for the quality factor $Q = 5$.

(b) What will be the value of the subsequent extreme deflection that occurs in the direction opposite to the first? Verify your answers using the simulation.

**1.3.2.3** **Complex Initial Conditions.**

(a) Let the initial deflection of the torsion pendulum be 155 degrees, and the initial angular velocity be $2\omega_0$. The quality factor $Q = 5$. Calculate the maximal deflection of the flywheel.

(b) With the same initial deflection (155 degrees) and the same quality factor $Q = 5$ as in the preceding item (a), calculate the maximal deflection of the flywheel, if the initial angular velocity equals $-2\omega_0$.

(c) Let the initial deflection of the torsion pendulum be $-155$ degrees. What initial angular velocity would ensure the maximal deflection of 155 degrees (to the opposite side), if the quality factor $Q = 20$?

**1.3.2.4*** **The Phase Trajectory of Damped Oscillations.** The phase trajectory of damped free oscillations for $Q > 0.5$ is a spiral that makes an infinite number of gradually shrinking loops around the focus located at the origin of the phase plane. This focus corresponds to the state of rest in the equilibrium position, and the phase trajectory approaches it asymptotically.

(a) How does the size of these loops change while the curve approaches the focus?

(b) Does the time interval during which the representative point makes one revolution along the spiral change as the loops of the curve shrink?

**1.3.2.5*** **The Dissipation of Energy.** Compare the transformation of potential energy into kinetic energy (and vice versa) for free undamped oscillations in the absence of friction with that for free damped oscillations in the presence of viscous friction.

(a) Show, using a simulation experiment, that if $Q = 18.1$, the amplitude is halved during four complete oscillations and the total energy is halved during two complete oscillations.

(b) Why is the dissipation of mechanical energy nonuniform during one cycle of oscillations? At what instants during a cycle is the time-rate of energy dissipation greatest and at what instants is it smallest?

## 1.3.3 Non-Oscillatory Motion of the System

When viscous friction is strong ($Q \leq 0.5$), a disturbed system returns to the equilibrium position without oscillating. In the computer simulation, the needle asymptotically approaches the zero point from one side.

**1.3.3.1*** **Non-Oscillatory Motion at Critical Damping.** Consider the case of critical damping, $\gamma = \omega_0$.

(a) Why is critical damping preferable in measuring instruments using a needle as an indicator? How might your answer apply to the suspension system in an automobile?

Show that the value of $Q$ in the case of critical damping is 0.5.

(b) Calculate the maximal angle of deflection if the system, with $Q = 0.5$, receives an initial velocity $\Omega = 5\omega_0$ in the equilibrium position.

(c) In what lapse of time does the needle move towards this extreme point?

Verify your answers by simulating the experiment on the computer. Note that the needle approaches the equilibrium position from one side – it does not cross the zero point of the dial.

### 1.3.3.2  Critical Damping.

(a) Prove that the value $Q = 0.5$ ($\gamma = \omega_0$) is really critical. Do so by showing that at slightly greater values of $Q$, the needle of a perturbed oscillator executes heavily damped oscillations, slowly moving to and fro across the zero point of the dial.

(b) For a critically damped system, express the constants $C_1$ and $C_2$ in the general solution $\varphi(t) = (C_1 t + C_2) \exp(-\gamma t)$ of the differential equation, Eq. (1.5), in terms of the initial displacement $\varphi(0) = \varphi_0$ and the initial angular velocity $\dot{\varphi}(0) = \Omega_0$.

(c) Is it possible for a critically damped system to move after an initial disturbance according to a pure exponential law? If so, what initial conditions give rise to such motion? What is the phase trajectory of this motion? Prove your answers experimentally.

(d) At what initial conditions will the flywheel of a disturbed critically damped system cross the equilibrium position? For a given initial displacement $\varphi_0$, what initial angular velocity $\Omega$ should you impart to the flywheel of the critically damped oscillator in order to have it cross the equilibrium position after a lapse of time $t = 3T_0$, where $T_0 = 2\pi/\omega_0$ is the natural period (the period of oscillations in the absence of friction)?

### 1.3.3.3*  Motion of an Overdamped System.

(a) For arbitrary initial conditions ($\varphi(0) = \varphi_0$, $\dot{\varphi}(0) = \Omega_0$), express the values of $C_1$ and $C_2$ in the general solution (1.14) of the differential equation for an overdamped system in terms of $\varphi_0$ and $\Omega_0$.

(b) At what initial conditions will the motion of an overdamped system be described by a monoexponential function of time? What are the phase trajectories that correspond to such motions?

(c) Explain why, at arbitrary initial conditions, non-oscillatory motion of the flywheel towards the equilibrium position occurs more slowly and requires more time than at critical damping. Is it possible for an overdamped system to return to the equilibrium position faster than for the critically damped system with the same $\omega_0$? If so, what conditions of excitation ensure the motion?

(d) What is the principal difference between the phase trajectories corresponding to a non-oscillatory motion and those corresponding to damped oscillations?

(e) Is it possible for an overdamped system ($\gamma > \omega_0$) to cross the equilibrium position after excitation? If so, what initial conditions give rise to such motion? Is it possible for the oscillator to cross the equilibrium position more than once?

# Chapter 2

# Torsion Spring Oscillator with Dry Friction

**Annotation.** Chapter 2 deals with natural oscillations in a simple mechanical system with dry (Coulomb) friction. It includes a description of the simulated physical system and a summary of the relevant theoretical material for students as a prerequisite for the virtual lab "Torsion Spring Oscillator with Dry Friction" from the software package *Physics of Oscillations* (Part I, Simple Systems). Peculiarities of damping of oscillations under dry friction are discussed in detail. Chapter 2 also includes a set of theoretical and experimental problems to be solved by students, as well as various assignments that the instructor can offer students for possible individual work on their own.

## 2.1 Summary of the Theory

### 2.1.1 General Concepts

Chapter 2 and the relevant simulation program are aimed at investigation of free oscillations of a torsion spring pendulum damped by dry (Coulomb) friction. An idealized mathematical model of dry friction described by the so-called z-characteristic is assumed. In this model, the force of kinetic friction does not depend on speed and equals the limiting force of static friction. The physical system modeled here allows us to understand the origin of accidental errors in reading some measuring instruments.

### 2.1.2 The Physical System

The rotating component of the torsion spring oscillator is a balanced flywheel (a rigid rod with two equal weights) whose center of mass lies on the axis of rotation. Figure 2.1 shows a schematic image of the simulated system. A spiral spring, one

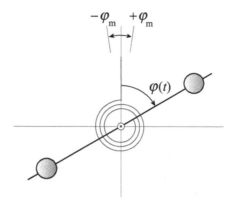

Figure 2.1: The image of the torsion spring oscillator with dry friction.

end fixed and the other end attached to the flywheel, flexes when the flywheel is turned. The system is in equilibrium when the rod of the flywheel is vertical ($\varphi = 0$). The spring provides a restoring torque whose magnitude is proportional to the angular displacement $\varphi$ of the flywheel from the equilibrium position.

The dynamical behavior of such a system under the influence of viscous friction (for which the torque is proportional to the angular velocity) is discussed in Chapter 1, "Free oscillations of a linear torsion pendulum." When friction is viscous, free oscillations of a spring pendulum are described by a *linear* differential equation. The amplitude of such oscillations decreases exponentially with time. That is, the consecutive maximal deflections of the oscillator from its equilibrium position are in a diminishing geometric progression because their ratio is constant. In principle such oscillations continue indefinitely, their amplitude asymptotically approaching zero. However, it is convenient to characterize the duration of exponential damping by a *decay time* $\tau$. This conventional time of damping $\tau$ is the lapse of time during which the amplitude of free oscillations decreases by a factor of $e \approx 2.72$.

The exponential character of damping caused by viscous friction follows from the proportionality of friction to velocity. Some other relationship between friction and velocity produces damping with different characteristics.

The case of *dry* or *Coulomb* friction has important practical applications. In this case, as long as the system is moving, the magnitude of dry friction is very nearly constant and its direction is opposite that of the velocity. An idealized simplified characteristic of dry friction (called the z-characteristic) is shown in Figure 2.2. The graph shows dependence of the frictional torque $N$ on the angular velocity $\dot{\varphi}$ of rotation. Here the magnitude of friction is constant, but its direction changes each time the direction of the velocity changes. When the system is at rest, the torque of static dry friction takes on any value from $-N_{\max}$ to $N_{\max}$. The actual value depends on the friction needed to balance the other forces exerted

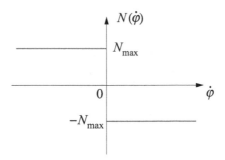

Figure 2.2: An idealized characteristic of dry friction (z-characteristic).

on the system. The magnitude of the torque of kinetic dry friction is assumed in this model to be equal to the limiting torque $N_{\mathrm{max}}$ of static friction.

In real physical systems dry friction is characterized by a more complicated dependence on velocity. The limiting force of static friction is usually greater than the force of kinetic friction. When the speed of a system increases from zero, kinetic friction at first decreases, reaches a minimum at some speed, and then gradually increases with a further increase in speed. These peculiarities are ignored in the idealized z-characteristic of dry friction. Nevertheless, this idealization helps us to understand many essential properties of oscillatory processes in real physical systems.

Because the magnitude of the torque of static friction can assume any value up to $N_{\mathrm{max}}$, there is a range of values of displacement called the *stagnation interval* or *dead zone* in which static friction can balance the restoring elastic force of the strained spring. The stagnation interval extends equally to either side of the point at which the spring is unstrained. The stronger the dry friction in the system, the more extended the stagnation interval. The boundaries $\pm\varphi_{\mathrm{m}}$ of the interval are determined by the limiting torque $N_{\mathrm{max}}$ of static friction: $N_{\mathrm{max}} = D\varphi_{\mathrm{m}}$. If the velocity becomes zero at some point of the dead zone, the system remains at rest there. The boundaries $\pm\varphi_{\mathrm{m}}$ of the dead zone are indicated in Figure 2.1. At any point within the stagnation interval the system can be at rest in a state of neutral equilibrium, in contrast to a single position of stable equilibrium provided by the spring in the case of viscous friction.

An important feature of oscillations damped by dry friction is that motion ceases after a *finite* number of cycles. As the system oscillates, the sign of its velocity changes periodically, and each subsequent change occurs at a smaller displacement from the mid-point of the stagnation interval. Eventually the turning point of the motion occurs within the stagnation interval, in which static friction can balance the restoring force of the spring, and so the motion abruptly stops. The exact position in the stagnation interval at which this event occurs depends on the initial conditions, which may vary from one situation to the next.

These characteristics are typical of various mechanical systems with dry friction. For example, dry friction may be encountered in measuring instruments, such as a moving-coil galvanometer, in which readings are taken with a needle. In the galvanometer, a light coil of wire is pivoted between the poles of a magnet. When a current flows through the coil, it turns against a spiral return spring. If the coil axis is fixed in unlubricated bearings and hence experiences dry friction, the needle of the coil may come to rest and show at any point of the stagnation interval on either side of the dial point, which gives the true value of the measured quantity. So we can now understand one of the reasons that random errors inevitably occur in the readings of moving-coil measuring instruments. The larger the dry friction, the larger the errors of measurement.

### 2.1.3   The Differential Equation of the Oscillator

The rotating flywheel of the torsion oscillator is simultaneously subjected to the action of the restoring torque $-D\varphi$ produced by the spring, and of the torque $N_{\mathrm{fr}}$ of kinetic dry friction. The differential equation describing the motion of the flywheel, whose moment of inertia is $J$, is thus

$$J\ddot{\varphi} = -D\varphi + N_{\mathrm{fr}}. \qquad (2.1)$$

According to the idealized z-characteristic of dry friction, the torque $N_{\mathrm{fr}}$ is directed oppositely to the angular velocity $\dot{\varphi}$, and is constant in magnitude while the flywheel is moving, but may have any value in the interval from $-N_{\max}$ up to $N_{\max}$ while the flywheel is at rest:

$$N_{\mathrm{fr}}(\dot{\varphi}) = \begin{cases} -N_{\max} & \text{for} \quad \dot{\varphi} > 0, \\ N_{\max} & \text{for} \quad \dot{\varphi} < 0, \end{cases} \qquad (2.2)$$

or $N_{\mathrm{fr}} = -N_{\max}\,\mathrm{sign}(\dot{\varphi})$. Here $N_{\max}$ is the limiting value of the static frictional torque. It is convenient to express the value $N_{\max}$ in terms of the maximal possible deflection angle $\varphi_{\mathrm{m}}$ of the flywheel at rest:

$$N_{\max} = D\varphi_{\mathrm{m}}. \qquad (2.3)$$

The angle $\varphi_{\mathrm{m}}$ corresponds to the boundary of the stagnation interval.

The differential equation, Eq. (2.1), in the general case of an oscillator with dry friction, is *nonlinear* because the torque $N_{\mathrm{fr}}(\dot{\varphi})$ abruptly changes when the sign of $\dot{\varphi}$ changes at the extreme points of oscillation, and because when the flywheel moves, the torque is usually not constant. But in the idealized case of the z-characteristic we may consider the following two *linear* equations instead of Eq. (2.1):

$$J\ddot{\varphi} = -D(\varphi + \varphi_{\mathrm{m}}) \qquad \text{for} \quad \dot{\varphi} > 0, \qquad (2.4)$$

$$J\ddot{\varphi} = -D(\varphi - \varphi_{\mathrm{m}}) \qquad \text{for} \quad \dot{\varphi} < 0. \qquad (2.5)$$

Whenever the sign of the angular velocity $\dot{\varphi}$ changes, the pertinent equation of motion also changes. The nonlinear character of the problem reveals itself in alternate transitions from one of the linear equations, Eqs. (2.4)–(2.5), to the other.

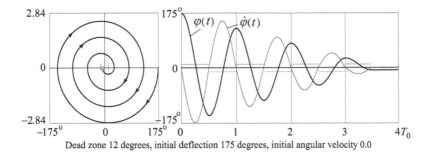

Dead zone 12 degrees, initial deflection 175 degrees, initial angular velocity 0.0

Figure 2.3: Phase trajectory (left) and graphs of $\varphi(t)$ and $\dot{\varphi}(t)$ (right) for oscillations whose damping occurs under dry friction.

The solution to Eqs. (2.4)–(2.5), which corresponds to a set of given initial conditions, can be found by using the method of the stage-by-stage integration of each of the linear equations for the half-cycle during which the direction of motion is unchanged. These solutions are then joined together at the instants of transition from one equation to the other in such a way that the displacement at the end point of one half-cycle becomes the initial displacement at the beginning of the next half-cycle. This array of solutions continues until the end point of a half-cycle lies within the dead zone.

If in addition to dry friction the oscillator also experiences viscous friction, we must add to the equations of motion, Eqs. (2.4)–(2.5), one more term proportional to the angular velocity $\dot{\varphi}$:

$$\ddot{\varphi} = -\omega_0^2(\varphi + \varphi_m) - 2\gamma\dot{\varphi} \qquad \text{for} \quad \dot{\varphi} > 0, \tag{2.6}$$

$$\ddot{\varphi} = -\omega_0^2(\varphi - \varphi_m) - 2\gamma\dot{\varphi} \qquad \text{for} \quad \dot{\varphi} < 0. \tag{2.7}$$

Here $\omega_0^2 = D/J$ is the squared natural frequency of the oscillator (the frequency of oscillations in the absence of friction), and $\gamma$ is the damping constant. It is convenient to characterize viscous friction by the dimensionless quality factor, $Q = \omega_0/2\gamma$.

## 2.1.4 Damping Caused by Dry Friction

In order to discover the basic characteristics of oscillations that are damped under the action of dry friction, we shall first assume that viscous friction is absent, that is, the damping constant $\gamma = 0$.

At the initial instant $t = 0$, let the flywheel be displaced to the right (clockwise) from the equilibrium position so that $\varphi(0) > 0$. If this displacement exceeds the boundary of the stagnation interval, i.e., if $\varphi(0) > \varphi_m$, the flywheel, being released without a push, begins moving to the left ($\dot{\varphi} < 0$), and its motion is described by Eq. (2.5). The solution to Eq. (2.5) with the given initial conditions ($\varphi(0) = \varphi_0$, $\dot{\varphi}(0) = 0$) is simple harmonic motion whose frequency is $\omega_0$. The midpoint of the motion is $\varphi_m$. This point coincides with the right-hand boundary of the stagnation interval. The displacement $\varphi_m$ of the midpoint from zero is caused by the constant torque of kinetic friction. This torque is directed to the right side (clockwise) while the flywheel is moving to the left. The amplitude of this oscillation about the midpoint $\varphi_m$ equals $\varphi_0 - \varphi_m$. The first segment of the graph in the right-hand panel of Figure 2.3 (the first half-cycle of the sine curve, whose midpoint is at a height of $\varphi_m$ above the abscissa axis) is a plot of this portion of the motion. Since the amplitude of the first half-cycle is $\varphi_0 - \varphi_m$, the extreme left position of the flywheel at the end of the half-cycle is $\varphi(0) - 2\varphi_m$. When the flywheel reaches this position, its velocity is momentarily zero, and it starts to move to the right. Since its angular velocity $\dot{\varphi}$ is subsequently positive, we must now consider Eq. (2.4). The values of $\varphi$ and $\dot{\varphi}$ at the end of the preceding half-cycle are taken as the initial conditions for this half-cycle. Thus the subsequent motion is again a half-cycle of harmonic oscillation with the same frequency $\omega_0$ as before but with the midpoint $-\varphi_m$ displaced to the left, i.e., with the midpoint at the left-hand boundary of the stagnation interval. This displacement is caused by the constant torque of kinetic friction, whose direction was reversed when the direction of motion was reversed. The amplitude of the corresponding segment of the sine curve is $\varphi_0 - 3\varphi_m$.

Continuing this analysis half-cycle by half-cycle, we see that the flywheel executes harmonic oscillations about the midpoints alternately located at $\varphi_m$ and $-\varphi_m$. The frequency of each cycle is the natural frequency $\omega_0$, and so the duration of each full cycle equals the period $T_0 = 2\pi/\omega_0$ of free oscillations in the absence of friction.

The joining together of these sinusoidal segments, whose midpoints alternate between the boundaries of the stagnation interval, produces the curve that describes oscillatory motion damped by dry friction (Figure 2.3). The maximal deflection decreases after each full-cycle of these oscillations by a constant value equal to the doubled width of the stagnation interval (i.e., by the value $4\varphi_m$). The oscillation continues until the end point of some next-in-turn segment of the sine curve occurs within the dead zone $(-\varphi_m, \varphi_m)$.

Thus, in the case of dry friction, consecutive maximal deflections diminish linearly in a decreasing arithmetic progression, and the motion stops after a final number of cycles, in contrast to the case of viscous friction, for which the maximal displacements decrease exponentially in a geometric progression, and for which the motion continues indefinitely.

### 2.1.5   The Phase Trajectory

The character of oscillations in the presence of dry friction is given clearly by the phase trajectory shown in the left-hand panel of Figure 2.3. The system is initially at rest $(\dot{\varphi}(0) = 0)$ and displaced to the right $(\varphi(0) = \varphi_0 > \varphi_m)$. This initial state is represented by the point on the curve that lies to the extreme right on the horizontal axis, the $\varphi$-axis. The portion of the phase trajectory lying below the horizontal axis represents the motion during the first half-cycle, when the flywheel is moving to the left. This curve is the lower half of an ellipse (or of a circle if the scales have been chosen appropriately) whose center is at the point $\varphi_m$ on the horizontal axis. This point corresponds to the right-hand boundary of the stagnation interval.

The second half-cycle, when the flywheel is moving to the right, is represented by half an ellipse lying above the $\varphi$-axis, where angular velocities are positive. The center of this second semi-ellipse is at the point $-\varphi_m$, on the $\varphi$-axis. The complete phase trajectory is formed by such increasingly smaller semi-ellipses, alternately centered at $\varphi_m$ and $-\varphi_m$. The diameters of these consecutive semi-ellipses lie along the $\varphi$-axis and decrease each half-cycle by $2\varphi_m$. The phase trajectory terminates on the $\varphi$-axis at the point at which the curve meets the $\varphi$-axis inside the dead zone (the portion of the $\varphi$-axis lying between $\varphi_m$ and $-\varphi_m$).

This phase trajectory is to be compared with that of the oscillator acted upon by viscous friction (see Figure 1.4 of Chapter 1). In the latter case, the curve spirals around a focal point located at the origin of the phase plane. The curve consists of an infinite number of turns, which gradually become smaller and approach the focus asymptotically. In the present case of dry friction, the loops of the phase curve are equidistant. The phase trajectory consists of a finite number of cycles and terminates at the point at which it meets the dead zone – the segment of the $\varphi$-axis between the points $-\varphi_m$ and $\varphi_m$.

If dry friction in the system is accompanied by a rather weak viscous friction ($\gamma < \omega_0$), the semi-ellipses become distorted and their axes shrink during the motion. The loops of the phase trajectory are no longer equidistant. Nevertheless their shrinking does not last indefinitely: the phase trajectory in this case also terminates after some finite number of turns around the origin when it reaches the stagnation interval on the $\varphi$-axis.

## 2.1.6 Energy Transformations

While the flywheel is rotating in one direction, the torque $N_{max}$ of kinetic friction, independent of the velocity, is constant, and the total energy of the oscillator decreases linearly with the angular displacement, $\varphi$, of the flywheel. This linear dependence of the total energy on $\varphi$ is clearly indicated in the left-hand panel of Figure 2.4, where the parabolic potential well of the elastic spring is shown. The representing point whose ordinate gives the total energy $E(\varphi)$ and whose abscissa gives the angular displacement of the flywheel travels in the course of time between the slopes of this well, gradually descending to the bottom of the well. The trajectory of this point consists of rectilinear segments lying between the sides of the well. These segments are straight because the negative work done by the force of dry friction is proportional to the angle of rotation, $\Delta\varphi$. The amount of this work $|N_{max}\Delta\varphi|$ equals the decrease $-\Delta E$ of total energy. However, the dependence of total energy on time, $E(t)$, is not linear because the rotation of the flywheel is nonuniform. The time rate of dissipation of the total energy, $-dE/dt$, is proportional to the magnitude of the angular velocity, $|\dot{\varphi}(t)|$. Thus, the greatest rate of dissipation of mechanical energy through friction occurs when the mag-

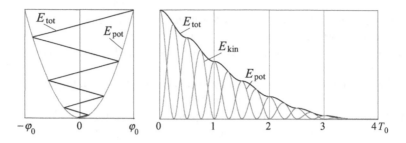

Figure 2.4: Energy transformations in oscillations damped by dry friction.

nitude of the angular velocity, $\dot{\varphi}$, is greatest, that is, when the flywheel crosses the boundaries of the dead zone. Near the points of extreme deflection, where the angular velocity is near zero, the time rate of dissipation of mechanical energy is smallest. Typical plots of energy transformation are shown in the right-hand panel of Figure 2.4.

Unlike the case of viscous friction, the oscillator with dry friction may retain some mechanical energy $E_f$ at the termination of the motion. Such occurs if the final angular displacement (within the dead zone) is not at the midpoint of the stagnation interval. Then the spring remains strained, and its elastic potential energy is not zero. The remaining energy does not exceed the value $D\varphi_{\mathrm{m}}^2/2 = N_{\mathrm{max}}\varphi_{\mathrm{m}}/2$.

When the initial excitation is large enough, that is, when the initial energy is much greater than $D\varphi_{\mathrm{m}}^2/2$, the oscillator executes a large number of cycles before the oscillations cease. In this case it is reasonable to consider the total energy averaged over the period of an oscillation, $\langle E(t)\rangle$. The decrease of $\langle E(t)\rangle$ during a large number of cycles depends quadratically on the lapse of time because the amplitude of oscillation decreases linearly with time and because the averaged total energy is proportional to the square of the amplitude.

If we let $t_f$ be the final moment when oscillations cease, then at the time $t$ the averaged total energy $\langle E(t)\rangle$ is proportional to $(t - t_f)^2$. This statement (which clearly applies only for $t < t_f$) is exactly true only when the flywheel comes to rest at the center of the stagnation interval. However, even if such is not the case and there is a residual potential energy stored in the spring after the motion ceases, the statement is approximately true.

## 2.1.7   The Role of Viscous Friction

In real systems, dry friction is always accompanied to some extent by viscous friction. The damping of oscillations in this case can also be investigated by the above-described method, namely by the stage-by-stage solving of Eqs. (2.6) and (2.7) and by using the final mechanical state (the angular displacement and velocity) of every half-cycle as the initial conditions for the next half-cycle. That is, the solutions are joined by equating their angular displacements and angular velocities (always zero) at the boundaries.

The clearest representation of the mechanical evolution of the system experiencing both dry and viscous friction is given by a phase diagram. Unlike the case of pure dry friction, the path in phase space is no longer a series of diminishing semi-ellipses (or semicircles) with alternating centers. Instead the phase trajectory consists of the shrinking alternating halves of spiral loops that are characteristic of a linear damped oscillator. The focal points of these spirals alternate between the boundaries of the stagnation interval.

To compare the relative importance of viscous versus dry friction, we consider below the decrease in amplitude caused by each of these effects during one complete cycle.

It was established above that under the action of dry friction this decrease equals the constant value of the doubled width of the stagnation interval $4\varphi_m$. On the other hand, viscous friction decreases the amplitude of the oscillation during a complete cycle by an amount that is not constant but rather is proportional to the amplitude. Indeed, for $\gamma T_0 \ll 1$, i.e., for rather large values of the quality factor $Q$, expression for the decrease $\Delta a$ during one period $T_0$ in the momentary amplitude $a$ due to viscous friction can be expanded in a series:

$$\Delta a = a(1 - e^{-\gamma T_0}) \approx a\gamma T_0 = a\gamma \frac{2\pi}{\omega_0} = \frac{\pi a}{Q}. \tag{2.8}$$

Equating $\Delta a$ to the doubled width $4\varphi_m$ of the stagnation interval, we find the amplitude $\tilde{a}$, which delimits the predominance of one type of friction over the other:

$$\tilde{a} = \frac{4\varphi_m}{\gamma T_0} = \frac{4}{\pi}\varphi_m Q \approx \varphi_m Q. \tag{2.9}$$

If the actual amplitude is greater than $\tilde{a}$, the effect of viscous friction dominates. Conversely, if the actual amplitude is less than $\tilde{a}$, the effect of dry friction dominates.

When the initial excitation of the oscillator is great enough, the amplitude may exceed the value $\tilde{a} \approx Q\varphi_m$. In this instance, the initial damping of the oscillations is influenced mainly by viscous friction. This case may be illustrated in the phase diagram. The decrement in the width of several initial loops of the phase trajectory (caused by viscous friction) is greater than the separation of the centers of adjoining half-loops (i.e., the decrement exceeds the width of the stagnation interval). It is clear that in this case the shrinking of the spiral caused by viscous friction is more influential in showing the effects of damping than is the alternation of the centers of half-loops caused by dry friction.

When the value of $a$ falls below that of $\tilde{a}$ (when $a < \tilde{a} = Q\varphi_m$), the effects of dry friction dominate. In the phase plane this dominance produces a trajectory of consecutive half-loops whose centers alternately jump between the ends of the stagnation interval, $-\varphi_m$ and $\varphi_m$, until the phase trajectory reaches the segment of the $\varphi$-axis in the stagnation interval.

When viscous friction is strong, that is, when values of the quality factor $Q$ are less than the critical value of $0.5$ (when $\gamma > \omega_0$), and when the initial displacement of the flywheel $\varphi(0)$ lies beyond the boundaries of the stagnation interval, $|\varphi(0)| > \varphi_{\mathrm{m}}$, the needle of the released flywheel moves without oscillating toward the point of the dial that corresponds to the nearest boundary of the stagnation interval. At this point the flywheel stops turning.

## 2.2   Review of the Principal Formulas

The differential equation of motion of an oscillator acted upon by dry friction:

$$J\ddot{\varphi} = -D(\varphi + \varphi_{\mathrm{m}}) \qquad \text{for} \qquad \dot{\varphi} > 0, \qquad (2.10)$$

$$J\ddot{\varphi} = -D(\varphi - \varphi_{\mathrm{m}}) \qquad \text{for} \qquad \dot{\varphi} < 0, \qquad (2.11)$$

where $\varphi_{\mathrm{m}}$ is the angle corresponding to the boundaries of the dead zone. If in addition, viscous friction is present, a term proportional to the angular velocity is also present:

$$\ddot{\varphi} = -\omega_0^2(\varphi + \varphi_{\mathrm{m}}) - 2\gamma\dot{\varphi} \qquad \text{for} \qquad \dot{\varphi} > 0, \qquad (2.12)$$

$$\ddot{\varphi} = -\omega_0^2(\varphi - \varphi_{\mathrm{m}}) - 2\gamma\dot{\varphi} \qquad \text{for} \qquad \dot{\varphi} < 0, \qquad (2.13)$$

where $\omega_0$ is the natural frequency of oscillations in the absence of friction:

$$\omega_0^2 = \frac{D}{J}. \qquad (2.14)$$

The damping factor $\gamma$ that characterizes the viscous friction is related to the quality factor $Q$ by the equation:

$$Q = \frac{\omega_0}{2\gamma}. \qquad (2.15)$$

The boundary value of the amplitude that delimits the two cases in which the effects either of viscous friction or of dry friction predominate:

$$a = \frac{4\varphi_{\mathrm{m}}}{\gamma T} = \frac{4}{\pi}\varphi_{\mathrm{m}}Q \approx \varphi_{\mathrm{m}}Q. \qquad (2.16)$$

## 2.3   Questions, Problems, Suggestions

The preceding analysis of the behavior of the oscillator under the influence of dry friction is based upon the method of the stage-by-stage analytic integration of the differential equations that describe the system. These equations are linear for the time intervals occurring between consecutive extreme deflections. These intervals are bounded by the instants at which the angular velocity is zero. The complete solution is obtained by joining together these half-cycle solutions for consecutive time intervals. On the other hand, the computer simulation of the torsion pendulum in this software package is based on the numerical integration of differential equations (the Runge – Kutta method to fourth order). To answer the questions below, you may apply the analysis described above. Then you can verify your analytic results by simulating the experiment on the computer.

## 2.3.1  Damping Caused by Dry Friction

The strength of dry friction in the system is characterized by the width of the dead zone. This interval is defined in the program when you input the value of the angle $\varphi_m$, which sets the limits of the dead zone on both sides of the middle position at which the spring is unstrained. Total width of this dead zone is $2\varphi_m$. The value of $\varphi_m$ must be expressed in degrees.

**2.3.1.1  Oscillations without Dry Friction.** Begin with the value $\varphi_m = 0$ corresponding to the absence of dry friction. Show that in this case the system displays the familiar behavior of a linear oscillator, i.e., simple harmonic oscillations with a constant amplitude in the absence of friction and with an exponentially decaying amplitude in the presence of viscous friction. The strength of viscous friction is characterized by the quality factor $Q$.

**2.3.1.2  Dry Friction after an Initial Displacement.** To display the role of dry friction clearly, choose a large value of the angle $\varphi_m$ that determines the limits of the dead zone (say, 15 to 20 degrees), and let viscous friction be zero. Such conditions are somewhat unrealistic. They are far unlike the situation characteristic of measuring instruments using a needle, such as moving-coil galvanometers. These instruments are constructed so that the dead zone is as small as possible, and critical viscous damping is deliberately introduced in order to avoid taking a reading from an oscillating needle. When an instrument is critically damped, its moving system just fails to oscillate, and it comes to rest in the shortest possible time. If the dead zone is narrow, the needle stops at a position very close to the dial point, which gives the true value of the measured quantity. Here, on the other hand, conditions are chosen to clarify the role of dry friction.

(a) What can you say about the succession of maximal deflections if damping is caused only by dry friction with the ideal z-characteristic? What is the law of their diminishing? How is the difference of consecutive maximal deflections related to the half-width of the dead zone?

(b) Let the angle $\varphi_m$ that defines the boundaries of the stagnation zone be, say, 15°, let the initial angle of deflection $\varphi_0$ be 160°, and let the initial angular velocity be zero. Calculate the point of the dial at which the needle eventually comes to rest. How many semi-ellipses form the phase trajectory of this motion, from its initial point to the point at which the motion stops? Verify your predictions by simulating the motion on the computer.

(c) In the graph of the time dependence of the deflection angle, where are the midpoints of the half-cycles of the sinusoidal oscillations located? Note how these individual segments of the sine curves are joined to form a continuous plot of damped oscillations.

(d) In the graph of the angular velocity versus time, note the abrupt bends in the curve at the instants at which the midpoints abruptly replace one another. What is the reason for these bends? Prove that these instants are separated by half the period of harmonic oscillations in the absence of dry friction. (Note that points on the time scale of the graphs correspond to integral multiples of the period.)

**2.3.1.3\* Dry Friction after an Initial Push.** Choose different initial conditions: Let the initial deflection be zero, and the initial angular velocity be, say, $2\omega_0$ (where $\omega_0$ is the natural frequency of oscillations). Use the same value $\varphi_m = 15°$ as above.

(a) Calculate the maximal deflection of the needle.

(b) To what position on the dial does the needle point when oscillations cease? How many turns are present in the complete phase trajectory of this motion? Verify your answer using a simulation experiment on the computer.

**2.3.1.4\* Damping by Dry Friction at Various Initial Conditions.** Assuming the same width of the dead zone as above, calculate the maximal angle of deflection and the final position on the dial to which the needle points when oscillations cease, for the more complicated initial conditions:

(a) The initial deflection angle $\varphi(0) = 135°$, and the initial angular velocity $\dot{\varphi}(0) = 1.5\omega_0$ ($\omega_0$ is the natural frequency of the oscillator).

(b) The initial deflection angle $\varphi(0) = -135°$, and the initial angular velocity $\dot{\varphi}(0) = 1.5\omega_0$.

Verify your calculated values by simulating an experiment on the computer.

**2.3.1.5\* Energy Dissipation at Dry Friction.**

(a) The graph of the total mechanical energy versus the angle of deflection consists of rectilinear segments joining the slopes of the parabolic potential well (when you work in the section "Energy transformations" of the computer program). Suggest an explanation.

(b) Letting the initial angular velocity $\dot{\varphi}(0) = 2\omega_0$, where $\omega_0$ is the natural frequency, and using energy considerations, calculate the entire angular path of the flywheel, excited from the midpoint of the dead zone by an initial push if the half-width of the dead zone $\varphi_m = 10°$.

**2.3.1.6 Oscillations in the Case of a Narrow Dead Zone.** Choose a small value for the angle $\varphi_m$ (less than 5°), and set the initial angular displacement to be many times the width of the dead zone, $2\varphi_m$.

(a) How many cycles does the flywheel execute before stopping?

(b) When the number of cycles is large, the plots clearly demonstrate the linear decay of the amplitude and the equidistant character of the loops in the phase diagram. What can you say about the time dependence of the total energy, averaged over a cycle?

## 2.3.2 Influence of Viscous Friction

**2.3.2.1\* Transition of the Main Role from Viscous to Dry Friction.** When damping is caused both by dry and viscous friction, it is interesting to observe the change in the character of damping when the main contribution passes from viscous to dry friction.

Let the angle $\varphi_m$ that determines the width of the dead zone be about 1° and let the quality factor $Q$, which characterizes the strength of viscous friction, be about

30. Let the initial angular deflection be $120°$ and let the initial angular velocity be zero.

(a) Does dry or does viscous friction determine the initial damping effects?

(b) At what value of the amplitude does the character of damping change? How does this change manifest itself on the plots of time dependence of the angle of deflection and of the angular velocity? On the phase trajectory?

**2.3.2.2\* Both Viscous and Dry Friction.** Let the boundaries of the stagnation interval be at $\varphi_m = 10°$ and the quality factor $Q = 5$. Let the initial velocity be $2\omega_0$ and let the initial deflection be zero.

(a) Calculate the maximal angular deflection of the needle at these initial conditions. Verify your answer experimentally.

(b) What kind of friction, dry or viscous, initially dominates the damping of oscillations?

(c)\*\* Let the boundaries of the stagnation zone be determined by the angle $\varphi_m = 10°$. Let the quality factor $Q$ be 3, the initial deflection $65°$, and the initial angular velocity $-2\omega_0$. Calculate the maximal angular deflection of the needle in the direction opposite the initial deflection. Verify your answer experimentally.

**2.3.2.3 Dry Friction and Critical Viscous Damping.**

(a) Choose the quality factor $Q$ to be near the critical value 0.5 and investigate the character of damping experimentally. Where within the limits of the dead zone is the needle most likely to stop if the quality $Q$ is slightly greater than the critical value? Give some physical explanation of your observations.

(b) Where would the needle stop if the quality factor $Q$ is less than 0.5 (that is, if the system is overdamped)? Does the answer depend on the initial conditions?

# Chapter 3

# Forced Oscillations in a Linear System

**Annotation.** Chapter 3 deals with a linear torsion pendulum driven by a sinusoidal external torque and its electromagnetic analogue – $LCR$-circuit with applied alternate voltage. The chapter includes a description of the simulated physical system and a summary of the relevant theoretical material for students as a prerequisite for the virtual lab "Forced Oscillations of Linear Torsion Pendulum" from the software package *Physics of Oscillations* (Part I, Simple Systems).

A preliminary study of natural oscillations (Chapter 1, "Free Oscillations of a Linear Torsion Pendulum") is strongly recommended. In Chapter 3 steady-state forced oscillations and transient processes at different driving frequencies are investigated. This chapter also includes a set of theoretical and experimental problems to be solved by students on their own, as well as various assignments that the instructor can offer students for possible individual mini-research projects.

## 3.1   Summary of the Theory

This chapter is concerned with forced oscillations of a torsion spring pendulum driven by a periodic external force with *sinusoidal* dependence on time. In the model of the physical system adopted here, a *kinematic mode* of excitation is used: One part of the system (a driving rod) is constrained to execute simple harmonic motion.

We consider both steady-state sinusoidal forced oscillations, which take place later than a sufficiently large time interval elapsed after the driving force started to operate, and transient processes, the latter depending on the initial conditions. The decomposition of a transient process onto the sum of sinusoidal oscillations of a constant amplitude, which have the frequency of the external force, and damped free oscillations with the natural frequency, are examined.

## 3.1.1    Basic Concepts

In the conventional classification of oscillations by their mode of excitation, oscillations are called *forced* if an oscillator is subjected to an external periodic influence whose effect on the system can be expressed by a separate term, a periodic function of the time, in the differential equation of motion. We are interested in the response of the system to the periodic external force.

The behavior of oscillatory systems under periodic external forces is one of the most important topics in the theory of oscillations. A noteworthy distinctive characteristic of forced oscillations is the phenomenon of *resonance,* in which a small periodic disturbing force can produce an extraordinarily large response in the oscillator. Resonance is found everywhere in physics and so a basic understanding of this fundamental problem has wide and various applications. The phenomenon of resonance depends upon the whole functional form of the driving force and occurs over an extended interval of time rather than at some particular instant.

In the case of unforced (free, or natural) oscillations of an isolated system, motion is initiated by an external influence acting *before* a particular instant. This influence determines the mechanical state of the system, that is, the displacement and the velocity of the oscillator, at the *initial instant.* These in turn determine the amplitude and phase of subsequent free oscillations. Frequency and damping of such oscillations are determined by the physical properties of the system. On the other hand, the characteristics of forced oscillations generated by a periodic external influence depend not only on the initial conditions and physical properties of the oscillator but also on the nature of the external disturbance, that is, on its amplitude and (primarily) on frequency.

## 3.1.2    Discussion of the Physical System

To study forced oscillations in a linear system excited by a sinusoidal external force, we consider here the same torsion spring pendulum described in Chapter 1 (which is devoted to free oscillations), namely, a balanced flywheel attached to one end of a spiral spring. The flywheel turns about its axis of rotation under the restoring torque of the spring, much like the devices used in mechanical watches. However, unlike the situation of free oscillations in which the other end of the spring is fixed, now this end is attached to an exciter, which is a rod that can be turned back and forth about an axis common with the axis of rotation of the flywheel. A schematic diagram of the driven torsion oscillator is shown in the left-hand panel of Figure 3.1.

The right-hand panel of Figure 3.1 shows an oscillatory *LCR*-circuit with alternate input voltage. This circuit can be regarded as an electromagnetic analog of the mechanical device. Both systems are described by identical differential equations and thus are dynamically isomorphic. However, the mechanical system has a definite didactic advantage for exploration of forced oscillations because it allows us to observe a direct visualization of motion.

Indeed, a mechanical system such as this one is ideal for the study of resonance

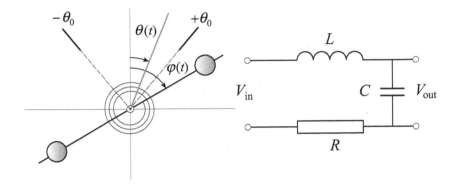

Figure 3.1: The torsion spring oscillator excited by a given sinusoidal motion of the driving rod attached to the spiral spring (left), and its electromagnetic analogue – $LCR$-circuit under sinusoidal input voltage (right).

because it is possible to see directly what is happening. When the driving rod (the exciter) is turned through a given angle $\theta$, the equilibrium position of the flywheel is displaced through the same angle, alongside the rod. The flywheel can execute free damped oscillations about this displaced position. For weak and for moderate friction the angular frequency of these oscillations is close to the natural frequency $\omega_0$ of the flywheel. This frequency depends on the torsion spring constant $D$ and the moment of inertia $J$ of the flywheel: $\omega_0 = \sqrt{D/J}$.

If the rod is forced to execute a periodic oscillatory motion, the flywheel is subjected to the action of a periodic external torque. This action is an example of the *kinematic excitation* of forced oscillations. This method of excitation is characterized by a given periodic motion of some part of the system. The kinematic mode of excitation is chosen here for the computer simulations of forced oscillations because the motion of the exciting rod can be displayed directly on the computer screen. Computer experiments with the system can show clearly, among other things, the phase shift between the exciter and the flywheel, and the ratio of their amplitudes.

Another possible mode of excitation of forced oscillations is characterized by a given periodic external force whose value does not depend on the position and velocity of the excited oscillator. This mode of excitation is called *dynamic*. Such excitation is difficult to display on the screen because it does not arise from the mechanical *motion* of the external source. Moreover, this mode of excitation of a mechanical system is not easy to realize experimentally. Nevertheless, in most textbooks forced oscillations are treated under the assumption that an oscillatory system is excited by a given periodic force.

The differential equations describing forced oscillations are the same for both modes of excitation. The physical differences appear primarily in the character of energy transformations. When the excitation is kinematic, the equilibrium position of the flywheel moves alongside the moving rod. The corresponding parabolic potential well then moves also as a whole, to and fro, alongside the rod. On the other hand, when the excitation is dynamic, the potential well is stationary. We discuss these differences below (see Section 3.2.4, p. 42).

### 3.1.3   The Differential Equation for Forced Oscillations

We assume that the exciting rod is constrained by some external source to execute simple harmonic motion about a middle position (the vertical in Figure 3.1). The amplitude of the rod is $\theta_0$ and the angular frequency is $\omega$. The angular displacement of the rod $\theta(t)$ varies with time $t$ sinusoidally:

$$\theta(t) = \theta_0 \sin \omega t. \tag{3.1}$$

If at some instant $t$ the flywheel is displaced through an angle $\varphi(t)$ from the central position (which is the origin of the scale in Figure 3.1), and the rod of the exciter is simultaneously displaced through an angle $\theta$, the spring exerts a torque $-D(\varphi - \theta) = -D\varphi + D\theta_0 \sin \omega t$ on the flywheel because the spring is strained through the angle $\varphi - \theta$. (Compare this torque with the torque $-D\varphi$ for the case of free oscillations.) Hence, in the absence of friction, the differential equation of rotation of the flywheel with the moment of inertia $J$ is:

$$J\ddot{\varphi} = -D\varphi + D\theta_0 \sin \omega t. \tag{3.2}$$

This equation is also the differential equation of forced oscillations excited by a given external sinusoidally varying torque $D\theta_0 \sin \omega t$ whose constant amplitude is $D\theta_0$. That is, the equation of motion is the same for forced oscillations excited by these two modes (an oscillating motion of the rod and a sinusoidal external torque).

Dividing both sides of this equation by the moment of inertia $J$ of the flywheel and introducing the notation $\omega_0 = \sqrt{D/J}$ for the natural frequency ($\omega_0^2 = D/J$), we rewrite the above equation in a canonical form:

$$\ddot{\varphi} + \omega_0^2 \varphi = \omega_0^2 \theta_0 \sin \omega t. \tag{3.3}$$

In the presence of viscous friction whose torque is proportional to the angular velocity $\dot{\varphi}$ of the flywheel, we must add the appropriate frictional term to the differential equation of motion describing forced oscillations:

$$\ddot{\varphi} + 2\gamma\dot{\varphi} + \omega_0^2 \varphi = \omega_0^2 \theta_0 \sin \omega t. \tag{3.4}$$

The damping constant $\gamma$ characterizes the strength of viscous friction. As in the case of free oscillations, it is related to the dimensionless quality factor $Q$ by the expression $2\gamma/\omega_0 = 1/Q$.

Forced oscillations of the electric charge $q$ stored in a capacitor of a resonant series $LCR$-circuit (see the right-hand panel of Figure 3.1) excited by a sinusoidal input voltage $V_{in}(t)$ obey the same differential equation as does the forced oscillation of a mechanical torsion spring oscillator excited by sinusoidal variations in position of the driving rod:

$$\ddot{q} + 2\gamma\dot{q} + \omega_0^2 q = \omega_0^2 C V_{in}(t). \tag{3.5}$$

In this equation $\omega_0$ is the natural frequency of oscillations of charge in the circuit in the absence of resistance. It depends on the capacitance $C$ of the capacitor and the inductance $L$ of the coil: $\omega_0 = 1/\sqrt{LC}$. The damping constant $\gamma = R/(2L)$ characterizes the dissipation of electromagnetic energy occurring in a resistor whose resistance is $R$.

Because of this similarity, the mechanical system described above enables us to give a very clear explanation for transformation of the input voltage $V_{in}(t) = V_0 \sin \omega t$ into the output voltage $V_{out}(t) = V_C(t) = q/C$ (voltage across the capacitor $C$). The output voltage $V_C(t)$ is analogous to the deflection angle $\varphi(t)$ of the rotor, while the alternating electric current $I(t) = \dot{q}(t)$ in the circuit is analogous to the rotor angular velocity $\dot{\varphi}(t)$ of the mechanical model. However, some caution is necessary in interpreting the analogy between the mechanical oscillator and the electric $LCR$-circuit with respect to the energy transformations.

The momentary position of the modeled physical system at any time instant is determined by two angular coordinates, namely, by the angles $\varphi(t)$ and $\theta(t)$. But the coordinate $\theta(t)$ is entirely determined by external conditions. (The physical meaning of $\theta(t)$ is the instantaneous position of the rod that executes a given motion.) The angle $\theta$ it is not a "free" coordinate, and so the system does not actually have a second degree of freedom. The only "free" coordinate (that is, the coordinate whose functional dependence on time is yet to be determined) is the angle $\varphi$, which gives the deflection of the flywheel from the central position. To find this unknown function $\varphi(t)$, we need only one differential equation, Eq. (3.4). The differential equation corresponding to the second coordinate $\theta$ can be used to find the external torque that must be exerted on the rod in order to provide its given sinusoidal motion. The external source of such torque inevitably experiences the reaction of the oscillator.

### 3.1.4 The Principle of Superposition

Our investigation of forced oscillations in a system whose differential equation of motion is *linear* is facilitated by the *principle of superposition*. This principle states that if several external forces act simultaneously on a linear system, the forced oscillations caused by each force acting separately are to be added together (superimposed) to get the complete solution. In other words, in linear systems there is no interaction (no mutual influence) of the individual oscillations excited by several external forces acting simultaneously.

It follows from the principle of superposition that, in addition to the forced oscillations caused by a given external force, a linear oscillator can simultaneously execute free damped oscillations. These free (or natural) oscillations may be thought of as arising from a null external force on the right-hand side of the differential equation of forced oscillations, Eq. (3.4). We can always infer that along with a given driving force a null force is also present. These natural oscillations are excited when an external force is switched on (or when its amplitude or initial phase is changed).

A superposition of such damped natural oscillations and driven forced oscillations of constant amplitude occurs during a *transient process,* when forced oscillations, over a period of time, acquire the frequency of the external force and a constant amplitude. The duration $\tau$ of this transient process of establishing forced *steady-state oscillations* equals (in the general case) the duration of damping of free oscillations: $\tau = 1/\gamma$.

## 3.2  Steady-State Forced Oscillations

During some time after the external force has been activated (after the rod has begun its given periodic motion), the transient natural oscillations inevitably damp out. Since only these oscillations depend on the initial conditions, we can say figuratively that the oscillator eventually "forgets" its initial state, and its forced oscillations become steady: The flywheel executes harmonic oscillations of a constant amplitude with the frequency of the external driving force. These steady-state oscillations are described by the *periodic* particular solution of the inhomogeneous differential equation of motion, Eq. (3.4):

$$\varphi(t) = a \sin(\omega t + \delta). \tag{3.6}$$

The steady-state oscillations are characterized by definite values of amplitude $a$ and phase lag $\delta$. The phase lag $\delta$ is the angular difference between the instants at which the flywheel and the driving rod cross the zero point of the dial (or reach their maximal displacements). Both $a$ and $\delta$ depend on $\omega$ (the frequency of the external action), $\omega_0$ (the natural frequency), and $\gamma$ (the damping constant of the oscillator).

The dependencies of $a$ and $\delta$ on the external driving frequency, $\omega$, are called the amplitude—frequency and the phase—frequency characteristics of the oscillator. When friction is relatively weak ($\gamma \leq \omega_0$ or $Q \geq 1$), the dependence of the amplitude on the frequency has a *resonance character*: the amplitude increases sharply as $\omega$ approaches $\omega_0$.

The graph of the dependence of the steady-state amplitude on the frequency $\omega$ is called the *resonance curve* (see Figure 3.2). The greater the quality factor $Q$, the sharper the peak of the resonance curve, that is, the more pronounced the resonance in the system.

### 3.2.1  Forced Oscillations in the Absence of Friction

When the driving frequency is sufficiently far from the resonant frequency, we may neglect the influence of friction on the amplitude $a$ and the phase lag $\delta$ of steady-state oscillations. That is, we may employ here the idealized frictionless model to describe the behavior of a real system in which there is some friction. (Note that the applicability of a physical model to a real system depends not only on the properties of the system, but also on the problem that we are solving.) Thus, in order to describe steady-state oscillations in the case when $|\omega - \omega_0| \gg \gamma$, we can use Eq. (3.3), which is valid for a forced linear oscillator in the absence of friction.

As a particular periodic solution describing steady-state oscillations, we try the expression:

$$\varphi(t) = a \sin \omega t. \tag{3.7}$$

Substituting this expression into Eq. (3.3), we find that Eq. (3.7) actually gives a solution of Eq. (3.3) if the amplitude $a(\omega)$ as a function of frequency $\omega$ is:

$$a(\omega) = \frac{\omega_0^2 \theta_0}{\omega_0^2 - \omega^2}. \tag{3.8}$$

If the driving frequency $\omega$ is set equal to zero, Eq. (3.8) yields $a = \theta_0$: The flywheel is at rest in the displaced equilibrium position. If $\omega \ll \omega_0$, $a \approx \theta_0$: In the case of a very slow motion of the driving rod, the flywheel follows the rod quasistatically. That is, the flywheel remains in the equilibrium position, which itself moves slowly alongside the slowly moving rod. At very low frequencies of the external action, kinematically excited steady-state forced oscillations of the flywheel occur with almost the same amplitude and the same phase as the compelled motion of the driving rod.

Equation (3.8) shows that as the driving frequency $\omega$ is increased, the amplitude of forced oscillations of the flywheel becomes greater. For $\omega \to \omega_0$ the value $a$ of the amplitude tends to infinity. Consequently, it is inadmissible to ignore friction in the vicinity of resonance (at $\omega \approx \omega_0$). This case is considered below.

We note that according to Eq. (3.8), the value of $a$ becomes negative if $\omega > \omega_0$. The negative sign of $a$ means here that for $\omega > \omega_0$, steady-state oscillations occur with a phase opposite that of the external force: When the rod turns in one direction, the flywheel turns in the other, both reaching their opposite extreme deflections simultaneously.

We can write the solution for $\omega > \omega_0$ in the form of Eq. (3.6), retaining the positive amplitude $a$ for all frequencies, if we assume $a$ to be equal to the absolute value of the right-hand side of Eq. (3.8), and the phase shift $\delta$ to be equal to $-\pi$.

When $\omega < \omega_0$, the phase shift $\delta$ in the absence of friction (and also for relatively weak friction, as we shall see later) is zero, so that the flywheel and the rod oscillate in phase. That is, moving in the same direction, they pass the mid-point at the same time, and reach the extremes in their deflections simultaneously. However, as indicated by Eq. (3.8), the extreme deflection of the flywheel is greater than that of the rod and increases infinitely when the driving frequency approaches the natural frequency (when $\omega \to \omega_0$).

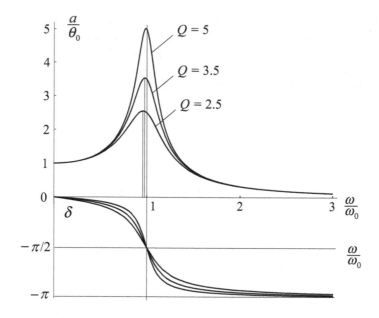

Figure 3.2: Resonance curves of a linear oscillator.

## 3.2.2   The Resonance Curve

In the vicinity of resonance (at driving frequencies $\omega$ that satisfy the condition $|\omega - \omega_0| \leq \gamma$), it is necessary to take friction into account in the differential equation of forced oscillations. That is, we need to solve Eq. (3.4). Steady-state forced oscillations are described by its particular periodic solution. We can write this solution in the form $\varphi(t) = a\sin(\omega t + \delta)$ (see Eq. (3.6)). Substituting that solution into Eq. (3.4), we can search for the values of $a$ and $\delta$ for which the function $a\sin(\omega t + \delta)$ satisfies Eq. (3.4). Leaving the derivations as an exercise, we give here the final expressions for the amplitude $a$ and the phase shift $\delta$:

$$a(\omega) = \frac{\omega_0^2 \theta_0}{\sqrt{(\omega_0^2 - \omega^2)^2 + 4\gamma^2\omega^2}}, \quad \tan\delta = -\frac{2\gamma\omega}{\omega_0^2 - \omega^2}. \tag{3.9}$$

The graphs of the dependence of the amplitude on the frequency $a(\omega)$ (the resonance curves) and the phase lag on the frequency $\delta(\omega)$ for different values of the quality factor $Q$ are shown in upper and lower parts of Figure 3.2, respectively. The frequency $\omega$ of excitation is measured in this figure in units of the natural frequency $\omega_0$, and the amplitude $a$ of the flywheel – in units of the exciter amplitude $\theta_0$. The maximum value of the amplitude of steady oscillations $a_{\max}$ occurs at the *resonant frequency* $\omega_{\text{res}}$:

$$\omega_{\text{res}} = \sqrt{\omega_0^2 - 2\gamma^2}. \tag{3.10}$$

This expression for $\omega_{\text{res}}$ is valid if friction is not too large, that is, if $\omega_0 > \sqrt{2}\,\gamma$. When friction is sufficiently small, that is, if $\gamma \ll \omega_0$ or $Q \gg 1$, we can use instead of Eq. (3.10) the following approximate expression:

$$\omega_{\text{res}} = \omega_0 \sqrt{1 - \frac{2\gamma^2}{\omega_0^2}} \approx \omega_0 \left(1 - \frac{\gamma^2}{\omega_0^2}\right) = \omega_0 \left(1 - \frac{1}{4Q^2}\right). \qquad (3.11)$$

That is, the resonant frequency nearly coincides with the natural frequency $\omega_0$: The value $\omega_{\text{res}}$ differs from $\omega_0$ only by a term of the second order in the small parameter $\gamma/\omega_0$. For example, if $Q = 10$ (moderate friction), the resonant frequency differs from the natural frequency only by 0.25%.

The amplitude of steady-state oscillations at resonance $a_{\max}$ is determined by the following expression:

$$a_{\max} = \frac{\omega_0^2 \theta_0}{2\gamma \sqrt{\omega_0^2 - \gamma^2}} \approx \frac{\omega_0 \theta_0}{2\gamma} = Q\theta_0. \qquad (3.12)$$

We see from Eq. (3.12) that the amplitude $a_{\max}$ of steady-state oscillations at resonance is approximately $Q$ times greater than the amplitude $\theta_0$ of the driving rod (provided that the quality factor $Q$ is not too low). In other words, the amplitude $a_{\max}$ of steady-state oscillations at resonance is $Q$ times greater than the amplitude $a(0)$ of steady-state oscillations at a very low driving frequency $\omega$ (at slow oscillations of the rod). We note that the resonant properties of a linear oscillator under forced oscillations and the damping of its natural free oscillations are characterized by the same quantity, the quality factor $Q$.

When there is no friction, Eq. (3.9) shows that the amplitude of the flywheel during steady-state forced oscillations is greater than the amplitude $\theta_0$ of the rod at all frequencies $\omega$ between zero and the boundary value $\sqrt{2}\,\omega_0$. When the frequency $\omega$ of the driving force exceeds the natural frequency $\omega_0$ by more than $\sqrt{2}$ times, the amplitude of steady-state forced oscillations is smaller than $\theta_0$ and approaches zero as the frequency $\omega$ increases further. In this range of frequencies the dynamic effect of an external driving force is less than the static effect of a constant force of the same magnitude. The physical cause of such behavior is the inertia of the flywheel: When the driving frequency of the rod is considerably greater than the natural frequency of the flywheel, the massive flywheel cannot follow the rapid motion of the rod. The same is true also in the presence of moderate friction, except that the bounding frequency is slightly smaller than $\sqrt{2}\,\omega_0$, as can be seen from the graphs of $a(\omega)$ plotted in Figure 3.2 using Eq. (3.9).

Equation (3.9) for the phase shift $\delta$ and the corresponding graphs in the lower panel of Figure 3.2 show that steady-state forced motion always lags behind the driving force since $\delta$ is always negative. Far from resonance at $\omega < \omega_0$ this lag is nearly zero, and the flywheel oscillates nearly in phase with the exciting rod. When $\omega = \omega_0$, steady-state oscillations of the flywheel lag in phase behind oscillations of the exciting rod by a quarter of the period ($\delta = -\pi/2$) for all values of friction. In this case the displacement of the flywheel is greatest when the displacement of the rod is zero, and vice versa. When $\omega$ is much greater than $\omega_0$, the phase shift $\delta$

approaches $-\pi$. That is, the lag is nearly $180°$, which means that in this case the flywheel and the exciting rod always rotate in opposite directions.

If there is no friction, Eq. (3.9) indicates that the phase lag is either 0 (for $\omega < \omega_0$) or $180°$ (for $\omega > \omega_0$). That is, when $\omega = \omega_0$, there is an abrupt change from motion of the flywheel exactly in phase with that of the rod to motion in which they oscillate exactly in opposite phase. (In the absence of friction, the amplitude of the flywheel at the transition is infinite.) In the presence of friction, the transition from in-phase steady-state oscillations of the flywheel and the rod to opposite-phase steady-state oscillations takes place gradually over a range of frequencies centered about $\omega_0$. The width of this range, as can be seen from Figure 3.2, is proportional to the damping constant $\gamma$.

### 3.2.3   Resonance of the Angular Velocity

In steady-state oscillations under a sinusoidal force, the angular velocity of the flywheel $\dot{\varphi} = a\omega \cos(\omega t + \delta)$ changes with time harmonically with the frequency $\omega$ of the external driving force. The expression for the amplitude of the angular velocity $\Omega = \dot{\varphi}$ differs from Eq. (3.9) for the amplitude $a(\omega)$ of these oscillations by an additional factor $\omega$:

$$\Omega(\omega) = \omega a(\omega) = \frac{\omega_0^2 \theta_0}{\sqrt{(\omega_0^2/\omega - \omega)^2 + 4\gamma^2}}. \tag{3.13}$$

Dependence of the velocity amplitude on the frequency is shown in the upper panel of Figure 3.3.

As we can see from Eq. (3.13), the maximum of the resonance curve for the angular velocity is located at $\omega = \omega_0$ independently of the damping factor $\gamma$. Therefore resonance of the angular velocity occurs at the value of the driving frequency $\omega$, which exactly equals the natural frequency $\omega_0$ of the oscillator for both weak and strong friction. On the other hand, resonance of the angular displacement occurs at $\omega_{res} = \sqrt{\omega_0^2 - 2\gamma^2}$.

The lower panel of Figure 3.3 shows the dependence of the phase shift between the driving rod and the angular velocity. At resonance (at $\omega = \omega_0$) this phase shift equals zero: The driving rod oscillates in phase with the velocity. This means that at resonance the energy is transmitted in one direction – from the exciter to the oscillator – during the whole period.

### 3.2.4   Energy Transformations

Though the amplitude is constant in steady-state forced oscillations, the total energy of the oscillator is constant only on the average. During one quarter of a cycle, energy is transmitted from the driving rod to the oscillator, and during the next quarter cycle, energy is transferred back from the oscillator to the external source driving the rod. In contrast to free oscillations, not only do the kinetic and potential energies oscillate, but so does their sum, the total mechanical energy. The total energy oscillates with double the frequency of the external force.

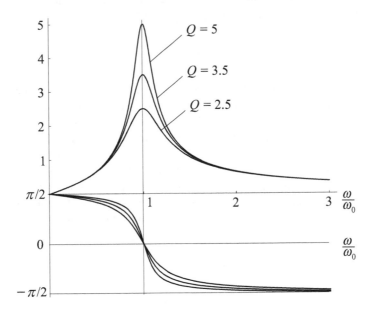

Figure 3.3: Resonance curves for the velocity of a linear oscillator.

In our discussion of steady-state energy transformations, we pay special attention to the distinguishing characteristics of the energy transfer for kinematic excitation of forced oscillations modeled in these computer simulations, and for dynamic excitation, for which the external torque is a given time-dependent quantity. We have already mentioned these distinctions above. Here we discuss them in detail.

When the dynamic mode of excitation is used, an external torque is applied directly to the flywheel. With one end of the spiral spring fixed, the deformation of the spring (the amount of twisting from its unstrained state) is determined by the angular displacement $\varphi$ of the flywheel from the midpoint (from the equilibrium position). Thus the potential energy of the spring is given by

$$E_{\text{pot}} = \frac{1}{2}D\varphi^2(t) = \frac{1}{2}Da^2 \sin^2(\omega t + \delta) = \frac{1}{4}J\omega_0^2 a^2[1 - \cos 2(\omega t + \delta)]. \quad (3.14)$$

The kinetic energy of the oscillating flywheel is independent of the mode of excitation and is given by the following expression

$$E_{\text{kin}} = \frac{1}{2}J\dot{\varphi}^2(t) = \frac{1}{2}J\omega^2 a^2 \cos^2(\omega t + \delta) = \frac{1}{4}J\omega^2 a^2[1 + \cos 2(\omega t + \delta)]. \quad (3.15)$$

It is seen that, while the oscillator executes steady-state forced oscillations with the frequency $\omega$, the values of its potential and kinetic energies are oscillating harmonically with the frequency $2\omega$ in opposite phases with respect to one another. The ratio of their maximal (and average) values is equal to the squared ratio of the natural frequency to the driving frequency:

$$\frac{\langle E_{\text{pot}} \rangle}{\langle E_{\text{kin}} \rangle} = \frac{\omega_0^2}{\omega^2}. \tag{3.16}$$

Hence, when $\omega < \omega_0$, the potential energy predominates on the average over the kinetic energy. In particular, when $\omega \ll \omega_0$, the spring is twisted quasistatically, and nearly all the energy of the oscillator is the elastic potential energy of the strained spring. On the other hand, for external frequencies that exceed the natural one ($\omega > \omega_0$), the kinetic energy predominates over the potential energy.

The peculiarities of energy transformations for the kinematic excitation of oscillations are related to the fact that the equilibrium position of the flywheel (and its potential well as a whole) is displaced when the driving rod is turned. The deformation of the spring in this case is determined by the difference in the angles $\varphi(t)$ and $\theta(t)$. The expression for its potential energy then takes the form:

$$E_{\text{pot}} = \frac{1}{2} D (\varphi - \theta)^2. \tag{3.17}$$

When the frequency of the rod is much less than the resonant frequency, the flywheel moves as though it were attached to the slowly moving rod. The flywheel remains close to its equilibrium position, which is displaced by the driving rod. The spring remains nearly unstrained and its potential energy is nearly zero. In other words, the oscillator is always located near the bottom of its potential well, which slowly oscillates alongside the rod. Therefore, at low frequencies of the driving mechanism, the kinetic energy predominates over the potential energy, in direct contrast to the case of dynamic excitation.

When the driving frequency is large compared to the natural frequency, the inertia of the flywheel diminishes the response the flywheel is able to make to the displacement of the equilibrium position by the external source, and so the flywheel oscillates with a relatively small amplitude. At high driving frequencies, the amplitude of the steady-state deflections of the flywheel from its central position is much smaller than the amplitude $\theta_0$ of the angular displacement of the driving rod. The spring is twisted back and forth through approximately the angle $\theta_0$, while the angular velocity of the flywheel remains relatively small. Hence the elastic potential energy arising from the deformation of the spring predominates over the kinetic energy of the flywheel, again in direct contrast to the case of dynamic excitation.

When friction is small ($\gamma \ll \omega_0$), the ratio of the average value of the potential energy to the average value of the kinetic energy for kinematic excitation can be found from Eq. (3.17) and Eq. (3.6) for $\varphi(t)$ and Eq. (3.1) for $\theta(t)$:

$$\frac{\langle E_{\text{pot}} \rangle}{\langle E_{\text{kin}} \rangle} = \frac{\omega^2}{\omega_0^2}. \tag{3.18}$$

Comparing Eq. (3.18) with Eq. (3.16), we see that the ratios of the average value of the potential energy to the average value of the kinetic energy for dynamic and kinematic modes of excitation of forced oscillations are inverses of one another, in agreement with the general discussion above.

## 3.3 Transient Processes

The amplitude and the phase of steady-state oscillations do not depend on the initial conditions. Loosely speaking, during the transient process the oscillator eventually "forgets" them. We should keep in mind that steady-state oscillations are described by Eq. (3.6), which is the *periodic particular solution* to the inhomogeneous differential equation, Eq. (3.4). The graphs of the amplitude and phase versus the driving frequency (the amplitude—frequency and the phase—frequency characteristics of a linear oscillator), displayed in Figure 3.2, refer to this particular solution and are valid only for steady-state oscillations.

The initial conditions, namely the initial angle of deflection $\varphi(0)$ and the initial angular velocity $\dot{\varphi}(0)$, are influential only during the *transient process*. During the transition, natural damped oscillations are superimposed on the steady-state forced oscillations. The effects of the natural oscillations disappear once the steady-state oscillations have been established.

Mathematically a transient process is represented by the *general solution* to the inhomogeneous equation, Eq. (3.4). This complete solution is given by:

$$\varphi(t) = a \sin(\omega t + \delta) + C e^{-\gamma t} \cos(\omega_1 t + \alpha). \tag{3.19}$$

The first term on the right is the periodic particular solution, Eq. (3.6), to the inhomogeneous equation, Eq. (3.4). The second term on the right, called the *transient term*, is the general solution to the corresponding homogeneous equation, namely Eq. (3.4) in which the right-hand side is zero. This solution of the homogeneous equation is the contribution of damped natural oscillations to the transient process. The frequency $\omega_1$ of this term nearly equals the natural frequency $\omega_0$ provided the friction is not too large:

$$\omega_1 = \sqrt{\omega_0^2 - \gamma^2} = \omega_0 \sqrt{1 - \frac{\gamma^2}{\omega_0^2}} \approx \omega_0 \left(1 - \frac{\gamma^2}{2\omega_0^2}\right) = \omega_0 \left(1 - \frac{1}{8Q^2}\right). \tag{3.20}$$

The fractional difference $(\omega_0 - \omega_1)/\omega_0$ in most cases of practical importance is so small that we can neglect it and assume that $\omega_1 = \omega_0$. Indeed, if $Q = 5$, the fractional difference is only 0.5%: $(\omega_0 - \omega_1)/\omega_0 = 0.005$.

The transient term of the general solution contains two arbitrary constants, $C$ and $\alpha$. Their values depend on the initial conditions, namely on the angular displacement and the angular velocity of the flywheel at the instant the external force begins to act.

Thus the transient process is described by a superposition of two oscillations: A sinusoidal steady oscillation with a constant amplitude $a$ and a frequency $\omega$ of

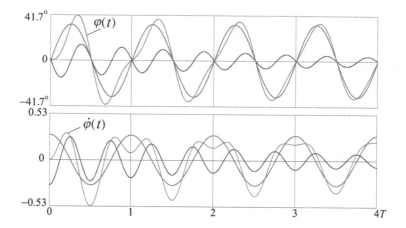

Figure 3.4: Decomposition of a transient process onto the sum of steady-state forced oscillations and damped natural oscillations (time-dependent graphs of the angular displacement and angular velocity) at $\omega = 0.5\omega_0$, $\varphi(0) = 0$, $\dot{\varphi}(0) = 0$.

the driving force, and a damped natural oscillation with the frequency $\omega_1 \approx \omega_0$ and a decaying amplitude. In principle, the transient process continues indefinitely, but in practice it is considered completed after the transient term essentially dies out, that is, after about $Q$ cycles of natural oscillations. (We recall that $Q$ is the dimensionless quality factor, given by $Q = \omega_0/2\gamma$.) Generally, the smaller the friction, the longer the transient process lasts. However, as we shall see in the next section, it is possible to choose initial conditions such that there is no transient term and no transient process.

There is an option in the suggested simulation program that allows the user to display the plots of the two simple oscillations, which constitute the transient process, while simultaneously plotting their superposition. An example of the decomposition of a transient process onto the two separate simple components—the steady-state sinusoidal oscillation with a constant amplitude and the frequency of the external driving action (periodic particular solution of the differential equation), and the decaying natural transient oscillation (solution of the corresponding homogeneous equation)—is shown in Figure 3.4. The graphs correspond to $\omega = 0.5\omega_0$ (the driving frequency equals one half the natural one), and zero initial conditions ($\varphi(0) = 0$, $\dot{\varphi}(0) = 0$).

## 3.3.1    Initial Conditions That Eliminate a Transient

As noted above, it is possible to choose the initial conditions in such a way that there is no transient process, that is, so that steady-state oscillations appear immediately upon activation of the external force. To find these conditions, we observe that if we take as the angular displacement and the angular velocity at $t = 0$

the corresponding values that characterize the steady-state oscillations, the initial conditions are satisfied by the steady-state itself, without the addition of natural oscillations.

It follows from Eq. (3.6) that the required initial angular deflection $\varphi(0)$ is $a \sin \delta$, and the required initial angular velocity $\dot{\varphi}(0)$ is $a\omega \cos \delta$, where $a$ and $\delta$ are the amplitude and the phase of steady-state oscillations given by Eq. (3.9). Since the steady-state term in Eq. (3.19) satisfies the initial conditions by itself, the transient term vanishes, leaving only the steady-state, so that $C$ in Eq. (3.19) must be zero. In other words, if $\varphi(0) = a \sin \delta$ and $\dot{\varphi}(0) = a\omega \cos \delta$, no transient natural oscillation arises when the external force begins to act, and there is no contribution to the motion from the homogeneous equation.

Transient processes make the phenomenon of forced oscillations much more complicated than are simple harmonic steady-state oscillations. In many cases these transient processes are important and interesting in themselves. They are worth considerable attention in your work with the simulation computer program.

### 3.3.2  Forced Oscillations from Rest at Resonance

We next restrict our study of transient processes to the case in which the initial conditions are zero. That is, we consider forced oscillations that correspond to the flywheel at rest in the equilibrium position at the instant the driving force begins to operate:

$$\varphi(0) = 0, \qquad \dot{\varphi}(0) = 0. \tag{3.21}$$

At $t = 0$, the driving rod starts moving according to:

$$\theta(t) = \theta_0 \sin \omega t. \tag{3.22}$$

Let us first consider the case of an oscillator damped by relatively weak friction ($\gamma \ll \omega_0$). Because there is little friction, the resonant frequency is very nearly the natural frequency $\omega_0$. For the case in which the driving frequency $\omega$ is set equal to $\omega_0$, it follows from Eqs. (3.9) that the periodic particular solution that describes the steady-state oscillations is given by:

$$\varphi(t) \approx \frac{\omega_0}{2\gamma} \theta_0 \sin \left( \omega_0 t - \frac{\pi}{2} \right) = -Q\theta_0 \cos \omega_0 t. \tag{3.23}$$

In this case the amplitude of oscillation of the flywheel is greater than the amplitude of oscillation of the rod by the factor $Q$, and the phase lag is $-\pi/2$. That is, the oscillations of the flywheel are one quarter of a cycle behind the oscillations of the driving rod.

The transient term in Eq. (3.19) with $\omega_1 = \omega_0$ is next added to Eq. (3.23) to obtain a complete solution:

$$\varphi(t) = -Q\theta_0 \cos \omega_0 t + C e^{-\gamma t} \cos(\omega_0 t + \alpha). \tag{3.24}$$

The arbitrary constants $C$ and $\alpha$ are determined by satisfying the initial conditions, Eqs. (3.21). In the case of weak friction, when $\gamma \ll \omega_0$, the exponential factor $e^{-\gamma t}$ in Eq. (3.24) changes little over many oscillations and when taking the time derivative of Eq. (3.24), can be considered approximately constant for long durations:

$$\dot{\varphi}(t) \approx Q\theta_0\omega_0 \sin\omega_0 t - Ce^{-\gamma t}\omega_0 \sin(\omega_0 t + \alpha). \qquad (3.25)$$

Requiring that $\dot{\varphi}(0)$ be zero implies that $\alpha = 0$, and requiring that $\varphi(0)$ be zero implies that $C = Q\theta_0$. Hence, if $Q \gg 1$ and $\omega = \omega_0$, the solution of differential equation of motion (3.6), satisfying the zero initial conditions defined by Eq. (3.21), is:

$$\varphi(t) = -Q\theta_0(1 - e^{-\gamma t})\cos\omega_0 t = -b(t)\cos\omega_0 t, \qquad (3.26)$$

where

$$b(t) = Q\theta_0(1 - e^{-\gamma t}). \qquad (3.27)$$

This superposition of forced and slowly decaying natural oscillations, each with the frequency $\omega_0$, can be considered as a single nearly harmonic oscillation with the frequency $\omega_0$ and an amplitude $b(t)$, which slowly increases with time, asymptotically approaching the steady-state value, $Q\theta_0$. The graph of $\varphi(t)$ for such a transient process at resonance in the case of zero initial conditions is shown in the upper panel of Figure 3.5 (the angular displacement of the flywheel $\varphi(t)$ together with the angular displacement $\theta(t)$ of the exciter). The lower panel of this figure shows decomposition of this process onto the sum of steady-state forced oscillations of constant amplitude and damped natural oscillations.

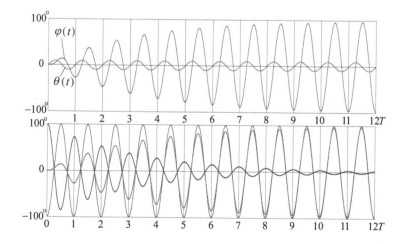

Figure 3.5: The transient process of resonant excitation from the state of rest in the equilibrium position (upper panel), and decomposition of this process onto the sum of steady-state forced oscillations and damped natural oscillations ($\omega = \omega_0$, $\theta_0 = 10°$, $Q = 10$, initial conditions $\varphi(0) = 0$, $\dot{\varphi}(0) = 0$).

By convention, the duration of this transient process is assumed to be equal to the time of damping $\tau$ ($\tau = 1/\gamma$), which is the time during which natural free oscillations of the transient term fade away.

This gradual growth of the amplitude $b(t)$ and its gradual asymptotic approach to a maximum value during the transient process at the resonant frequency can be easily explained on the basis of energy transformations. At resonance a definite phase relation establishes between the oscillation of the flywheel and that of the rod. Namely, the rod oscillates in phase with the angular velocity of the flywheel. This phase relation provides the conditions that are favorable for the transfer of energy from the rod to the oscillator. For large values of the quality factor $Q$, the amplitude of the flywheel eventually increases to the value $Q\theta_0$, which considerably exceeds (by a factor of $Q$) the amplitude $\theta_0$ of the rod. The greater the quality factor $Q$, the greater the energy eventually stored by the oscillator and the greater the number of cycles required to transfer this energy to the oscillator by a weak driving force. The growth of the amplitude decreases to zero when the velocity-dependent friction dissipates any further addition of energy from the driving rod. During steady-state oscillations, the energy dissipated over a cycle equals the energy transmitted to the oscillator from the external source that drives the exciting rod.

In the case of weak friction the duration of the transient process is large compared with the period of oscillations: $\tau \gg T_0$. The amplitude of oscillation of the flywheel, initially at rest, increases over many cycles. The early growth of the amplitude occurs almost linearly with time. This behavior is evident from Eq. (3.27), in which we let $\gamma t \ll 1$. Expanding the exponential in a power series and keeping only the linear term, we have:

$$b(t) = Q\theta_0(1 - e^{-\gamma t}) = \frac{\omega_0}{2\gamma}\theta_0(1 - e^{-\gamma t}) \approx \frac{1}{2}\theta_0\omega_0 t. \tag{3.28}$$

Cancelation of the damping constant $\gamma$ in the latter expression means that the linear growth of the amplitude during the early stage of the transient process (while $\gamma t \ll 1$, or $t \ll QT_0$) occurs just as though friction were absent. In the idealized case of the complete absence of friction, such linear growth of the amplitude would continue indefinitely.

Thus, steady-state oscillations are impossible for a frictionless oscillator at the resonant frequency. In this case, the particular solution of the inhomogeneous differential equation of motion (Eq. (3.4), with $\gamma = 0$ and $\omega = \omega_0$), rather than being periodic, increases without limit: $\varphi(t) = \frac{1}{2}\theta_0\omega_0 t \cos\omega_0 t$. To obtain the general solution containing two arbitrary constants, we must also include the general solution of the corresponding homogeneous equation, which in this case describes natural oscillations of constant amplitude. Determining the constants from the zero initial conditions, we find:

$$\varphi(t) = \frac{1}{2}\theta_0(\omega_0 t \cos\omega_0 t - \sin\omega_0 t). \tag{3.29}$$

The amplitude of this oscillation with the frequency $\omega_0$ is $\frac{1}{2}\theta_0\sqrt{(\omega_0 t)^2 + 1} \approx \frac{1}{2}\theta_0\omega_0 t$. Thus, during the resonant transient process in the absence of friction the amplitude of the flywheel, initially at rest, increases steadily at almost a constant rate. This rate is the same as the early rate in the presence of weak friction, because during the early stage of the transient process the growth of mechanical energy is almost unimpeded by the velocity-dependant friction.

The unlimited growth of the amplitude at resonance predicted by Eq. (3.29) means, as noted above, that in this frictionless system steady-state oscillations are not possible when $\omega = \omega_0$. In fact this growth means that in conditions of resonance the idealized frictionless model does not work; that is, we cannot use the model for description of resonance in a real system no matter how small the friction.

In a real system, at sufficiently large amplitudes, either friction dissipates the added energy (so that in the model of the system it is necessary to take friction into account) or the amplitude of the oscillation increases beyond the linear limit of the restoring force of the spring so that Hooke's law fails. In the latter case, the nonlinear dependence of the restoring force on the angle of deflection changes the period of natural oscillations as the amplitude grows, and resonance is destroyed. For a nonlinear system, the growth of the amplitude is restricted even in the absence of friction: The resonant conditions become violated at large amplitudes. To say which of these reasons (friction or nonlinearity) restricts the growth of the amplitude in a real physical system, we need to know more about the properties of the system.

### 3.3.3   Mechanical Analogue of the Stimulated Emission of Radiation

As we have seen above, in conditions of exact tuning to resonance, that is, when the external frequency is equal to the natural frequency of the oscillator, the amplitude of forced oscillations grows monotonically. In the absence of friction, as follows from Eq. (3.29), the amplitude grows indefinitely almost linearly with the time $t$: $b(t) \approx \frac{1}{2}\theta_0\omega_0 t$. The growth of energy is provided by the external source that drives the rod of the exciter. Now let us think about the following question: Is the backward transfer of energy from the oscillator to the source of external excitation possible in conditions of exact tuning to resonance?

The process of resonant growth of the amplitude from the state of rest in the equilibrium position occurs at definite relations between the phases of the exciter and the flywheel. These relations are favorable for the transfer of energy from the external source to the oscillator: As follows from Eq. (3.29), the angular velocity of the flywheel depends on time according to $\dot{\varphi}(t) \approx \frac{1}{2}\theta_0\omega_0^2 t \sin\omega_0 t$ almost from the very beginning, that is, the velocity changes in the same phase with the external torque $D\theta_0\sin\omega_0 t$ in the right-hand side of Eq. (3.4). This means that the torque is always directed in the same way as the angular velocity, pushing the flywheel in the direction of its rotation. We note that such phase relationships are characteristic for zero initial conditions.

But what will happen if at the moment at which the external force is switched on, the oscillator is already excited, i. e., it already has energy and is executing natural oscillations? It is clear that the subsequent process depends on the phase relations between the existing natural oscillations of the flywheel and oscillations of the exciter. If the angular velocity varies in the same phase with the external torque, the energy again will be transferred from the external source to the oscillator, and the amplitude of existing oscillations will be increasing immediately after the external source is switched on. On the other hand, if the angular velocity of the flywheel varies in the opposite phase with respect to the external torque, that is, according to $\dot{\varphi}(t) \sim -\sin \omega_0 t$, the external torque will be directed all the time against the velocity of the flywheel, and will slow down its rotation. This means that the energy will be transferred from the oscillator to the external source. Certainly, this process will continue only until the amplitude becomes zero, that is, until all the energy of the oscillator is transferred to the external source. After this moment all will happen in the way that corresponds to zero initial conditions: The amplitude will grow because the phase relations become favorable for the transfer of energy from the external source to the oscillator.

We can easily see that if the exciting rod moves according to $\theta(t) = \theta_0 \sin \omega_0 t$, the phase of the flywheel's natural oscillation that is favorable for the transfer of energy from the oscillator to the external source will be provided, for example, if at the initial moment the flywheel is at rest and is displaced from the equilibrium position in the positive direction, that is, at the initial conditions of the type $\varphi(0) = \varphi_0$ (where $\varphi_0 > 0$), $\dot{\varphi}(0) = 0$. The solution to differential equation of motion, Eq. (3.4), which corresponds to such initial conditions, has the form:

$$\varphi(t) = (\varphi_0 - \frac{1}{2}\theta_0\omega_0 t) \cos \omega_0 t + \frac{1}{2}\theta_0 \sin \omega_0 t$$

$$\approx (\varphi_0 - \frac{1}{2}\theta_0\omega_0 t) \cos \omega_0 t. \tag{3.30}$$

The approximate expression in (3.30) is valid when it is possible to ignore the contribution of natural oscillations into $\varphi(t)$, i. e., when the constant amplitude $\frac{1}{2}\theta_0$ of the $\sin \omega_0 t$ term is small compared to the time-dependent amplitude $|\varphi_0 - \frac{1}{2}\theta_0\omega_0 t|$ of the $\cos \omega_0 t$ term in Eq. (3.30). It follows from Eq. (3.30) that the amplitude first diminishes linearly with time to almost the zero value during the interval $t = 2\varphi_0/(\omega_0\theta_0) = (\varphi_0/\pi\theta_0)T_0$. Then the time-dependent term of the amplitude in Eq. (3.30) changes its sign. This means that the phase relation between the flywheel and the exciting rod becomes the same as in the above considered case of resonant swinging from rest; that is, it becomes favorable for the transfer of energy to the oscillator from the external source. As a result, the amplitude grows indefinitely.

The time-dependent graph of the angle $\varphi(t)$ and angular velocity $\dot{\varphi}(t)$ for this process is shown in Figure 3.6. We note that during the initial stage of the process the external torque that is proportional to $\theta(t)$ varies in the opposite phase with respect to $\dot{\varphi}(t)$ and thus slows down the flywheel.

During the second stage the torque $\theta(t)$ varies in phase with the angular velocity $\dot{\varphi}(t)$ and thus accelerates the flywheel – the energy is transferred from the external source to the flywheel.

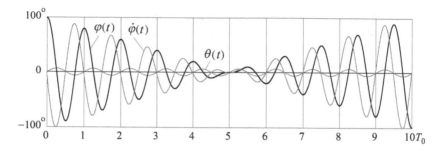

Figure 3.6: The transfer of energy from the excited oscillator to the external source and back in conditions of exact tuning to resonance.

We can see here an analogue between the above considered processes of resonant interaction of the mechanical oscillator with the periodic external source, and the familiar optical phenomena of resonant absorption and stimulated emission of radiation by atoms. The mechanical oscillator can be considered as an atom, or, more precisely, as an optical electron linked to the atom by a quasi elastic force. Such an electron can execute natural oscillations at a definite frequency. The external source that excites forced oscillations of the mechanical oscillator can be regarded as an analog of the electromagnetic field of the light wave that interacts with the atom. The energy can be transferred either to the atom from the electromagnetic field whose frequency equals the natural frequency of the electron in the atom (more precisely, with the frequency corresponding to a transition of the optical electron between atomic energy levels), or from the excited atom to the electromagnetic wave which interacts with the atom. The first possibility corresponds to the resonant absorption of light by the atom; the second possibility corresponds to the stimulated emission of radiation by the atom. Which one of the two possibilities actually takes place depends on the phase relationships between oscillations of the electromagnetic field in the light wave and oscillations of the optical electron in the atom.

When the energy is transferred from the atom to the electromagnetic field, the amplitude of the light wave increases with conservation of all other characteristics of the wave (including its phase). This explains the coherent character of the stimulated radiation emitted by a lot of excited atoms in the field of one and the same electromagnetic wave. By virtue of this high coherence of the stimulated radiation, lasers have their remarkable properties: The possibility to concentrate the energy in the spectrum of radiation (to get a highly monochromatic emitted light); the possibility to concentrate the emitted energy in space and in the direction of propagation (to focus the radiation); and the possibility to concentrate the energy in time (to generate extremely short light impulses).

This analogue is very useful for understanding optical phenomena, but we should not take it too literally. The processes of absorption and emission of light by atoms obey quantum laws, which can predict only the probabilities of different

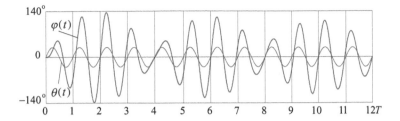

Figure 3.7: Gradually fading beats during a transient process near the resonant frequency ($\omega = 0.8\,\omega_0$). The graphs show time dependencies for angular displacements $\varphi(t)$ of the flywheel and $\theta(t)$ of the exciter at the zero initial conditions.

results of interaction of light with an atom. Therefore the deterministic processes in a mechanical oscillator can be compared to those in a statistical group of atoms interacting with radiation, rather than to those in an isolated atom.

### 3.3.4 Transient Processes Near Resonance

If the driving frequency $\omega$ is near the frequency $\omega_1$ of damped natural oscillations, then during the transient process, while the natural oscillations have not yet damped away, we observe the addition of two oscillations with slightly different frequencies, $\omega$ and $\omega_1$. As already mentioned, the frequency $\omega_1$ nearly equals the natural frequency $\omega_0$ provided the friction is not too large (see Eq. (3.20), p. 45), so that we need not distinguish between them here.

The oscillation resulting from this addition is *modulated*: its amplitude slowly alternately increases and decreases with a *beat frequency* equal to the difference $|\omega - \omega_0|$ between the driving and natural frequencies. The external force first drives the oscillator to amplitudes that exceed the steady-state value; then the accumulated phase shift between the oscillations of the flywheel and those of the driving rod causes energy to flow back from the oscillator to the source of the external action, and the amplitude of the flywheel decreases. These cycles of *transient beats* (of slow variations of the amplitude) are repeated over and over until the damped natural oscillations die out.

In the presence of viscous friction, the modulation of the amplitude is gradually diminished as damping decreases the contribution of the transient oscillations with the natural frequency.

Figure 3.7 displays graphs of such fading transient beats for the case in which the flywheel is initially at rest. In this example, four cycles of the driving force at the frequency $\omega = 0.8\,\omega_0$ occur during five cycles of natural oscillations at the frequency $\omega_0$. (On the graph, the vertical hatch marks correspond to the external period $T$.) Hence one period, $T_b = 2\pi/|\omega - \omega_0|$, of the beat cycle occurs during four periods, $T = 2\pi/\omega$, of the driven (steady-state) cycle and during five periods, $T_0 = 2\pi/\omega_0$, of natural oscillations.

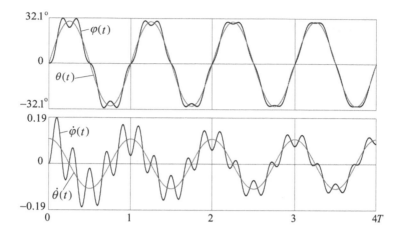

Figure 3.8: The angular displacement and angular velocity during a transient process at a low driving frequency ($\omega = 0.2\omega_0$, $\theta_0 = 30°$, $Q = 40$, initial conditions $\varphi(0) = 0$, $\dot{\varphi}(0) = 0$).

In the absence of friction, steady-state oscillations with the frequency $\omega$ occur in phase with the external force at $\omega < \omega_0$, or $180°$ out of phase at $\omega > \omega_0$. Their amplitude $a$ is determined by Eq. (3.8). The other contribution in the transient process is given by oscillations with the natural frequency $\omega_0$. At $\gamma = 0$ these natural oscillations are not damped. Thus, in this case an addition of two harmonic oscillations with slightly different frequencies $\omega$ and $\omega_0$ and *constant* amplitudes occurs after the external force is activated. For the zero initial conditions the amplitude of the natural oscillation equals $-(\omega/\omega_0)a$. The resulting beat oscillation is *pure* in the sense that the beats do not decay. We can consider the beat oscillation as an oscillation with a mean frequency $(\omega + \omega_0)/2$ and a slowly periodically varying amplitude. The envelope of the oscillations changes sinusoidally, periodically equaling zero. The transient process lasts indefinitely, and so there is no steady-state oscillation in the absence of friction (at $\gamma = 0$).

### 3.3.5   Transient Processes Far from Resonance

Here we consider non-resonant cases in which the external frequency $\omega$ is much different from the natural frequency $\omega_0$.

If the external driving frequency is much less than the natural frequency of the oscillator ($\omega \ll \omega_0$), the equilibrium position of the flywheel (in which the spring is unstrained) slowly moves back and forth alongside the rod. Simultaneously, the flywheel executes relatively rapid damped oscillations at its natural frequency about this slowly moving equilibrium position. As a result, these gradually fading rapid natural oscillations, superimposed on the slow steady-state forced oscillations of a constant amplitude, produce a pattern of motion like that shown by the

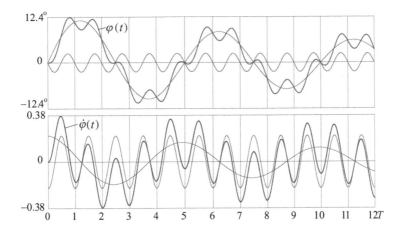

Figure 3.9: The angular displacement and angular velocity during a transient process at a high driving frequency ($\omega = 5.0\omega_0$, $\theta_0 = 60°$, $Q = 10$, initial conditions $\varphi(0) = 0$, $\dot{\varphi}(0) = 0$).

graph in Figure 3.8. When these natural oscillations die out, the plot evolves into a pure, undistorted sine wave corresponding to steady-state oscillations whose frequency is the slow driving frequency $\omega$.

In the opposite case in which the exciting rod oscillates with a high frequency ($\omega \gg \omega_0$), relatively rapid forced oscillations at the frequency $\omega$ and a constant amplitude occur about a middle position, which, during the transient process, slowly executes damped oscillations at the natural frequency $\omega_0$. After these slow oscillations have died out, only the rapid forced oscillations of a constant amplitude remain. These rapid steady-state oscillations occur symmetrically about the value $\theta = 0$, i.e., about the central position of the driving rod. This case is illustrated in Figure 3.9. Together with the graphs of $\varphi(t)$ and $\dot{\varphi}(t)$, the graphs of damped slow natural oscillations and of steady-state rapid oscillations that constitute the transient are also shown. We note that the steady-state oscillations have opposite phase with respect to oscillations of the driving rod.

### 3.3.6 Transient Processes and the Phase Trajectory

The equation of motion describing forced oscillations, Eq. (3.4), is explicitly time-dependent. In this equation a given function of time, $\theta(t) = \theta_0 \sin \omega t$, describes the forcing periodic motion of the driving rod. Thus the mechanical state of the system under consideration is determined by the three quantities: $\varphi$, $\dot{\varphi}$, and $t$.

In order to display all of the characteristics of the mechanical state of an oscillator acted upon by a given time-dependent external force, we add a temporal dimension to the phase plane ($\varphi$, $\dot{\varphi}$). This temporal third dimension is introduced by erecting a time axis perpendicular to the phase plane. In plotting the solution

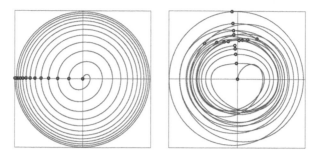

Figure 3.10: Phase trajectories with Poincaré sections for the transient at resonance swinging of the oscillator (left), and for the gradually fading transient beats at $\omega = 0.8\,\omega_0$ (right).

to Eq. (3.4), the computer simulation program traces the projection onto the plane $(\varphi,\ \dot{\varphi})$ of a twisted three-dimensional phase trajectory arising from forced oscillations.

To obtain a clear graphic representation of the entire transient process, we mark positions of the representative point on this phase trajectory at equal time intervals, at the moments when the rod, moving from left to right, crosses the zero point of the dial.

These points on the phase trajectory show the mechanical state of the system at times equal to integer multiples of the period of oscillation, $T = 2\pi/\omega$, of the external force. Such points are called *Poincaré sections*. The simulation program displays the phase trajectory with Poincaré sections (Figure 3.10) when you open the window "Phase diagram" by using a relevant button on the control panel of the program.

Since steady-state oscillations have the period of the external force and a constant amplitude, the corresponding three-dimensional phase trajectory intersects all the planes $t = T,\ 2T,\ \ldots,\ nT$ at the same values of $\varphi$ and $\dot{\varphi}$. Thus the projections of Poincaré sections onto the plane $(\varphi,\ \dot{\varphi})$ at the times $t_n = nT$ coincide.

However, for a transient process, projected Poincaré sections form a set of points in the plane $(\varphi,\ \dot{\varphi})$ that condense gradually to the point $\varphi = a \sin \delta$, $\dot{\varphi} = a\omega \cos \delta$, the projected Poincaré section for the steady-state oscillations.

At the resonant frequency ($\omega = \omega_0$), the oscillation of the flywheel lags behind that of the rod by a quarter cycle ($\delta = \pi/2$), and the coordinates of the limiting condensing point of the Poincaré sections for the transient process are $\varphi = -a$, $\dot{\varphi} = 0$. If the initial angular deflection and angular velocity of the flywheel are zero, the rod starts to move when the flywheel is at rest in the equilibrium position. Then all the projections of the Poincaré sections lie on the abscissa axis of the phase plane, starting at the origin and gradually approaching the above-mentioned condensing point, $\varphi = -a$, $\dot{\varphi} = 0$ (left-hand panel of Figure 3.10). The right-hand panel of this figure corresponds to the process of transient beats at $\omega = 0.8\,\omega_0$, whose graph is shown in Figure 3.7, p. 53.

## 3.4 Review of the Principal Formulas

The differential equation of forced oscillations for the kinematic excitation:

$$\ddot{\varphi} + 2\gamma\dot{\varphi} + \omega_0^2\varphi = \omega_0^2\theta_0\sin\omega t. \tag{3.31}$$

Steady-state forced oscillations are described by the particular periodic solution of this inhomogeneous differential equation:

$$\varphi(t) = a\sin(\omega t + \delta). \tag{3.32}$$

The amplitude $a$ and the phase shift $\delta$ of steady-state forced oscillations:

$$a(\omega) = \frac{\omega_0^2\theta_0}{\sqrt{(\omega_0^2 - \omega^2)^2 + 4\gamma^2\omega^2}}, \quad \tan\delta = -\frac{2\gamma\omega}{\omega_0^2 - \omega^2}. \tag{3.33}$$

The driving frequency $\omega_{res}$ at which the amplitude of steady-state forced oscillations is a maximum is given by:

$$\omega_{res} = \sqrt{\omega_0^2 - 2\gamma^2} \approx \omega_0\left(1 - \frac{\gamma^2}{\omega_0^2}\right) = \omega_0\left(1 - \frac{1}{4Q^2}\right). \tag{3.34}$$

The amplitude of steady-state oscillations at resonance (for $\gamma < \sqrt{2}\,\omega_0$) and its approximate value for $\gamma \ll \omega_0$ are given by:

$$a_{max} = \frac{\omega_0^2\theta_0}{2\gamma\sqrt{\omega_0^2 - \gamma^2}} \approx \frac{\omega_0\theta_0}{2\gamma} = Q\theta_0. \tag{3.35}$$

Here the quality factor $Q$ is the same dimensionless quantity, $Q = \omega_0/(2\gamma)$, that characterizes the damping of free oscillations.
The amplitude $\Omega$ of the angular velocity of steady-state oscillations:

$$\Omega = \omega a(\omega) = \frac{\omega_0^2\theta_0}{\sqrt{(\omega_0^2/\omega - \omega)^2 + 4\gamma^2}}. \tag{3.36}$$

The maximal amplitude $\Omega$ of the angular velocity of steady-state oscillations at resonance (for which $\omega = \omega_0$) is given by:

$$\Omega_{max} = \omega a(\omega_0) = \frac{\omega_0^2\theta_0}{2\gamma} = \omega_0 Q\theta_0. \tag{3.37}$$

The initial conditions that eliminate the transient process are:

$$\varphi_0 = a\sin\delta, \qquad \dot{\varphi}(0) = a\omega\cos\delta, \tag{3.38}$$

where $a$ and $\delta$ are the amplitude and the phase of steady-state oscillations.
The transient process at the resonant frequency and zero initial conditions:

$$\varphi(t) = -Q\theta_0(1 - e^{-\gamma t})\cos\omega_0 t. \tag{3.39}$$

# 3.5  Questions, Problems, Suggestions

The simulation of forced oscillations of a torsion spring pendulum in the computer program is based on the numerical integration of the differential equation of motion, Eq. (3.4), by means of the Runge–Kutta method of the fourth order. Although this linear differential equation can be solved analytically, the analytic solution is not used in this simulation program. The agreement between the analytic theoretical predictions and the observed behavior of the linear oscillator in these simulations is an indicator of the reliability of the numerical method. Thus we get confidence in the reliability of the simulations of nonlinear systems in other programs in the package "Physics of Oscillations," because these simulations are based on the same numerical method. This confidence is not without value since analytic solutions are unavailable for nonlinear systems whose behavior in simulations is often hard to reconcile with common sense.

## 3.5.1  Steady-State Forced Oscillations

In order to display steady-state oscillations without a preliminary transient process (when the option "Show steady-state" is chosen), the simulation program automatically sets the initial conditions to be $\varphi(0) = a \sin \delta$ and $\dot\varphi(0) = a\omega \cos \delta$ (overriding any initial conditions you may have entered). These initial conditions provide steady-state oscillations immediately after the external force is activated. The values of the amplitude $a$ and the phase $\delta$ are calculated in the program from Eq. (3.9) using the values of the external frequency $\omega$ and the quality factor $Q$ you have entered.

### 3.5.1.1  Steady-State Forced Oscillations without Friction.

(a) Setting properties of the system, choose full absence of friction. In this case, the transient process lasts indefinitely, so that steady-state oscillations do not establish. How can you explain the physical sense of the periodic analytical solution that describes steady-state forced oscillations in the absence of friction? Is this solution applicable to a real system? If so, what conditions must be satisfied in order that it be possible in a real system to observe the motion described by this analytical solution?

(b) Convince yourself that for driving frequencies less than the natural frequency ($\omega < \omega_0$), the steady-state sinusoidal oscillations of the flywheel occur exactly in phase with the oscillations of the driving rod. At what frequency is the amplitude of the flywheel twice that of the rod? Calculate this frequency and verify your answer with a simulation experiment.

(c) Convince yourself that for driving frequencies greater than the natural frequency ($\omega > \omega_0$), the phase of the steady-state oscillations of the flywheel is exactly opposite that of the rod. At what value of the driving frequency ($\omega > \omega_0$) is the amplitude of the flywheel again twice that of the rod? At what driving frequency are these amplitudes equal? At what frequency is the amplitude of the flywheel one half that of the rod?

**3.5.1.2\*  Transformations of Energy for Steady-State Oscillations.**

(a) Using the plots of potential, kinetic, and total mechanical energy, find out during which parts of the driving cycle is energy transmitted from the rod to the oscillator. Give a physical explanation of this direction of energy transfer. During which parts of the cycle is energy transferred back from the oscillator to the external source?

(b) For the kinematic excitation of forced oscillations, what is the ratio of the average value of the potential energy to the average value of the kinetic energy for each of those values of frequency for which the amplitude of the flywheel is twice that of the rod? For the case when the amplitudes are equal? For the case when the amplitude of the flywheel equals one half that of the rod? Compare the values observed on the experimental plots with your calculated values.

**3.5.1.3  The Amplitude and Phase of Steady-State Oscillations.**

Examine steady-state oscillations in the presence of friction. Input some moderate value of the quality factor, say, $Q = 5$.

(a) Evaluate the percentage shift of the resonant frequency from the natural frequency.

(b) What is the ratio of the amplitude of steady-state oscillation of the flywheel to the amplitude of the exciting rod at resonance?

(c) What is the phase lag of the oscillation of the flywheel relative to the phase of the rod at the resonant frequency and at a driving frequency equal to 0.8 of the resonant value?

Answer the same questions for $Q = 20$.

**3.5.1.4\*\*  Peculiarities of the Kinematic Excitation.** In the case of the dynamic excitation of oscillations by a given force whose value is independent of the position of the flywheel, the ratio of the average potential energy to the average kinetic energy equals $(\omega_0/\omega)^2$, so that for low frequencies the potential energy predominates. For kinematic excitation, the ratio of the average energies is different.

(a) Analyze the variations with time of both kinds of energy, and of the total energy for the kinematic mode of excitation. Give a reasonable physical explanation for these energy variations. Calculate the ratio of the average values of potential energy to kinetic energy in this case.

(b) At what frequency of the external torque are the average values of the potential and kinetic energy equal to each other?

(c) In the case of dynamic excitation, mean values of both kinds of energy at resonance are equal to one another, and their changes occur exactly in opposite phase, so that total mechanical energy remains constant. However, in the kinematic mode of excitation of forced oscillations, total mechanical energy is subjected to variations even at resonance. Explain these variations. Calculate how much the maximal and minimal values of the total energy differ from its average value (in percent).

**3.5.1.5\*\*  Steady-State Oscillations at Various Frequencies.**

(a) Let the driving frequency $\omega$ of the rod be a little less than the natural frequency $\omega_0$, say, $\omega = 0.9\,\omega_0$, and let the value of $Q$ be 5. What is the ratio of the amplitude of the steady-state oscillations to the resonant amplitude? What is the phase lag of the oscillation of the flywheel relative to the phase of the rod (in fractions of a cycle)?

(b) At what values of the driving frequency (on either side of the resonant frequency) is the amplitude of steady-state oscillations one half of the resonant amplitude? What is the corresponding phase lag in each of these cases? What kind of energy (averaged over a cycle) predominates in each of these cases?

(c) At what driving frequency $\omega$ are the amplitudes of the flywheel and of the rod equal? What is the phase lag of the flywheel relative to the phase of the rod in this case?

### 3.5.1.6** Half-Width of the Resonance Curve.

(a) Examine analytically and experimentally the dependence of the resonant amplitude of steady-state oscillations on the value of the quality factor $Q$. How does the half-width of the resonance curve depend on the quality factor $Q$? (The *half-width* is the interval of driving frequencies within the limits of which the amplitude of steady-state oscillations exceeds one half of its maximal value.) In other words, how does the sharpness of the resonance peak change when damping is increased?

Make the necessary calculations and then verify your answers by simulating the appropriate experiments on the computer.

(b) How does the position of the resonance peak change as damping is increased? At what value of $Q$ is this maximum shifted to a frequency of zero? (In this case the effect of a static external force applied to the system exceeds that of a driving force oscillating at any frequency.)

### 3.5.1.7** Power Absorbed and Dissipated. Lorentzian.

(a) Prove analytically that for steady-state forced oscillations the power received by the oscillator from the external source, averaged over a period, equals the averaged value of energy dissipated by friction. Use the values of amplitude $a$ and phase $\delta$ for steady-state oscillations, expressed by Eq. (3.9).

(b) Show that the spectral distribution of power absorbed by the oscillator with weak friction for steady-state oscillations is described by the function

$$F(\omega) = \frac{1}{1 + (\omega - \omega_0)^2\tau^2},$$

where $\tau = 1/\gamma$. (This function is encountered in various problems of physics. It is called Lorentzian.)

## 3.5.2  Transient Processes

### 3.5.2.1* Initial Conditions That Eliminate a Transient.

(a) Under certain initial conditions there is no transient term. That is, immediately after the external driving force is activated, the oscillator executes steady-state oscillations with a constant amplitude at the driving frequency. What are these initial conditions? Express the angle of deflection and the angular velocity corresponding to these initial conditions in terms of the quality factor $Q$ of the oscillator and the parameters of the external action – the driving frequency $\omega$ and the amplitude $\theta_0$ of oscillations of the rod.

(b) Letting $Q = 5$, $\omega = \omega_0$, and $\theta_0 = 15°$, calculate the initial angle of deflection and the initial angular velocity for which there is no transient. Enter your calculated values and verify experimentally that the forced oscillations are purely sinusoidal from the beginning of the motion. Repeat the same procedure for different values of the system parameters: $Q = 25$, $\omega = 0.5\,\omega_0$, $\theta_0 = 15°$, and say, $Q = 25$, $\omega = 1.5\,\omega_0$, $\theta_0 = 15°$.

### 3.5.2.2* Transient Processes at Resonance.

Examine transient processes at resonance experimentally. Enter the following values for the parameters: $Q = 5$, $\omega = \omega_0$, $\theta_0 = 15°$, and enter the initial conditions $\varphi(0) = 0$, $\dot{\varphi}(0) = 0$.

(a) Calculate the lapse of time, measured in units of the period, during which the amplitude reaches 90% of its steady-state value. Verify your answer experimentally. Note the monotonic growth of the amplitude and its exponential asymptotic approach to its steady-state value.

Analyze the character of energy conversions using the graphs of the kinetic, potential, and total energy. At what instants of time is the growth in the total energy of the oscillator a maximum?

(b) Carefully examine the graphs of the decomposition of the resonant transient process into its simple component parts (the transient term and the steady-state oscillation). Note especially the exponential damping of the transient term. Why does the initial value of the amplitude of this natural oscillation equal the amplitude of the steady-state oscillation?

(c)* Taking into account the analytic expression, Eq. (3.26) for $\varphi(t)$ corresponding to the resonant case ($\omega = \omega_0$), predict the behavior of the Poincaré sections in the phase plane. Verify your prediction by a simulation experiment.

(d)* Consider a transient process at resonance in the absence of friction. How does the amplitude of oscillations increase with time during the transient process that begins from the state of rest in the equilibrium position?

(e)** During a transient process, is it possible for the amplitude to decrease if the frequency of the external force is exactly the resonant frequency? Give physical arguments for your answer. Can you prove your answer experimentally?

### 3.5.2.3* Transient Processes Near Resonance.

Explore transient processes near resonance. Let the frequency of the external force $\omega$ be equal to, say, $1.2\,\omega_0$ ($\omega_0$ is the natural frequency of the oscillator). Consider first of all behavior of the system in the absence of friction when the initial conditions are zero. The left-hand panel of Figure 3.11 shows the phase diagram with Poincaré sections, and the right-hand panel shows the graphs of $\dot{\varphi}(t)$ and $\varphi(t)$.

(a)* Calculate the amplitude of the transient term, that is, the amplitude of oscillations with the natural frequency contributing to the transient process. Also

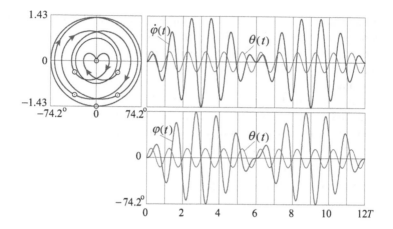

Figure 3.11: Beats in the absence of friction during forced oscillations at driving frequency $\omega = 1.2\,\omega_0$.

calculate the amplitude of the angular velocity of these oscillations. What is the ratio of the amplitude of the transient term to the amplitude of steady-state forced oscillations? Verify your answers experimentally using the option "Decomposition of the Transient Process."

(b) Through how many cycles does the rod oscillate before the amplitude of oscillation of the flywheel reaches its maximal value? What is the lapse of time between successive moments at which the amplitude is zero? In other words, what is the beat period?

(c)** Calculate the maximal values of the angular velocity and of the angular displacement during the beats. What is the ratio of the maximal amplitude of the flywheel to the amplitude of the rod?

(d)** Note the distribution of the Poincaré sections of the phase trajectory for this process of beats in the absence of friction (Figure 3.12). Explain this distribution.

(e)** Consider the effect of friction by entering a moderate value of the quality factor $Q$ $(20-25)$, while keeping the previous values of the remaining parameters. What changes in the behavior of the system do you expect? What is the corresponding distribution of the Poincaré sections in the phase plane? (Several initial loops of the phase trajectory are shown in Figure 3.12.) Follow behavior of the Poincaré sections for as long a time as is needed for the steady-state oscillations to establish. Explain the distribution of the Poincaré sections displayed.

(f)** Change the driving frequency by a small amount. For example, let $\omega$ be $1.19\,\omega_0$ or $1.21\,\omega_0$ instead of $1.20\,\omega_0$. What changes in the transient process does this change in the frequency cause? What are the corresponding distributions of the Poincaré sections in the phase plane if there is no friction and if there is weak friction? What are the reasons for the differences in appearance of the display

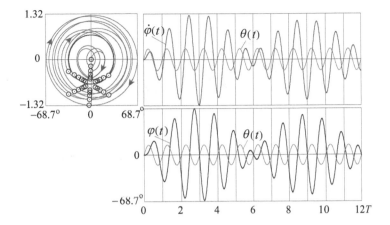

Figure 3.12: Fading of beats caused by weak friction ($Q = 50$) during forced oscillations at $\omega = 1.2\,\omega_0$.

from the case in which $\omega = 1.2\,\omega_0$?

(g)* Predict the distribution of Poincaré sections for $\omega = 1.25\omega_0$. Verify your prediction by the simulation.

### 3.5.2.4* Transient Processes Far from Resonance.

Investigate transient processes far from resonance. Let the frequency of the driving rod be, say, four times smaller than the natural frequency of the oscillator: $\omega = 0.25\,\omega_0$. Set the two initial conditions to zero.

(a) What are the oscillations like during the transient process if there is no friction? Calculate the amplitudes of the two superimposed oscillations, one with the natural frequency $\omega_0$ and the other with the driving frequency $\omega$. What are the the two amplitudes of the corresponding angular velocities?

(b) Observe and explain the shape of the phase trajectory and the distribution of Poincaré sections.

(c) Next introduce moderate friction by setting $Q$ to be approximately 15, keeping the values of the other parameters. Note the gradual fading of the contribution of the natural oscillations. (Use the option "Decomposition of the Transient Process" to see how the time dependencies of these contributing simple oscillations are plotted.) Observe how the complex phase trajectory plotted during the early stage of the transient process is transformed into the ellipse corresponding to steady-state oscillations. Explain the shape of the phase trajectory and the distribution of Poincaré sections (Figure 3.13a).

(d)** Consider the opposite case of a driving force with a high frequency. For example, let the driving frequency be, say, four times greater than the natural frequency of the oscillator: $\omega = 4\omega_0$. Set the two initial conditions to zero, and let friction be zero. What is the ratio of the amplitudes of the two superimposed oscillations with the frequencies $\omega$ and $\omega_0$? By how many times does the maxi-

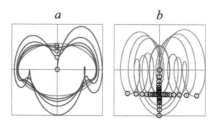

Figure 3.13: Several loops of the phase trajectory and the succession of Poincaré sections for transients at $\omega = 0.25\,\omega_0$ (a) and $\omega = 4\,\omega_0$ (b). The thick lines show the ellipses which correspond to steady-state oscillations.

mal deflection during the transient process exceed the amplitude of steady-state oscillations, contributing to the transient?

(e)** Compare the shapes of the phase trajectories and the distributions of the Poincaré sections for the two cases, $\omega = 0.25\,\omega_0$ and $\omega = 4\,\omega_0$, in the absence of friction. How can you explain the similarity in the phase trajectories for these cases?

(f)* Introduce weak friction for the case in which $\omega = 4\,\omega_0$. Observe the way in which frequent steady-state forced oscillations are established while the contribution of slow transient natural oscillation gradually fades away. How does the phase trajectory evolve in this case? What is the behavior of the Poincaré sections during the corresponding transient process?

(g)** Repeat the simulation experiments for other values of the driving frequency, say $\omega = 0.125\,\omega_0$, $\omega = 0.5\,\omega_0$, $\omega = 1.5\,\omega_0$, $\omega = 2\,\omega_0$, $\omega = 6\,\omega_0$). Explain the peculiarities of the transient processes for these cases.

# Chapter 4

# Square-Wave Excitation of a Linear Oscillator

**Annotation.** Chapter 4 deals with forced oscillations in a linear system driven by a non-sinusoidal external force, namely by a force with square-wave time dependence. It includes a description of the simulated mechanical physical system and its electromagnetic analogue, and a summary of the relevant theoretical material for students as a prerequisite for the virtual lab "Square-wave Excitation of Linear Oscillator." Chapter 4 also includes a set of theoretical and experimental problems to be solved by students on their own, as well as various assignments that the instructor can offer students for possible individual mini-research projects.

## 4.1  Theoretical Background

This chapter deals with forced oscillations of a torsion spring pendulum excited by an external square-wave driving torque. Two different ways of determining the steady-state response of the oscillator to a non-harmonic driving force are described and compared. Behavior of this familiar mechanical system can help a student to better understand why and how an electromagnetic oscillatory *LCR*-circuit transfers the square-wave voltage from input to output with a distortion of its shape.

### 4.1.1  Model of the Physical System

Forced oscillations in a linear system under a sinusoidal driving force are considered in Chapter 3. Most textbooks on general physics treat this case rather extensively. The general case of a periodic but non-sinusoidal excitation of a linear oscillator is usually only mentioned with reference to the principle of superposition and an expansion of an arbitrary periodic force as the Fourier series of sine

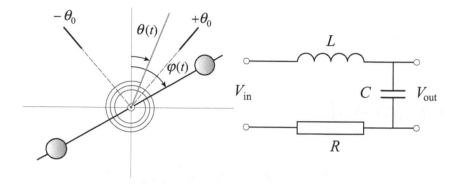

Figure 4.1: Schematic image of the torsion spring oscillator (left) and its electromagnetic analogue—*LCR*-oscillatory circuit excited by the square-wave input voltage (right).

and cosine functions. In this chapter an alternative approach to the problem of forced oscillations is suggested and compared with the traditional treatment.

To study forced oscillations caused in a linear system by a non-sinusoidal periodic external influence, we employ a simplified model of a torsion spring oscillator. Its schematic image is shown in the left panel of Figure 4.1. The oscillator is similar to the balance devices of ordinary mechanical watches—a balanced massive rotor (flywheel) attached to one end of an elastic spiral spring. A mechanical system such as this one is ideal for the study of resonance because it is possible to see directly what is happening. The spring provides a restoring torque proportional to the angular displacement of the flywheel from the equilibrium position. To provide an external excitation, the other end of the spiral spring is attached to a driving rod (exciter) that can be turned about the axis common with the axis of the flywheel. When the rod is constrained to move periodically to and fro about some middle position, an additional periodic torque is exerted on the flywheel. This mode of excitation is called *kinematical* because it is characterized by a given motion of some part of the system rather than by a given external torque.

When the driving rod is turned, the equilibrium position of the flywheel is displaced alongside the rod through the same angle. The flywheel can execute damped natural oscillations about this new displaced equilibrium position. An external piecewise constant torque whose shape is that of a periodic square-wave can be realized by abruptly displacing the driving rod alternately in opposite directions through the same angle in equal time intervals. We suppose that the displacements of the rod and thus of the equilibrium position of the flywheel occur so quickly that there is no significant change in either the angular position or velocity of the flywheel during the abrupt displacement of the driving rod.

The right-hand panel of Figure 4.1 shows an *LCR*-oscillatory circuit that can be regarded as an electromagnetic analogue of the mechanical device. Both systems are described by identical differential equations and thus are dynamically isomorphic. However, the mechanical system has a definite didactic advantage for exploration of forced oscillations because it allows us to observe a direct visualization of motion.

## 4.1.2 The Differential Equation of Forced Oscillations

When the system is at rest, we let the needle attached to the flywheel be parallel to the driving rod. In other words, the spring is not strained when the needle points in the same direction as does the rod. The zero point of the dial indicates the central position of the exciting rod (the vertical position in Figure 4.1). The angle of deflection $\varphi$ of the needle from this zero point indicates the position of the flywheel. When the rod is deflected from the vertical position through an angle $\theta$, the spiral spring is twisted from its unstrained state through the angle $\varphi - \theta$. The spring then exerts a torque $-D(\varphi - \theta)$ on the flywheel, where $D$ is the torsion spring constant. Thus, the differential equation of rotation of the flywheel, whose moment of inertia about the axis of rotation is $J$, is given by

$$J\ddot{\varphi} = -D(\varphi - \theta). \tag{4.1}$$

We transfer $-D\varphi$ (the part of the elastic torque that is proportional to $\varphi$) to the left side of Eq. (4.1), divide the resulting equation by $J$, and introduce the value $\omega_0 = \sqrt{D/J}$, whose physical meaning is the frequency of natural oscillations in the absence of friction. Thus we obtain:

$$\ddot{\varphi} + \omega_0^2\varphi = \omega_0^2\theta. \tag{4.2}$$

The right-hand side $\omega_0^2\theta$ of this equation can be treated as an external torque (divided by $J$) caused by the displacement of the rod from its central position through an angle $\theta$. We let the instantaneous displacements of the rod occur alternately to the right and to the left after the lapse of equal time intervals $T/2$, so that during the interval $(0, T/2)$ the equilibrium position of the flywheel is displaced to the right through a fixed angle $\theta_0$, and during the next interval $(T/2, T)$ the equilibrium position is displaced to the left through the same angle (the angle $\theta$ in Eq. (4.2) equals $-\theta_0$). Thus, $T$ is the full period of the external non-sinusoidal action, repeated indefinitely. In the presence of viscous friction, a term $2\gamma\dot{\varphi}$ proportional to the angular velocity $\dot{\varphi}$ should be added to Eq. (4.2), in which the damping constant $\gamma$ characterizes the strength of viscous friction in the system:

$$\ddot{\varphi} + 2\gamma\dot{\varphi} + \omega_0^2\varphi = \begin{cases} \omega_0^2\theta_0, & (0, T/2), \\ -\omega_0^2\theta_0, & (T/2, T). \end{cases} \tag{4.3}$$

Forced oscillations of the electric charge $q$ stored in a capacitor of a resonant series $LCR$-circuit (see the right-hand panel of Figure 4.1) excited by a square-wave input voltage $V_{in}(t)$ obey the same differential equation as does the forced oscillation of a mechanical torsion spring oscillator excited by periodic abrupt changes of position of the driving rod:

$$\ddot{q} + 2\gamma\dot{q} + \omega_0^2 q = \omega_0^2 C V_{in}(t). \tag{4.4}$$

In this equation, $\omega_0$ is the natural frequency of oscillations of charge in the circuit in the absence of resistance. It depends on the capacitance $C$ of the capacitor and the inductance $L$ of the coil: $\omega_0 = 1/\sqrt{LC}$. The damping constant $\gamma = R/(2L)$ characterizes the dissipation of electromagnetic energy occurring in a resistor whose resistance is $R$.

Because of this similarity, the mechanical system described above enables us to give a clear explanation for transformation of the square-wave input voltage $V_{in}(t) = \pm V_0$ into the output voltage $V_{out}(t) = V_C(t) = q/C$ (voltage across the capacitor $C$), whose time dependence differs considerably from the piecewise constant input voltage. The output voltage $V_C(t)$ is analogous to the deflection angle $\varphi(t)$ of the rotor, while the alternating electric current $I(t) = \dot{q}(t)$ in the circuit is analogous to the rotor angular velocity $\dot{\varphi}(t)$ of the mechanical model. However, some caution is necessary in interpreting the analogy between the mechanical oscillator and the electric $LCR$-circuit with respect to the energy transformations. We note that in using the mentioned analogy it would be incorrect to exactly associate the electric potential energy of a charged capacitor with the elastic potential energy of a strained spring because the latter depends directly on the angle $(\varphi \pm \theta_0)$, while the energy of a capacitor depends directly on the charge $q$ or on the corresponding voltage $V_C = q/C$ (*not* on the voltage $(V_C \pm V_0)$). In contrast to the spring oscillator, for which a jump in the position of the rod causes an abrupt change in the elastic potential energy of the spring, a jump of the input voltage across an electric circuit does not abruptly change the charge and the energy of a capacitor.

# 4.2   Steady-State Forced Oscillations under the Square-Wave Torque

Because of friction, natural oscillations of the flywheel gradually damp out, and a while after the external force began to act, a steady-state periodic motion of the flywheel is eventually established with a period equal to the period $T$ of the driving force. The greater the decay time, $\tau = 1/\gamma$, of natural oscillations, the longer the duration of this transient process.

In the case of a sinusoidal driving torque, the steady-state oscillations of the flywheel acquire not only the period of the external action but also the sinusoidal time dependence whose frequency equals the driving frequency. However, a periodic driving force, whose time dependence is something other than a pure sinusoid, produces a steady-state response that has the same period but whose time dependence differs from that of the driving force.

## 4.2.1   Harmonics of the Driving Force and of the Steady-State Response

We consider below two different ways of determining the steady-state response of the oscillator to a non-harmonic driving force, such as the square-wave time-dependent force discussed above. One (traditional) way is based on expansion of the time dependence of the external force in a Fourier series, i.e., on the representation of this force as a superposition of sinusoidal components (harmonics). Because the differential equation of motion for the spring oscillator is linear, the influence of each of these harmonic components of the external force can be considered independently. Each sinusoidal component of the driving torque produces its own sinusoidal response of the same frequency in the motion of the flywheel. The amplitude and phase of each sinusoidal response can be calculated separately. The corresponding formulas are the same as for the familiar case of monoharmonic excitation.

The net steady-state forced motion of the flywheel can be found as the superposition of these individual responses. Thus, to each sinusoidal component of the periodic driving force (to the *input harmonic*), there corresponds a sinusoidal component of the same frequency in the steady-state motion of the responding oscillator (we can call it the *output harmonic*). Since the relative contributions of harmonic components to this response differ from the corresponding contributions to the driving force, the graph of the net motion of the flywheel has a different shape than does the graph of the motion of the driving rod.

In particular, it may occur that one of the input harmonics with relatively small amplitude induces especially large amplitude in the output oscillations. Such is the case when the frequency of this harmonic is close to the natural frequency $\omega_0$ of the oscillator since forced oscillations caused by this sinusoidal force occur under conditions of resonance. On the other hand, the relative contributions of the input harmonics whose frequencies lie far from the maximum of the resonance curve are considerably attenuated in the output oscillations. In the output steady-state oscillations such harmonics are appreciably suppressed. The oscillator responds *selectively* to sinusoidal external forces of different frequencies. The phenomenon of resonance depends upon the whole functional form of the driving force and occurs only if its spectrum contains the component whose frequency is close to the natural frequency of the oscillator.

Differences between the time dependence of output steady oscillations and that of the input driving force (distortions of the signal from input to output) are caused not only by changes in the relative amplitudes of different harmonics but also by changes in their phases from input to output. In the case of weak damping the resonance curve (amplitude versus frequency) is very sharp, and the dependence of phase on frequency is nearly a step-function. Specifically, all harmonic components whose frequencies $\omega_k = k\omega = (2\pi/T)k$ ($T$ – driving period) are lower than the natural frequency $\omega_0$ contribute to the output oscillations of the flywheel nearly in the same phases as they do to the input driving force. But harmonics whose frequencies $\omega_k$ are higher than the natural frequency contribute to the output oscillations with nearly inverted phases: in the Fourier expansion of the output steady-state oscillations, their phases are almost opposite to the phases of the corresponding harmonics of the driving force. If there is a sinusoidal component in the driving force whose frequency lies close to $\omega_0$, this harmonic produces a significantly increased relative contribution to the output oscillations of

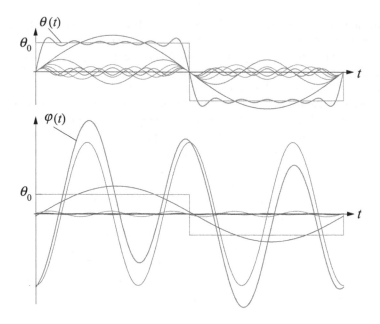

Figure 4.2: Transformation of the spectrum of the input square-wave external torque into the spectrum of steady-state output oscillations (see text for detail).

the flywheel. This output harmonic component lags in phase by $\pi/2$ behind the corresponding input harmonic in the spectrum of the driving force.

The analytic expression for Eq. (4.3) for which the square-wave shape of its right-hand side has been Fourier decomposed has the following form:

$$\ddot{\varphi} + 2\gamma\dot{\varphi} + \omega_0^2\varphi = \sum_{k=1,\,3,\,5\ldots}^{\infty} \frac{4\theta_0\omega_0^2}{\pi k} \sin\omega_k t. \tag{4.5}$$

The Fourier series of the piecewise-constant external force in Eq. (4.3) contains only odd-number harmonics with frequencies $\omega_k = k\omega$ ($k = 1, 3, 5, \ldots$), where $\omega = 2\pi/T$, the frequency of the driving force. We note that the amplitudes of harmonics of the square-wave function decrease rather slowly, as $1/k$, with the increase of their index $k$ and their frequency $\omega_k$. This case is a good example of a multi-harmonic external excitation of the oscillator since the frequency spectrum of the square-wave driving force is rich in harmonics.

Figure 4.2 illustrates the transformation of the input spectrum of an external square-wave periodic force into the output spectrum of the steady-state response of the oscillator. The upper panel of Figure 4.2 shows the square-wave graph of $\theta(t)$ that corresponds to the exciter's motion, and harmonics of $\theta(t)$ from the first up to the eleventh. The graph of the sum of all these harmonics, which approximates the square-wave motion of the exciter, is also shown.

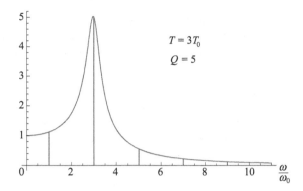

Figure 4.3: The resonance curve, which shows how the amplitudes of separate odd-numbered harmonics of the input square-wave motion of the exciter (whose period $T$ equals $3T_0$) are transformed by the oscillator with $Q = 5$ into the amplitudes of the output harmonics of the flywheel steady-state oscillations (see text for details).

The lower panel of Figure 4.2 shows the same harmonics in the output steady-state motion of the flywheel $\varphi(t)$. For each sinusoidal term (each input harmonic) in the right-hand side of Eq. (4.5), the amplitude and phase of the periodic sinusoidal particular solution (of the output harmonic) are given by the commonly known expression (see Eq. (3.9), p. 40 in Chapter 3). Figure 4.3 illustrates how the amplitudes of separate odd-numbered harmonics of the input square-wave motion of the exciter (whose period $T$ equals $3T_0$) are transformed by the oscillator with $Q = 5$ into the the amplitudes of the output harmonics of the flywheel steady-state oscillations. The resonance curve shows how the oscillator with certain given parameters responds to the individual harmonic components of the external force. Since for the case shown in Figure 4.3 the driving period equals three natural periods, the third harmonic of the square-wave force occurs under the maximum of the resonance curve, and this harmonic component dominates the output spectrum.

Adding solutions that correspond to all input harmonics, we get the following time dependence of the angular displacement, $\varphi(t)$, for steady-state forced oscillations under the square-wave excitation:

$$\varphi(t) = \sum_{k=1,\,3,\,5\ldots}^{\infty} \frac{4\theta_0}{\pi k} \frac{\omega_0^2}{\sqrt{(\omega_0^2 - \omega_k^2)^2 + 4\gamma^2\omega_k^2}} \sin(\omega_k t + \alpha_k), \qquad (4.6)$$

where the phases $\alpha_k$ of the individual harmonics are determined by

$$\tan \alpha_k = \frac{2\gamma\omega_k}{\omega_k^2 - \omega_0^2}. \qquad (4.7)$$

Equations (4.6) and (4.7) display clearly the above-discussed peculiarities of the oscillator response to the square-wave driving action of the exciting rod. A resonant response from the oscillator occurs each time the denominator in one of the terms of the sum in Eq. (4.6) is minimal, that is, when the frequency $\omega_k$ of one of the harmonics of the external force is equal to the resonant frequency $\omega_{\text{res}}$ of the oscillator:

$$\omega_{\text{res}} = \sqrt{\omega_0^2 - 2\gamma^2} \approx \omega_0 \left( 1 - \frac{\gamma^2}{\omega_0^2} \right).$$

The latter approximate expression for $\omega_{\text{res}}$ is valid for a weakly damped oscillator ($\gamma \ll \omega_0$), whose quality factor $Q$ is large ($Q = \omega_0/2\gamma \gg 1$). Since the fractional difference between $\omega_{\text{res}}$ and $\omega_0$ is of the second order in the small parameter $\gamma/\omega_0 = 1/(2Q)$, in most cases of practical importance we need not distinguish the resonant frequency from the natural one and can assume that $\omega_{\text{res}} = \omega_0$. For $\omega_k < \omega_0$ Eq. (4.7) yields $\alpha_k \approx 0$, which means that the corresponding harmonic contributes to the output oscillations in the same phase as to the input square-wave force. Conversely, for $\omega_k > \omega_0$ Eq. (4.7) yields $\alpha_k \approx -\pi$, and this harmonic component enters into the output oscillations with the inverted phase.

When the frequency of the *sinusoidal* external force is slowly varied, the resonant steady-state response of the oscillator can occur at only one value of the driving frequency $\omega$, namely $\omega = \omega_{\text{res}}$, the resonant frequency of the oscillator. In other words, in the case of sinusoidal excitation there is only one resonance, and it occurs when the driving period $T$ equals the natural period $T_0$ of the oscillator. However, in the case of the square-wave excitation, resonance occurs not only when the periods are equal, but each time the driving period $T$ is an odd-number multiple of the natural period $T_0$ of the oscillator, that is, when $T = (2n + 1)T_0$, where $n = 0, 1, 2, \ldots$ Resonances, for which $n \geq 1$, occur when the frequency of one of the odd harmonics of the square-wave driving torque approaches the resonant frequency of the oscillator. Each resonance corresponds to a definite harmonic in the input spectrum (spectrum of the square-wave force).

Generalizing, we note that a linear oscillator with a sharp resonance curve (and given resonant frequency) appreciably responds only to a certain single harmonic component of an arbitrarily complex external force. In this respect such an oscillator can be regarded as a spectral instrument, which selects a definite spectral component of an external action. That is, if we cause to "sweep" the natural frequency of an oscillator through a range of frequencies, such an oscillator responds resonantly to the complex external input each time its natural frequency coincides with one of the harmonic frequencies in the Fourier expansion of the external force. In other words, a sweep-frequency oscillator with a large quality factor provides us with a means by which a complex periodic input can be *physically* decomposed into its Fourier components.

The mathematical representation of the square-wave function by a series on the right-hand side of Eq. (4.3) is not unique. The function can be represented as a sum of other functions in many different ways. That is, it is possible to express the external action either as a Fourier series of sine and cosine functions or as a series of other complete sets of functions. From the mathematical point of view, all such expansions are equally valid. The usefulness of the Fourier expansion in the case under consideration is associated with *physics*. It is related to the capability of a linear harmonic oscillator to perform this expansion physically. When the phenomenon of resonance is used as a means of experimental investigation, only the Fourier representation of the analyzed complex process is adequate and expedient.

## 4.2.2 Forced Oscillations as Natural Oscillations about the Alternating Equilibrium Positions

Another way to obtain an analytic solution to the differential equation of motion (4.3) for steady-state oscillations forced by the square-wave external torque is based on viewing the steady-state motion as a sequence of free oscillations, which take place about an equilibrium position that periodically alternates between $+\theta_0$ and $-\theta_0$. During the first half-cycle, from $t = 0$ to $T/2$, the equilibrium position is located at $\varphi = +\theta_0$. For this half-cycle the general form of the dependence of $\varphi(t)$ on $t$ can be written as:

$$\varphi(t) = \theta_0 + Ae^{-\gamma t}\cos(\omega_1 t + \alpha), \qquad (0,\ T/2), \qquad (4.8)$$

where $\omega_1 = \sqrt{\omega_0^2 - \gamma^2}$ is the frequency of damped natural oscillations, $A$ and $\alpha$ are arbitrary constants of integration determined by conditions at the beginning of the half-cycle. The graphs of $\dot\varphi(t)$ and $\varphi(t)$ for this stage of motion (from $t = 0$ to $T/2$) are shown in the right-hand panel of Figure 4.4 by segments of curves between points marked $1$ and $2$. The corresponding segment of the phase trajectory is shown in the left-hand panel of this figure between points $1$ and $2$. The graphs correspond to the case for which $T = 3T_0$ (that is, when the period $T$ of the driving force is set equal to three natural periods $T_0$).

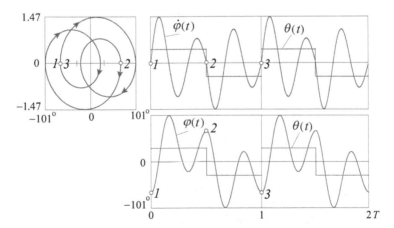

Figure 4.4: Phase diagram, graphs of the time dependence of the angular velocity $\dot\varphi(t)$ and the deflection angle $\varphi(t)$ at resonant steady-state oscillations for $T = 3T_0$, $Q = 5$, $\theta_0 = 30°$.

During the next half-cycle $(T/2,\ T)$, damped natural oscillations occur about an equilibrium position displaced through the same angle $\theta_0$ but in the opposite direction. For this half-cycle the time dependence of $\varphi(t)$ has the form:

$$\varphi(t) = -\theta_0 - Ae^{-\gamma(t-T/2)}\cos(\omega_1(t-T/2) + \alpha), \qquad (T/2,\ T), \qquad (4.9)$$

where the constants $A$ and $\alpha$ have the same values as they do in Eq. (4.8). These values follow from the fact that, in steady-state oscillations, the graph of time dependence during the second half-cycle (when the driving rod is displaced to the left) must be the mirror image of the graph for the first half-cycle, shifted by $T/2$ along the time axis. This relationship can clearly be seen from Figure 4.4, in which the graphs of $\dot{\varphi}(t)$ and $\varphi(t)$ for this stage of motion (from $t = T/2$ to $T$) are shown in the right-hand panel by segments of curves between points marked 2 and 3. The corresponding segment of the phase trajectory is shown in the left-hand panel of this figure between points 2 and 3 (the latter point coincides with the initial point 1).

The constants $A$ and $\alpha$ for any given values of $T$, $\theta_0$, and $\gamma$ can be calculated from the condition that during the instantaneous change in the positions of the driving rod at $t = T/2$, from one equilibrium position to the other, the angular deflection and the angular velocity of the flywheel do not change. In other words, we should equate the right-hand sides of Eqs. (4.8) and (4.9) and their time derivatives at $t = T/2$. These conditions give us two simultaneous equations for $A$ and $\alpha$. Solving the equations we find:

$$\tan\alpha = -\frac{e^{-\gamma T/2}[\omega_1\sin(\omega_1 T/2) + \gamma\cos(\omega_1 T/2)] + \gamma}{e^{-\gamma T/2}[\omega_1\cos(\omega_1 T/2) - \gamma\sin(\omega_1 T/2)] + \omega_1} \qquad (4.10)$$

and

$$A = -\frac{2\theta_0}{e^{-\gamma T/2}\cos(\omega_1 T/2 + \theta) + \cos\alpha}. \qquad (4.11)$$

Equations (4.8)–(4.11) describe the steady-state motion only during the time interval from 0 to $T$. That is, if we substitute a value of $t$ greater than $T$ into these equations, they do not give the correct value for $\varphi(t)$. Nevertheless, we can find the value of $\varphi(t)$ for an arbitrary $t$ by taking into account that $\varphi(t)$ is a periodic function of $t$: $\varphi(t + T) = \varphi(t)$. Thus, having obtained the graph of $\varphi(t)$ for the time interval $[0,\ T]$, we can simply translate the graph to the adjacent time intervals $[T,\ 2T]$, $[2T,\ 3T]$, and so on.

The treatment of forced oscillations as natural oscillations about alternating equilibrium positions provides especially clear explanation of a rather complex behavior of the oscillator under the square-wave force whose period is considerably longer than the natural period. Figure 4.5 shows the phase diagram (left), and the plots of $\dot{\varphi}(t)$ and $\varphi(t)$ (right) for the steady-state forced oscillations at $T = 7T_0$ and relatively strong friction ($Q = 3$). The segments of the phase trajectory are marked in the left-hand panel of this figure by points 1, 2, and 3 (the latter point coincides with the initial point 1), which correspond to the segments of the graphs of $\dot{\varphi}(t)$ and $\varphi(t)$ on the right-hand side, marked by the same figures.

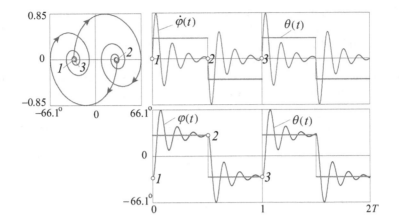

Figure 4.5: Damped oscillations about alternating displaced equilibrium positions at resonant steady-state oscillations for $T = 7T_0$, $Q = 3$, $\theta_0 = 30°$.

We see clearly how after each, in turn, abrupt displacement of the driving rod, the flywheel makes several natural oscillations of gradually diminishing amplitude about the new equilibrium position. These natural oscillations replace both abrupt fronts of each rectangular impulse, thus distorting its shape from input to output.

# 4.3  Transient Processes under the Square-Wave External Torque

The above treatment of forced oscillations excited by a square-wave external torque as natural oscillations about alternating equilibrium positions helps us to understand many characteristics of both steady-state oscillations and transient processes. In particular, it enables us to clearly understand the physical reason for the resonant growth of the amplitude when the period of the driving force equals the natural period of the oscillator or when it equals some odd-number multiple of that period.

Suppose that before the external square-wave torque is applied, the oscillator has been at rest in its equilibrium position, $\varphi = 0$. When, at $t = 0$, the driving rod abruptly turns into a new position, $\theta_0$, the flywheel, initially at rest, begins to execute damped natural oscillations about the new equilibrium position at $\theta_0$ with the frequency $\omega_1 \approx \omega_0$. This oscillation begins with an initial velocity of zero. As long as the rod remains at $\theta_0$, the time dependence of the angular displacement of the flywheel, $\varphi(t)$, is given by the following function:

$$\varphi(t) = \theta_0 - \theta_0 \exp(-\gamma t) \cos \omega_1 t.$$

That is, the flywheel, starting out with $\varphi = 0$ at $t = 0$, passes through the displaced equilibrium position $\varphi = \theta_0$ when $\omega_0 t = \pi/2$, and reaches its extreme deflection of nearly $\varphi = 2\theta_0$ at $\omega_0 t = \pi$. (Damping prevents it from quite reaching $\varphi = 2\theta_0$.) If the period $T$ of the driving rod equals $T_0$, the flywheel arrives at the extreme point $\varphi \approx 2\theta_0$ and its angular velocity becomes zero just at the moment $t = T/2$, when one half of the driving period has elapsed. At this moment the rod instantly moves to the new position $-\theta_0$, which becomes the new equilibrium position of the flywheel for the next time interval $(T_0/2, \ T)$. Hence, the next half-cycle of its natural oscillation starts again with an angular velocity of zero, but its initial angular displacement from the new central point is nearly $3\theta_0$. This value is nearly $2\theta_0$ greater than its value in the preceding half-cycle. It would be exactly $2\theta_0$ greater in the absence of friction, and the amplitude of oscillation would increase by the same value $4\theta_0$ during each full cycle of the external force, provided the driving period equals the natural period of the oscillator (or some odd-number multiple of that period).

In a real system such an unlimited growth of the amplitude linearly with time is impossible because of friction. The growth of the amplitude is approximately linear during the initial stage of the transient process. This resonant growth gradually decreases, and steady-state oscillations are eventually established, during which the increment of the amplitude occurring at every instantaneous displacement of the equilibrium position (at any jump of the driving rod) is nullified by an equal decrement caused by viscous friction during the intervals between successive jumps.

Such a process of gradual growth of the amplitude, which eventually results in oscillations of a constant amplitude, is depicted very clearly by the phase trajectory shown in the left-hand upper corner of Figure 4.6. Since the oscillator is at rest at the moment the external force is activated, the phase trajectory originates at the origin of the phase plane (point *1*). Its first section up to point *2* is a portion of a spiral that winds around a focus located at the point $(+\theta_0, \ 0)$. This focus corresponds to the equilibrium position displaced to the right. The next section between points *2* and *3*, a continuation of the phase trajectory, describes damped natural oscillation about the other equilibrium position after the driving rod has jumped to the left. This is a segment of a similar spiral that winds around the symmetrical point $(-\theta_0, \ 0)$ of the phase plane.

If the period of the square-wave external action equals an odd-numbered multiple of the natural period, the transition of the representative point from one spiral, centered say at $(+\theta_0, \ 0)$, to the adjoining spiral centered at the other focus $(-\theta_0, \ 0)$, occurs at a point of the $\varphi$-axis to the right of $\theta_0$, at a maximal distance from the new focus. As a result, the new loop of the phase trajectory turns out to be larger than the preceding one. Such untwisting of the phase trajectory continues at a decreasing rate until the expansion of loops due to the alternation of the foci is nullified by their contraction caused by viscous friction. Eventually a closed phase trajectory is formed that corresponds to steady-state oscillations. This curve has a central symmetry about the origin of the phase plane. It consists of two branches, each representing damped natural oscillations about one of the two alternating symmetrical equilibrium positions. For $T = 7T_0$ such a closed

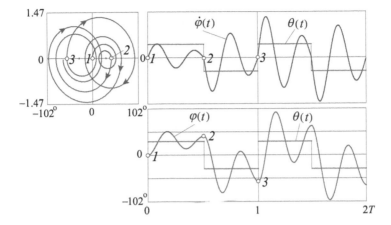

Figure 4.6: The phase diagram (left), graphs of the time dependence of the angular velocity and position angle (right) for the transient process of excitation from equilibrium for resonance occurring at $T = 3T_0$.

phase trajectory is shown in Figure 4.5 (left-hand upper corner).

Any transient process in a linear system can be represented as a superposition of the periodic solution to Eq. (4.3) that describes the steady-state oscillations, and a solution of the corresponding homogeneous equation (with the right-hand side equal to zero) that describes the damped natural oscillations. The simulation program displays such a decomposition of the transient process if the corresponding option is chosen. Figure 4.6 shows the above-discussed graphs for the transient process of swinging from the state of rest in the equilibrium position that takes place under the square-wave torque with a period $T = 3T_0$. A decomposition of this transient process onto periodic steady-state oscillations and fading natural oscillations is given by Figure 4.7, in which the graphs of damped natural oscillations with exponentially decreasing amplitude are singled out especially clearly. Curves 1 show the time-dependent graphs of the transient; curves 2 correspond to the contribution of natural oscillations, curves 3 show the periodic steady-state oscillations.

## 4.4 Estimation of the Amplitude of Steady-State Oscillations

Next we evaluate the maximal angular deflection, $\varphi_m$, attained in the steady-state oscillations when the oscillator is driven by a square-wave torque. The value $\varphi_m$ certainly can be found from Eqs. (4.8)–(4.11). However, such a calculation is rather complicated. Using the simple arguments suggested in the previous sections, we can avoid tedious calculations, at least for some special cases.

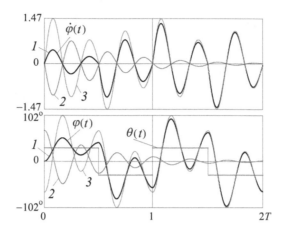

Figure 4.7: Graphs of the angular velocity (upper panel) and the angular position (lower panel) that show the decomposition of the transient (curves *1*) onto the damping natural oscillations (curves *2*) and the periodic steady-state oscillations (curves *3*).

### 4.4.1   Resonant Amplitude of Steady-State Oscillations

We consider first the main resonance in which the driving period equals the natural period: $T = T_0$. The estimation of $\varphi_m$ can be made in the following way. The closed phase trajectory for the steady-state oscillation consists in this case of a single loop intersecting the $\varphi$-axis at the points $-\varphi_m$ and $\varphi_m$, which are the points of extreme angular displacements of the oscillator. The angular separation of these points from the equilibrium position at $\theta_0$ are $\varphi_m + \theta_0$ (on the left side of $\theta_0$) and $\varphi_m - \theta_0$ (on the right side of $\theta_0$).

The upper part of the phase trajectory is a half-loop of a spiral whose focus is at the point $+\theta_0$, displaced to the right from the origin. While the representative point passes along this upper half-loop from $-\varphi_m$ to $\varphi_m$, the oscillator executes one half of a period of damped natural oscillation about the equilibrium position $\theta_0$, displaced to the right side: the flywheel passes from the extreme deflection $|\varphi_m + \theta_0|$ on the left side (measured from $\theta_0$) to the extreme deflection $\varphi_m - \theta_0$ on the right side.

When the oscillator reaches this extreme point, the equilibrium position instantly switches to the focus $-\theta_0$, and the representative point then passes along the lower half-loop, thus closing the phase trajectory of the steady-state motion.

The relative decrease of the amplitude because of the viscous friction during one half of the natural period ($t = T_0/2$) of oscillation equals $\exp(-\gamma T_0/2)$. So the left and right separations for the upper half-loop are related to one another through this exponential factor giving the frictional decay for a half-cycle:

$$(\varphi_m + \theta_0)\exp(-\gamma T_0/2) = \varphi_m - \theta_0. \tag{4.12}$$

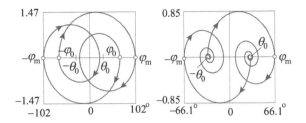

Figure 4.8: Phase trajectories for the resonance at $T = 3T_0$, $Q = 5$, $\theta_0 = 30°$ (left) and at $T = 7T_0$, $Q = 3$, $\theta_0 = 30°$ (right).

For the case in which $\gamma \ll \omega_0$, that is, $\gamma T_0 \ll 1$ (oscillator with relatively weak friction), we can assume $\exp(-\gamma T_0/2) \approx 1 - \gamma T_0/2$. Using this approximation in Eq. (4.12) and solving for $\varphi_m$, we obtain the desired estimate:

$$\varphi_m \approx \theta_0 \frac{2}{\gamma T_0/2} = \frac{4}{\pi} Q \, \theta_0. \tag{4.13}$$

The product of the damping constant $\gamma$ and the natural period $T_0$ is expressed here in terms of the quality factor $Q = \omega_0/2\gamma$.

Equation (4.13) shows that for resonance induced by the fundamental harmonic of the square-wave external torque ($T = T_0$) the amplitude of steady-state oscillation is $Q$ times greater than the amplitude $(4/\pi)\theta_0$ of this harmonic component in the square-wave motion of the exciting rod. (See Eq. (4.5).) The same conclusion can be reached from a spectral approach to the treatment of stationary forced oscillations.

Through a similar (though more complicated) calculation we can obtain an estimate of the maximal displacement, $\varphi_m$, attained in steady-state oscillations for any of the higher resonances when the period of the square-wave external torque is an odd multiple of the natural period. For example, for the resonance occurring at $T = 3T_0$ (three natural periods during one cycle of the exciting rod) we can consider a half of the closed phase trajectory (left-hand panel of Figure 4.8), which consists of three shrinking half-loops spiraling to the right-hand focus $(\theta_0, 0)$.

At the beginning of each cycle, when the exciter abruptly turns from left to right, the representative point occurs on the abscissa axis of the phase plane at some point $-\varphi_0$ (see the left-hand panel of Figure 4.8) to the left of the new equilibrium position $\theta_0$. After a half-period of natural oscillation about $\theta_0$ (after $T_0/2$) the flywheel reaches its maximal deflection $\varphi_m$ to the right-hand side, which we are going to evaluate. Then during the next period $T_0$ of natural oscillation the representative point makes a full revolution about the focus $(\theta_0, 0)$, and occurs again on the abscissa axis at the point $\varphi_0$, which is displaced to the right from the focus $(\theta_0, 0)$. At this moment the exciting rod turns abruptly to the left, and the remaining part of the closed phase trajectory is formed by the one-and-a-half loops spiraling toward the left focus $(-\theta_0, 0)$.

After each half-period $T_0/2$ of natural oscillations the maximal angular elongation of the flywheel from the equilibrium position diminishes by the same factor $q \approx \exp(-\gamma T_0/2) = \exp(-\pi/2Q)$. At the beginning of the cycle the representative point is displaced to the left from the equilibrium position $\theta_0$ through $|\varphi_0 + \theta_0$—. After $T_0/2$ later this point reaches its maximal elongation $\varphi_m - \theta_0$ to the right of $\theta_0$. Therefore

$$(\varphi_0 + \theta_0)q = \varphi_m - \theta_0. \tag{4.14}$$

During the next period $T_0$ the representative point traces a whole loop around $\theta_0$, and its maximal elongation is multiplied by $q^2$. Since now its displacement from the origin equals $\varphi_0$, this point is displaced from the equilibrium position through $\varphi_0 - \theta_0$. Therefore we can write:

$$(\varphi_m - \theta_0)q^2 = \varphi_0 - \theta_0. \tag{4.15}$$

Solving the system of Eqs. (4.14)–(4.15), we get the following expression for the desired maximal deflection $\varphi_m$:

$$\varphi_m = \theta_0 \left( 1 + \frac{2q}{1 - q^3} \right). \tag{4.16}$$

For example, at $Q = 5$ the factor $q = \exp(-\pi/2Q)$ equals 0.73, and Eq. (4.16) yields $\varphi_m = 3.39\, \theta_0$. For $\theta_0 = 30°$ we get $\varphi_m = 102°$ (see Figure 4.4).

We note that for this case it is not so simple to evaluate the maximal elongation of the flywheel on the basis of the spectral approach. Indeed, the shape of output oscillations now depends not only on the third harmonic of the input square-wave oscillation that is in resonance with the oscillator, but also on the first (fundamental) harmonic that produces in the output a contribution of the same order of magnitude. The shape of output oscillations in this case due to this contribution differs considerably from a pure sinusoid. This is clearly seen from the phase trajectory and the plots in Figure 4.4 of resonant steady-state oscillations for $T = 3T_0$, $Q = 5$, $\theta_0 = 30°$.

## 4.4.2  Amplitude of Steady Oscillations at Strong Friction

For the case of strong friction, the maximum elongation of the flywheel even in resonant conditions is only slightly greater than the amplitude of the exciting rod. The shape of the output pulses differs from the square-wave input impulses not so drastically as at weak friction. The principal distortion of the pulses' shape reveals itself in some smoothing of the abrupt leading front of input impulses. This is clearly seen from the graphs of $\varphi(t)$ in Figure 4.5 for $T = 7T_0$, $Q = 3$.

In order to evaluate the maximum elongation of the flywheel at steady-state oscillations in conditions of strong friction (quality factor $Q$ of the order $2 - 3$), we note that for $T/T_0 > Q$ the natural oscillations about a displaced equilibrium position almost damp out before the rod of the exciter makes its next-in-turn abrupt rotation to the new position. In these conditions the phase trajectory that spirals in towards the focus $(\theta_0, 0)$ approaches very close to this point before the transition to spiraling towards the other focus (see the right-hand panel of Figure 4.8). This means that at the beginning of each in-turn half-cycle of steady-state oscillations the representative point starts almost from one of the foci. This occurs at $T/T_0 > Q$ independently of the exact value of $T/T_0$: it does not matter whether the period of excitation has a resonant value (that is, equals an odd number of natural periods) or not.

Next we find the maximum elongation in such conditions. At abrupt rotation of the exciting rod from left to right, the representing point starts to move from the left focus $(-\theta_0, 0)$ (see the right-hand panel of Figure 4.8) along a shrinking spiral, winding around the right focus $(\theta_0, 0)$. This spiral intersects the abscissa axis at the point $\varphi_m$, which we are interested in after a half-period $T_1/2 \approx T_0/2$ of damped natural oscillations about the displaced equilibrium position $\theta_0$. The initial angular displacement from this position equals approximately $2\theta_0$, and after a half-period the displacement equals $\varphi_m - \theta_0$. These initial and final displacements differ by a factor $q \approx \exp(-\gamma T_0/2) = \exp(-\pi/2Q)$; therefore $2\theta_0 q = \varphi_m - \theta_0$. This yields the following final expression for the maximum elongation of the flywheel:

$$\varphi_m \approx \theta_0(1 + 2q) \approx \theta_0(1 + 2e^{-\pi/2Q}). \tag{4.17}$$

For example, at $Q = 3$ Eq. (4.17) gives for $\varphi_m$ the value $2.18\,\theta_0$, i.e., the maximum elongation of the flywheel is more than twice the displacement $\theta_0$ of the exciting rod from its middle position. For $\theta = 30°$ we have $\varphi_m \approx 66°$ (see Figure 4.5).

This approximate estimate for $\varphi_m$ is almost independent of the period of excitation. However, it is valid only under assumption that during a half of this period natural oscillations nearly fade away. To find the precision of the above estimate, we can consider the residual displacement of the flywheel from the focal point $\theta_0$. For example, at $T = 7T_0$ this residual displacement equals approximately $2\theta_0 \exp(-7\pi/2Q)$, which for $Q = 3$ yields $0.03\,\theta_0$. This means that at the beginning of each half-cycle the representing point starts not exactly from $\pm\theta_0$ as we assumed above, but rather from some point slightly displaced from $\pm\theta_0$ to one or the other side depending on the exact value of $T/T_0$.

## 4.4.3 Amplitude of Steady Oscillations at $T = 2nT_0$

When the driving period is an even-numbered multiple of the natural period, the maximal deflection of the flywheel $\varphi_m$ attained in steady-state forced oscillations at weak friction is close to $2\theta_0$, that is, about twice the amplitude of square-wave motion of the exciter. We can easily see this result from the shape of the corresponding phase trajectory: each of its two symmetrical halves consists of an integral number of shrinking loops of a spiral winding around one of the foci $\theta_0$ and $-\theta_0$. Figure 4.9 shows this kind of the phase trajectory and the time-dependent graphs for a special case in which $T = 4T_0$.

For $T = 2T_0$, one complete cycle of the natural oscillation occurs while the equilibrium position is displaced to one side. In the absence of friction both closed

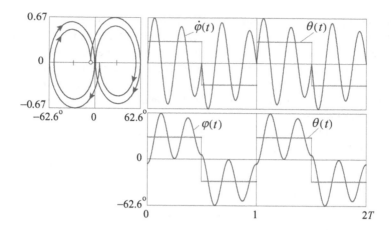

Figure 4.9: The phase trajectory (left) and graphs of the time dependence of the angular velocity and of the angular position for steady-state forced oscillations at $T = 4T_0$.

loops of the steady-state phase trajectory meet at the origin of the phase plane, and the magnitude $\varphi_m$ of the maximal displacement on each side of the zero point just equals $2\theta_0$. Friction causes the loops to shrink, and the maximal displacement $\varphi_m$ of the flywheel becomes slightly smaller than $2\theta_0$.

## 4.4.4    Steady-State Oscillations at High Frequencies of the Square-Wave Torque

For high frequencies of the external force, when the square-wave period $T$ of the driving rod is very short compared to the natural period $T_0$ of the oscillator, in steady-state motion the flywheel executes only small vibrations about the mid-point $\varphi = 0$. The period of these vibrations is the same as that of the driving rod. However, they occur in the opposite phase with respect to the exciter motion, and their amplitude is small compared to the amplitude $\theta_0$ of the exciter.

Since the flywheel moves little while the position of the rod is fixed at either $\theta_0$ or $-\theta_0$, we can consider the torque of the spring exerted on the flywheel as nearly constant during the intervals between successive jumps of the rod. Therefore the graph of the angular velocity in steady-state short-period oscillations consists of nearly rectilinear segments (Figure 4.10). They correspond to the rotation of the flywheel with a uniform angular acceleration $\omega_0^2\theta_0$ caused by the constant torque of the strained spring. Such rotation continues in one direction during time intervals between abrupt turns of the rod. After each succeeding turn the acceleration changes sign, remaining nearly the same in magnitude. In the graph of the angular velocity, the straight segments join to form a saw-toothed pattern of isosceles triangles. The corresponding graph of the angular deflection is formed by adjoining

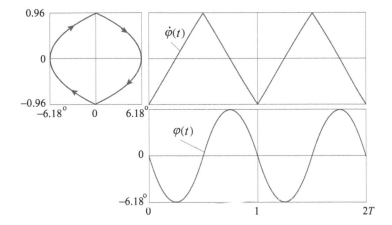

Figure 4.10: Steady-state oscillations at $T \ll T_0$ ($T = 0.25\,T_0$, $Q = 20$, $\theta_0 = 75°$)—phase diagram (left) and time dependent graphs (rectilinear segments of the saw-teeth graph of the angular velocity $\dot{\varphi}(t)$ and parabolic segments of the angle of deflection $\varphi(t)$).

parabolic segments alternating after each half of the driving period.

We can easily find the height of these segments, that is, the maximal displacement $\varphi_{\mathrm{m}}$ of the rotor in the case under consideration, by calculating the angular path of the rotor as it moves with a constant angular acceleration $\omega_0^2 \theta_0 = 4\pi^2 \theta_0 / T_0^2$ from the zero point of the dial to maximal elongation $\varphi_{\mathrm{m}}$ during a quarter driving period $T/4$:

$$\varphi_{\mathrm{m}} \approx \frac{\pi^2}{8} \left( \frac{T}{T_0} \right)^2 \theta_0. \tag{4.18}$$

## 4.5 Energy Transformations

The exchange of mechanical energy between the oscillator and the source of the external driving force occurs in the investigated system only at the instants when the driving rod turns abruptly from one position to the other. During the intervals between such instant turns, while the oscillator executes damped natural oscillations about one of the two displaced equilibrium positions, only an alternating partial conversion between the elastic potential energy of the strained spring and the kinetic energy of the flywheel occurs, accompanied by the gradual dissipation of mechanical energy because of friction.

To get a general idea of these energy transformations, we can refer to the section "Energy Transformations" in the corresponding simulation program, and consider the motion of the point representing the total energy in the graph of potential energy versus the angle of deflection.

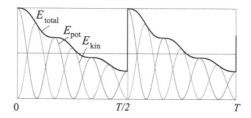

Figure 4.11: Energy transformations at steady-state oscillations with $T = 3T_0$, $Q = 8$. Curve $E_{\text{pot}}$ – potential energy, curve $E_{\text{kin}}$ – kinetic energy, curve $E_{\text{total}}$ – total mechanical energy.

A parabolic potential well corresponds to each of the two equilibrium positions of the oscillator. When the flywheel is located at the angle $\varphi$ from its central position, the corresponding potential energy of the spring is given by one of the two quadratic functions:

$$U(\varphi) = \frac{1}{2}k(\varphi \mp \theta_0)^2. \qquad (4.19)$$

We must take the upper sign in Eq. (4.19) if the equilibrium position is displaced to the right (to the point $+\theta_0$), and the lower sign if the equilibrium position is displaced to the left.

At an instantaneous turn of the driving rod from one position to the other the angular velocity of the massive flywheel and hence its kinetic energy do not change. An abrupt change occurs only in the value of the elastic potential energy of the spring. This causes the representative point to make an abrupt vertical transition from one of the parabolic potential wells to the other at a fixed value of the angle $\varphi$.

During the interval before the next jump, while the oscillator executes damped natural oscillations about a displaced equilibrium position, the point that represents the total energy travels back and forth between the walls of the corresponding potential well, descending gradually toward the bottom because of energy losses caused by friction.

The time-dependent graphs of the total mechanical energy $E_{\text{total}}(t)$, kinetic energy $E_{\text{kin}}(t)$, and potential energy $E_{\text{pot}}(t)$ during one period of steady-state oscillations are shown in Figure 4.11 for $T = 3T_0$, $Q = 8$.

It is important to note that in this simplified model of the physical system the deformation of the spiral spring is assumed to be quasistatic. In other words, we ignore the possibility that the spring vibrates as a system with distributed parameters, whose each portion has both elastic and inertial properties. For a light spring (attached to a comparatively massive flywheel) these vibrations are characterized by much higher frequencies than the frequencies of the torsional oscillations of the flywheel. Our simplified model is applicable to such a physical system because rapid vibrations of the spring quickly damp out.

# 4.6 The Electromagnetic Analogue of the Mechanical System

Forced oscillations of the electric charge $q$ stored in a capacitor in a resonant series $LCR$-circuit excited by a square-wave input voltage $V(t)$ (see the right-hand panel of Figure 4.1) obey the same differential equation as does the forced oscillation of a mechanical torsion spring oscillator excited by periodic abrupt changes of position of the driving rod:

$$\ddot{q} + 2\gamma\dot{q} + \omega_0^2 q = \omega_0^2 C V(t). \qquad (4.20)$$

In this equation $\omega_0$ is the natural frequency of oscillations of charge in the circuit in the absence of resistance. It depends on the capacitance $C$ of the capacitor and the inductance $L$ of the coil: $\omega_0 = 1/\sqrt{LC}$. The damping constant $\gamma = R/(2L)$ characterizes the dissipation of electromagnetic energy occurring in a resistor whose resistance is $R$.

Because of this similarity, the mechanical system simulated in the computer program enables us to give a very clear treatment of transformation of the square-wave input voltage $V(t) = \pm V_0$ into the output voltage $V_C(t) = q/C$ (voltage across the capacitor $C$). The output voltage $V_C(t)$ is analogous to the deflection angle $\varphi(t)$ of the rotor. The alternating electric current $I(t) = \dot{q}(t)$ in the circuit is analogous to the angular velocity $\dot{\varphi}(t)$ of the mechanical model.

The assumption concerning quasistatic character of the spring deformation in the mechanical model, i.e., the assumption of possibility to neglect rapid vibrations of the spring as a distributed parameter system, corresponds to the ordinary implicit assumption of quasistationary current in the circuit. According to this assumption, the momentary value of the current is the same along the whole circuit. The assumption is valid if the inductance and the capacitance of the wires are negligible compared with the inductance of the coil and the capacitance of the capacitor, respectively. In this case the oscillatory circuit can be treated as a system with lumped parameters (as a system with one degree of freedom), in which all the capacitance is concentrated in the capacitor and all the inductance is concentrated in the coil.

Some caution is necessary in interpreting the analogy between the mechanical oscillator and the electric $LCR$-circuit with respect to energy transformations. It is incorrect to identify exactly the electric potential energy of a charged capacitor with the elastic potential energy of a strained spring because the latter depends directly on the angle $(\varphi \pm \phi_0)$, while the energy of a capacitor depends directly on the charge $q$ or on the corresponding voltage $V_C = q/C$ (*not* on the voltage $(V_C \pm V_0)$). In contrast to the spring oscillator, for which a jump in the position of the rod causes an abrupt change of the elastic potential energy of the spring, a jump in the input voltage across an electric circuit does not abruptly change the charge and the energy of a capacitor.

The spectral treatment of the transformation of a square-wave driving force (an input) into the steady-state oscillations of the mechanical spring oscillator (the output) is equally valid for the transformation of an input square-wave voltage into the output oscillations of charge in the analogous electric circuit. Only those harmonics of the input signal that are near the resonant frequency of the circuit are noticeably present in the output voltage across the capacitor.

A resonant circuit selectively transmits to the output signal those harmonic components of the input signal whose frequencies are near the resonant frequency of the circuit. The greater the quality factor $Q$, the sharper the resonance curve and the finer the *selectivity* of the oscillatory circuit.

It is possible to vary the natural frequency $\omega_0 = 1/\sqrt{LC}$ of a resonant circuit by varying either the capacitance $C$ or the inductance $L$. Such tunable resonant circuits with high selectivity can serve as spectral instruments that are physically able to Fourier analyze a complex input signal.

The assumption concerning the quasistatic character of the spring deformation in the mechanical model, i.e., the assumption of possibility to neglect rapid vibrations of the spring as a distributed parameter system, corresponds to the ordinary implicit assumption of quasistationary current in the analogous oscillatory circuit. According to this assumption, the momentary value of the current is the same along the whole circuit. The assumption is valid if the inductance and capacitance of the wires are negligible compared with the inductance of the coil and the capacitance of the capacitor, respectively. In this case the oscillatory circuit can be treated as a system with lumped parameters (as a system with one degree of freedom), in which all the capacitance is concentrated in the capacitor and all the inductance is concentrated in the coil.

## 4.7    Concluding Remarks

We have considered in this chapter two different ways of determining the steady-state response of the linear oscillator to a non-harmonic driving force. The traditional approach based on the Fourier expansion of the input driving force is certainly quite general because it is applicable to an arbitrary periodic excitation. The second method based on representing forced oscillations as some sequence of natural oscillations about displaced equilibrium positions can be used only for piecewise constant excitations. Nevertheless, this approach is physically much more obvious, and also allows us to understand transient processes. Combining both approaches in teaching students, we can hope to give them better understanding of this important topic.

The obvious, intuitive treatment of the transformation of a square-wave driving force (an input) into the steady-state oscillations of the mechanical spring oscillator (the output) described in this chapter is equally valid for the transformation of an input square-wave voltage into the output oscillations of charge in the analogous electric circuit. Therefore, behavior of this familiar mechanical system can help a student to better understand why and how an electromagnetic oscillatory $LCR$-circuit transfers the square-wave voltage from input to output with a distortion of its shape. Mechanical analogies allow a direct visualization and thus can be very useful in gaining an intuitive understanding of complex phenomena.

This simple mechanical system helps us to understand both the complex shape of the output oscillations and also their spectral composition. Only those harmonics of the input signal, which are close to the resonant frequency of the circuit, are noticeably present in the output voltage across the capacitor. In other words, such a resonant circuit selectively responds to different harmonic components of the input signal. The greater the quality factor $Q$, the sharper the resonance curve, and the finer the *selectivity* of the oscillatory circuit.

It is possible to easily vary the natural frequency $\omega_0 = 1/\sqrt{LC}$ of a resonant circuit by varying either the capacitance $C$ or the inductance $L$. Such a tunable resonant circuit with high selectivity can serve as a spectral instrument that is able to accomplish the mathematical task of Fourier expansion of a complex input signal onto separate harmonic components on a physical level. The mechanical system described in this chapter provides a clear and plain way to understand this possibility.

## 4.8 Review of the Principal Formulas

The differential equation of motion for a torsion spring oscillator driven by a square-wave external force:

$$\ddot{\varphi} + 2\gamma\dot{\varphi} + \omega_0^2\varphi = \begin{cases} \omega_0^2\theta_0, & (0, T/2), \\ -\omega_0^2\theta_0, & (T/2, T). \end{cases} \tag{4.21}$$

The same equation in which the square-wave shaped right-hand side is represented as a Fourier series:

$$\ddot{\varphi} + 2\gamma\dot{\varphi} + \omega_0^2\varphi = \sum_{k=1, 3, 5...}^{\infty} \frac{4\theta_0\omega_0^2}{\pi k} \sin\omega_k t. \tag{4.22}$$

The particular periodic solution of the equation (describing steady-state oscillations):

$$\varphi(t) = \sum_{k=1, 3, 5...}^{\infty} \frac{4\theta_0}{\pi k} \frac{\omega_0^2}{\sqrt{(\omega_0^2 - \omega_k^2)^2 + 4\gamma^2\omega_k^2}} \sin(\omega_k t + \alpha_k), \tag{4.23}$$

where the phases $\alpha_k$ of the individual harmonics are determined by:

$$\tan\alpha_k = \frac{2\gamma\omega_k}{\omega_k^2 - \omega_0^2}. \tag{4.24}$$

The time dependence of $\varphi(t)$ during the interval $0 \le t \le T/2$, when the equilibrium position is located at $\varphi = \theta_0$:

$$\varphi(t) = \theta_0 + Ae^{-\gamma t}\cos(\omega_1 t + \theta), \quad (0, T/2), \tag{4.25}$$

where $\omega_1 = \sqrt{\omega_0^2 - \gamma^2}$, the frequency of natural damped oscillations, and $A$ and $\theta$ are some constants.

The time dependence of $\varphi(t)$ during the interval $T/2 \le t \le T$, when damped natural oscillations occur about the equilibrium position located at $-\theta_0$:

$$\varphi(t) = -\theta_0 - Ae^{-\gamma(t-T/2)} \cos(\omega_1(t - T/2) + \theta), \qquad (T/2,\ T). \quad (4.26)$$

For a steady-state process, the constants $A$ and $\theta$ here have the same values as they do for the interval $(0,\ T/2)$.

## 4.9  Questions, Problems, Suggestions

### 4.9.1  Swinging of the Oscillator at Resonance

**4.9.1.1  The Principal Resonance in the Absence of Friction.** Select the idealized case of no friction. Enter the period $T$ of the external force, letting this period be the period $T_0$ of natural oscillations. Choose null initial conditions. That is, let the oscillator be at rest in the equilibrium position at the moment the external force is activated.

(a) What must be the value $\theta_0$ of the angular amplitude of the square-wave motion of the driving rod in order that the amplitude reach $180°$ after the first 10 cycles? Verify your answer by a simulation experiment.

(b) What regularity does the growth of the amplitude exhibit? Explain the form of the phase trajectory displayed. How does the energy of the oscillator grow with time?

(c) If the oscillator is exactly tuned to resonance, is it possible for the amplitude to diminish? Give some physical justification for your answer. Can you test your answer by a simulation experiment?

**4.9.1.2  High Resonances in the Absence of Friction.** Examine the resonant excitation of the oscillator initially at rest in the equilibrium position when the period of the external square-wave force is three times longer than the natural period: $T = 3T_0$.

(a) At what value $\theta_0$ of the amplitude of a square-wave oscillation of the driving rod does the oscillator reach its maximal deflection of $180°$ in 10 first cycles of the external action? Verify your prediction experimentally.

(b) What are the differences in the graphs and the phase trajectories between this case and the previous case (Problem 1.1) for which $T = T_0$?

**4.9.1.3\*  Transient Process and Steady-State Oscillations at the Principal Resonance.** Letting the period of the square-wave motion of the driving rod be at the principal resonance, $T = T_0$, and letting the flywheel be initially at rest in the equilibrium position, examine the transient process and the steady-state oscillations in the presence of friction:

(a) Calculate the amplitude of steady-state oscillations for the values $\theta_0 = 10°$ and $Q = 10$. Verify your result experimentally.

(b) What regularity does the growth of the amplitude exhibit in this case? Explain the peculiarities of the phase trajectory.

(c) What is the initial amplitude of damped natural oscillations constituting the transient process for $\theta_0 = 10°$ and $Q = 10$?

(d) What initial conditions cause steady-state forced oscillations to appear from the moment the external force begins to act, thus eliminating the transient process? Verify your answer experimentally.

(e) Examine the spectrum of steady-state oscillations (the output) in this case. Why are these oscillations nearly purely harmonic in spite of the non-sinusoidal, square-wave shape of the input?

### 4.9.1.4* Steady-State Oscillations at High Resonances.

(a) Calculate the amplitude of steady-state oscillations for $\theta_0 = 25°$, $Q = 5$, and $T = 3T_0$. Explain the shape of the graphs displayed and of the phase trajectory.

(b) What energy transformations take place during steady-state oscillations? Compare the graphs of the time dependence of the kinetic, potential, and total energy with the corresponding graphs of the angular deflection and the angular velocity of the flywheel. Explain the shape of the graph of the total energy versus the angle of deflection, and explain its relationship to the parabolic potential wells shown in the same diagram.

(c) Which harmonic components determine the shape of output steady-state oscillations in this case? Why, in spite of the exact tuning of the oscillator to the frequency of the third harmonic of the input external force, does the first harmonic component of this force appreciably influence the shape of the output oscillations? How does this harmonic exhibit itself in the pattern of the output oscillations?

(d) Examine the influence of friction on the shape and on the spectral composition of steady-state oscillations at $T = 3T_0$. Note the relative reduction in the contribution of the first and fifth harmonics as the quality factor of the oscillator is increased.

(e) Explore the resonant oscillations of the flywheel when the frequency of the fifth or the seventh harmonic of the external square-wave driving force coincides with the natural frequency of the oscillator. Observe the transformation of the spectrum from input to output, and the dependence of the spectrum of steady-state oscillations on the quality factor of the oscillator. What is the shape of the phase trajectory in these cases? How can you estimate the value of the maximal displacement of the flywheel from the mid-point of its oscillations (from the zero point of the dial) when friction is large (when $Q$ ranges say from 1 to 3)?

## 4.9.2 Non-Resonant Forced Oscillations

### 4.9.2.1* Conditions That Eliminate a Transient at $T = 2T_0$.

(a) Predict the shape of the graphs of the angular deflection, angular velocity, and the phase trajectory of forced oscillations in the absence of friction, for $T = 2T_0$. Under what initial conditions is the steady-state oscillation established immediately after the force is activated? Verify your predictions in a simulation experiment.

(b) Why does the total energy remain constant in these oscillations? When the driving rod executes a jump, why does the oscillator neither gain energy from nor give back energy to the external source?

(c) Examine the spectral composition of steady-state oscillations. Note the contribution of the third harmonic and the influence of its alteration in phase on the shape of the output oscillations: The frequency of the third harmonic in this case is higher than the natural frequency of the oscillator. Therefore, its phase in the output oscillations is inverted. As a result, this harmonic component, super-imposed on the fundamental, produces a time-dependent graph with bulges at the positions of the flat parts of the square-wave input graph.

### 4.9.2.2  Steady-State Oscillations for $T = 2T_0$.

(a) Consider forced oscillations for the case in which $T = 2T_0$ in the presence of friction, by setting $Q \approx 5 - 10$. How do the time-dependent graphs and the phase trajectory differ from the preceding case (Problem 2.1), in which friction is absent? Calculate the maximal deflection of the flywheel attained in these steady-state oscillations. Note changes in the energy transformations.

(b) Why does friction not noticeably influence the spectral composition of steady-state oscillations in this case, in contrast to the case in which $T = 3T_0$?

### 4.9.2.3  Steady-State Oscillations for a Large External Period.

(a) Examine forced oscillations for a case in which the natural frequency of the oscillator lies somewhere between the frequencies of two consecutive high odd harmonics of the external action (e.g., let $5T_0 < T < 7T_0$). Which harmonic components dominate in the output steady-state oscillations? Compare the shape of the output steady-state oscillations with the shape of the input square-wave impulses. What is the main difference between the patterns of input and output oscillations?

(b) Investigate the influence of friction on the character of steady-state os-cillations. Why are the distortions of the output less prominent the stronger the damping? That is, why is the shape of the output curve for large friction nearly rectangular?

(c) Explain the energy transformations in these oscillations using the graph of the total energy versus the angle of deflection. What is the relationship of this graph with the parabolic potential wells shown in the same diagram?

### 4.9.2.4  Steady-State Oscillations Forced by Short-Period Impulses.

(a) Choose a value for the period $T$ of the square-wave external force to be a small fraction (say $0.2 - 0.3$) of the natural period $T_0$ of the oscillator. The graph of the angular velocity versus time for steady-state output oscillations has a saw-toothed pattern with teeth that are nearly rectilinear isosceles triangles. Suggest an explanation. What is the difference between the graph of the angular deflection versus time for this case and a sine curve?

(b) Evaluate theoretically the height of a tooth of the angular velocity graph. Also evaluate the maximal deflection angle for these non-sinusoidal steady-state oscillations. Consider the case of weak or moderate friction. Let, for example, the period $T$ be $T_0/4$ and the angle $\theta_0$ describing the instantaneous deflections of the driving rod be $30°$. What spectral composition is characteristic of such oscillations?

# Chapter 5

# Parametric Excitation of a Linear Oscillator

**Annotation.** Chapter 5 is concerned with the phenomenon of *parametric resonance* arising from a periodic square-wave modulation of the moment of inertia of a linear torsion spring oscillator. Computer simulations of oscillations show the motion of the system and are accompanied by plotting the time dependencies of the angle of deflection and the angular velocity. Phase trajectories and energy transformations for different values of the depth of modulation are analyzed, and conditions of parametric regeneration and of parametric resonance are discussed. Ranges of frequencies where parametric excitation is possible are determined. Stationary oscillations on the boundaries of these ranges are investigated both analytically and by computer simulation.

## 5.1 Summary of the Theory. General Concepts

### 5.1.1 Classification of Oscillations

In the conventional classification of oscillations by their method of excitation, oscillations are called *forced* if an oscillator is subjected to an external periodic influence whose effect on the system can be expressed by a separate term, a periodic function of the time, in the differential equation of motion describing the system. Forced oscillations are discussed in Chapter 3 and Chapter 4.

The investigation of *non-stationary, position-dependent forces*, i.e., the forces that are explicitly determined by both temporal and spatial coordinates, is more complicated. For example, let a restoring force $F = -kx$ arise when the system is displaced through some distance $x$ from the equilibrium position. But in contrast to the stationary case, the parameter $k$ changes with time because of some periodic influence: $k = k(t)$. In the differential equation of motion for the system,

$$m\ddot{x} = -k(t)x, \tag{5.1}$$

the coefficient of $x$ is not constant: It explicitly depends on time. Oscillations in such a system are essentially different from both free oscillations, which occur when $k$ is constant, and forced oscillations, which occur when $k$ is constant and an additional time-dependent forcing term is added to the right-hand side of the equation of motion, Eq. (5.1).

In the case of *periodic* changes of the parameter $k$, when $k(t + T) = k(t)$, where $T$ is the period, the corresponding differential equation, Eq. (5.1), is called Hill's equation. Oscillations in a system described by Hill's equation are called *parametrically excited* or simply *parametric oscillations*. When the amplitude of oscillation caused by the periodic modulation of some parameter increases steadily, we describe the phenomenon as *parametric resonance*. In parametric resonance, equilibrium becomes unstable and the system performs oscillations whose amplitude increases exponentially.

The characteristics and causes of parametric resonance are considerably different from those of the resonance occurring when the oscillator responds to a periodic external force. Specifically, the resonant relationship between the frequency of modulation of the parameter and the mean natural frequency of oscillation of the system is different from the relationship between the driving frequency and the natural frequency for the usual resonance in forced oscillations. And if there is friction, the amplitude of modulation of the parameter must exceed a certain threshold value in order to cause parametric resonance.

## 5.1.2   The Simulated Physical System

A physical system undergoes a parametric forcing if one of its parameters is modulated periodically with time. A common familiar example of parametric excitation of oscillations is given by the playground swing on which most people have played in childhood. The swing can be treated as a physical pendulum whose reduced length changes periodically as the child squats at the extreme points, and straightens when the swing passes through the equilibrium position.

It is easy to illustrate this phenomenon by the following simple experiment. Let a thread with a bob hanging on its end pass through a little ring fixed immovably in a support. The other end of the thread that you are holding in your hand you can pull by some small length each time when the swinging bob passes through the middle position and release the thread to its previous length each time the bob reaches the utmost deflection. These periodic variations of the pendulum length with the frequency twice the frequency of natural oscillation cause the amplitude to increase progressively. Another canonical example of parametric pumping is given by a pendulum whose support oscillates vertically.

However, such systems do not perfectly suit the initial acquaintance with the parametric excitation because the ordinary pendulum is a *nonlinear* physical system: The restoring torque of the gravitational force is proportional to the sine of the deflection angle.

That is why we suggest here to study the basics of parametric resonance by using the simplest *linear* mechanical system in which the phenomenon is possible,

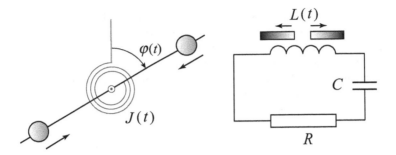

Figure 5.1: Schematic image of the torsion spring oscillator with a rotor whose moment of inertia is forced to vary periodically (left), and an analogous $LCR$-circuit with a coil whose inductance is modulated by moving periodically an iron core in and out of the coil (right).

namely, the torsion spring oscillator, similar to the balance device of a mechanical watch.

The suggested computer program simulates a simple physical system that perfectly suits the initial acquaintance with the basics of parametric resonance, namely, a torsion spring oscillator (Figure 5.1) similar to the balance device of a mechanical watch. It consists of a rigid rod that can rotate about an axis that passes through its center. Two identical weights are balanced on the rod. An elastic spiral spring is attached to the rod. The other end of the spring is fixed. When the rod is turned about its axis, the spring flexes. The restoring torque $-D\varphi$ of the spring is proportional to the angular displacement $\varphi$ of the rotor from the equilibrium position. After a disturbance, the rotor executes natural harmonic torsional oscillations.

We assume that the weights can be shifted simultaneously along the rod in opposite directions into other symmetrical positions so that the rotor as a whole remains balanced. However, its moment of inertia is changed by such displacements of the weights.

When the weights are shifted toward or away from the axis, the moment of inertia decreases or increases respectively. As the moment of inertia of the rotor is changed, so also is the natural frequency of its oscillation. Thus the moment of inertia of the rotor is the parameter to be modulated in this system.

This physical system is ideal for the study of parametric resonance and has several advantages in an educational context because it gives a very clear example of the phenomenon in a *linear* mechanical system. All peculiarities of parametric excitation in this linear system can be completely explained and exhaustively investigated by modest means even quantitatively.

Another similar mode of the parametric modulation – a smooth periodic variation of the moment of inertia by sinusoidal motion of the weights along the rod – is considered in Chapter 6.

In the case of the square-wave modulation, abrupt, almost instantaneous increments and decrements of the moment of inertia occur sequentially, separated by equal time intervals. We denote these intervals by $T/2$, so that $T$ equals the period of the variation in the moment of inertia (the *period of modulation*).

The square-wave variation of a parameter can produce considerable oscillation of the rotor if the period of modulation is chosen properly. For example, suppose that the weights are drawn closer to each other at the instant at which the rotor passes through the equilibrium position, when its angular velocity is almost maximal. While the weights are being moved, the angular momentum of the system remains constant since no torque is needed to effect this displacement. Thus the resulting reduction in the moment of inertia is accompanied by an increment in the angular velocity, and the rotor acquires additional energy. The greater the angular velocity, the greater the increment in energy. This additional energy is supplied by the source that moves the weights along the rod.

On the other hand, if the weights are instantly moved apart along the rotating rod, the angular velocity and the energy of the rotor diminish. The decrease in energy is transmitted back to the source.

In order that increments in energy occur regularly and exceed the amounts of energy returned, i.e., in order that, as a whole, the modulation of the moment of inertia regularly feed the oscillator with energy, the period of modulation must satisfy certain conditions.

For instance, let the weights be drawn closer to and moved apart from one another twice during one mean period of the natural oscillation. The angular velocity increases at the moment the weights come together, and vice versa. Furthermore, let the weights be drawn closer at the instant of maximum angular velocity, so that the rotor gains as much energy as possible. Then, after a quarter period of the natural rotary oscillation, the weights are moved apart, and this occurs almost at the instant of extreme deflection, when the angular velocity is nearly zero. Therefore this particular motion causes almost no change in the angular velocity and kinetic energy of the rotor. Thus, modulating the moment of inertia at a frequency twice the mean natural frequency generates the greatest growth of the amplitude, provided that the phase of the modulation is chosen in the way described above.

It is evident that the energy of the oscillator is increased most greatly not only when two full cycles of variation in the parameter occur during one natural period of oscillation, but also when two cycles occur during three, five, or any odd number of natural periods. We shall see later that the delivery of energy, though less efficient, is also possible if two cycles of modulation occur during an even number of natural periods.

If the changes of a parameter are produced with the above-mentioned periodicity but not abruptly, the influence of these changes on the oscillator is qualitatively quite similar, though the efficiency of the parametric delivery of energy (at the same amplitude of the parametric modulation) is a maximum for the square-wave time dependence, because this form of modulation provides optimal conditions for the transfer of energy to the oscillating system. The case of the sinusoidal modulation of some parameter is important for practical applications.

### 5.1.3 Electromagnetic Analogue of the Mechanical System

Parametric excitation is possible in various oscillatory systems. Electromagnetic oscillations in a series $LCR$-circuit containing a capacitor, an inductor (a coil), and a resistor can be excited by periodic changes of the capacitance if we periodically move the plates closer together and farther apart, or by changes of the inductance of the coil if we periodically move an iron core in and out of the coil, as shown in the right-hand panel of Figure 5.1. Such periodic changes of the inductance are quite similar to the changes of the moment of inertia in the mechanical system considered above. The strongest oscillations are excited when the cycle of such changes is repeated twice during one period of natural electromagnetic oscillations in the circuit, i.e., when the frequency of a parametric modulation is twice the natural frequency of the system. It is evident that parametric excitation can occur only if at least weak natural oscillations already exist in the system.

Parametric excitation is possible only with the modulation of one of the energy-consuming parameters, $C$ or $L$ ($D$ or $J$ in the mechanical system). Modulation of the resistance $R$ (or of the damping constant $\gamma$ in the mechanical system) can affect only the character of the damping of oscillations. It cannot generate an increase in their amplitude.

The mechanical system simulated in the relevant computer program has certain spectacular didactic advantages primarily because its motion is easily represented on the computer screen, and it is possible to see directly what is happening. Such visualization makes the simulation experiments very convincing and easy to understand, aiding a great deal in developing our physical intuition.

### 5.1.4 Conditions for Parametric Resonance

There are several important differences that distinguish parametric resonance from the ordinary resonance caused by an external force acting directly on the system. The growth of the amplitude and hence of the energy of oscillations during parametric excitation is provided by the work of forces that cause the modulation, that is, by the work of forces that periodically change the parameter.

The most efficient energy transfer to the oscillatory system occurs when the parameter is changed twice during one period of the excited natural oscillations. But the delivery of energy, though less efficient, is possible when the parameter changes once during one period, twice during three periods, and so on. That is, resonance is possible when one of the following conditions for the frequency $\omega$ (or for the period $T$) of a parameter modulation is fulfilled:

$$\omega = 2\omega_0/n, \qquad T = nT_0/2, \qquad n = 1, 2, \ldots \qquad (5.2)$$

For a given amplitude of parametric modulation, the higher the order $n$ of resonance, the less (in general) the amount of energy delivered to the oscillating system during one period.

One of the most interesting characteristics of parametric resonance is the possibility of exciting increasing oscillations not only at the frequencies $\omega_n$ given in

Eq. (5.2), but also in a range of frequencies $\omega$ lying on either side of the values $\omega_n$ (in the *ranges of instability.*) These intervals become wider as the degree (depth) of modulation is increased, that is, as the range of parametric variation is extended. By this range we mean, in the case of the rotor, the difference in the maximal and minimal values of its moment of inertia, and in the oscillating electrical circuit, the differences in the inductance of the coil.

An important difference between parametric excitation and forced oscillations is related to the dependence of the growth of energy on the energy already stored in the system. While for forced excitation the increment of energy during one period is proportional to the *amplitude* of oscillations, i.e., to the square root of the energy, at parametric resonance the increment of energy is proportional to the *energy* stored in the system.

Energy losses caused by friction (unavoidable in any real system) are also proportional to the energy already stored. In the case of direct forced excitation, an arbitrarily small external force gives rise to resonance. However, energy losses restrict the growth of the amplitude because these losses grow faster with the energy than does the investment of energy arising from the work done by the external force.

In the case of parametric resonance, both the investment of energy caused by the modulation of a parameter and the frictional losses are proportional to the energy stored (to the square of the amplitude), and so their ratio does not depend on the amplitude.

This means that parametric resonance is possible only when a *threshold* is exceeded, that is, when the increment of energy during a period (caused by the parametric variation) is larger than the amount of energy dissipated during the same time. To satisfy this requirement, the range of the parametric variation (the depth of modulation) must exceed some critical value.

The critical (threshold) value of the modulation depth depends on friction. However, if the threshold is exceeded, the frictional losses of energy cannot restrict the growth of the amplitude. In a linear system the amplitude of parametrically excited oscillations must grow infinitely.

In a nonlinear system the natural period depends on the amplitude of oscillations. If conditions for parametric resonance are fulfilled at small oscillations and the amplitude begins to grow, the conditions of resonance become violated at large amplitudes. In a real system the growth of the amplitude is restricted by nonlinear effects.

## 5.1.5   The Threshold of Parametric Excitation

We can use arguments employing the conservation laws to evaluate the modulation depth that corresponds to the threshold of parametric excitation of the torsion oscillator.

Let the changes in the moment of inertia $J$ of the rotor occur between maximal $J_1$ and minimal $J_2$ values that equal $J_0(1 + m)$ and $J_0(1 - m)$ respectively, where $J_0$ is a mean value of the moment of inertia, and $m$ is the dimensionless quantity called the *depth of modulation*, which equals fractional increments and decrements of the modulated parameter.

During abrupt radial displacements of the weights along the rod, the angular momentum $L = J\omega$ of the rotor is conserved. Therefore it is convenient to use the expression $E_{\mathrm{kin}} = L^2/(2J)$, which gives the kinetic energy of the rotor in terms of $L$ and $J$.

For the increment $\Delta E$ in the rotor kinetic energy that occurs during an abrupt shift of the weights toward the axis, when the moment of inertia decreases from the value $J_1 = J_0(1 + m)$ to the value $J_2 = J_0(1 - m)$, we can write:

$$\Delta E = \frac{L^2}{2J_0}\left(\frac{1}{1-m} - \frac{1}{1+m}\right) \approx 2m\frac{L^2}{2J_0} \approx 2mE_{\mathrm{kin}}. \qquad (5.3)$$

The approximate expressions in (5.3) are valid for small values of the modulation depth ($m \ll 1$). If the event occurs near the equilibrium position of the rotor, when the total energy $E$ of the pendulum is approximately its kinetic energy $E_{\mathrm{kin}}$, Eq. (5.3) shows that the fractional increment in the total energy $\Delta E/E$ approximately equals twice the value of the modulation depth $m$: $\Delta E/E \approx 2m$.

When the frequencies and phases have those values that are favorable for the most effective delivery of energy, the abrupt displacement of the weights toward the ends of the rod occurs at the instant when the rotor attains its greatest deflection (more precisely, when the rotor is very near it). At this instant the angular velocity of the rotor is almost zero, and so this radial displacement of the weights into their previous positions causes nearly no decrement in the energy.

For the principal resonance ($n = 1$) the investment in energy occurs twice during the natural period $T_0$ of oscillations. That is, the fractional increment in energy $\Delta E/E$ during one period approximately equals $4m$.

A process in which the increment in energy $\Delta E$ during a period is proportional to the energy stored $E$ (in the case under consideration $\Delta E \approx 4mE$) is characterized on the average by the exponential growth of the energy with time:

$$E(t) = E_0\exp(\alpha t). \qquad (5.4)$$

In this case the index of growth $\alpha$ is proportional to the depth of modulation $m$ of the moment of inertia: $\alpha = 4m/T_0$. When the modulation is exactly tuned to the principal resonance ($T = T_0/2$), the decrease of energy is caused almost only by friction.

Dissipation of energy due to viscous friction during an integral number of natural cycles (for $t = nT_0$) is described by the following expression:

$$E(t) = E_0\exp(-2\gamma t). \qquad (5.5)$$

Comparing Eqs. (5.4) and (5.5), we obtain the following estimate for the threshold (minimal) value $m_{\mathrm{min}}$ of the depth of modulation corresponding to the excitation of the principal parametric resonance:

$$m_{\mathrm{min}} = \frac{1}{2}\gamma T_0 = \frac{\pi}{2Q}. \qquad (5.6)$$

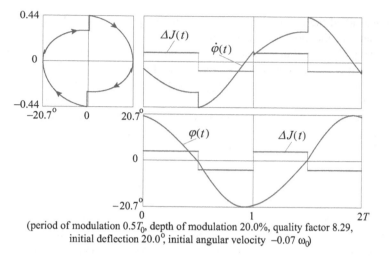

(period of modulation $0.5T_0$, depth of modulation 20.0%, quality factor 8.29,
initial deflection 20.0°, initial angular velocity $-0.07\,\omega_0$)

Figure 5.2: The phase trajectory (left) and the time-dependent graphs of $\dot\varphi(t)$ and $\varphi(t)$ (right) of stationary oscillations at the threshold condition $m \approx \pi/2Q$ for $T = T_0/2$.

Here we introduced the dimensionless quality factor $Q = \omega_0/(2\gamma)$ to characterize friction in the system.

The phase trajectory and the time-dependent plots of the angular velocity and the angle of deflection for parametric oscillations occurring at the threshold conditions, Eq. (5.6), are shown in Figure 5.2. This mode of steady oscillations (which have a constant amplitude in spite of the dissipation of energy) is called *parametric regeneration*.

For the third resonance ($T = 3T_0/2$) the threshold value of the depth of modulation is three times greater than its value for the principal resonance: $m_{\min} = 3\pi/(2Q)$. In this instance two cycles of the parametric variation occur during three full periods of natural oscillations. In conditions of the third parametric resonance, radial displacements of the weights again happen at the most favorable moments for pumping energy to the oscillator just like in conditions of the principal resonance. Therefore, at the same depth of modulation, the same investment in energy occurs during an interval that is three times longer than the interval for the principal resonance.

When the depth of modulation exceeds the threshold value, the energy of oscillations increases exponentially with time. The growth of the energy again is described by (5.4). However, now the index of growth $\alpha$ is determined by the amount by which the energy delivered through parametric modulation exceeds the simultaneous losses of energy caused by friction: $\alpha = 4m/T_0 - 2\gamma$. The amplitude of parametrically excited oscillations also increases exponentially with time (Figure 5.3): $a(t) = a_0\exp(\beta t)$. The index $\beta$ in the growth of amplitude

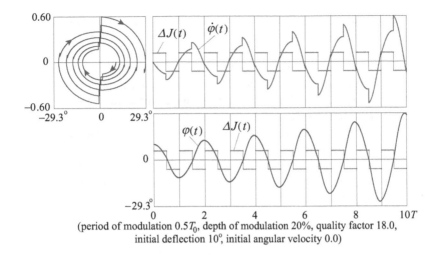

(period of modulation $0.5T_0$, depth of modulation 20%, quality factor 18.0, initial deflection 10°, initial angular velocity 0.0)

Figure 5.3: Exponential growth of the amplitude of oscillations at conditions of the first order parametric resonance ($n = 1$).

is one half the index of the growth in energy. For the principal resonance, when the investment in energy occurs twice during one natural period of oscillation, we have $\beta = 2m/T_0 - \gamma = m\omega_0/\pi - \gamma$.

## 5.1.6 Differential Equation for Parametric Oscillations

We next consider a more rigorous mathematical treatment of parametric resonance under a square-wave modulation of the parameter. This treatment is based on the differential equations governing the phenomenon.

During the time intervals $(0,\ T/2)$ and $(T/2,\ T)$, the value of the moment of inertia is constant, and the motion of the rotor can be considered as a free oscillation described by a linear differential equation. However, the coefficients in this equation are different for the adjacent time intervals $(0,\ T/2)$ and $(T/2,\ T)$:

$$\ddot{\varphi} = -\frac{1}{1+m}(\omega_0^2\varphi + 2\gamma\dot{\varphi}) \qquad \text{for} \qquad 0 < t < T/2, \tag{5.7}$$

$$\ddot{\varphi} = -\frac{1}{1-m}(\omega_0^2\varphi + 2\gamma\dot{\varphi}) \qquad \text{for} \qquad T/2 < t < T. \tag{5.8}$$

Here $\omega_0 = \sqrt{D/J_0}$ is the natural frequency of the oscillator and $\gamma$ is the damping constant characterizing the strength of viscous friction. Both these quantities correspond to the mean value $J_0 = \frac{1}{2}(J_1 + J_2)$ of the moment of inertia. For small and moderate values of $m$, the moment of inertia equals $J_0$ when the weights are near the half-way point between their extreme positions on the rod. For large $m$ this is not the case because the moment of inertia depends on the square of the distance of the weights from the axis of rotation.

At each instant $t_n = nT/2$ $(n = 1, 2, \ldots)$ of an abrupt change in the moment of inertia we must make a transition from one of the linear Eqs. (5.7)–(5.8) to the other. During each half-period $T/2$ the motion of the oscillator is a segment of some harmonic (or damped) natural oscillation. An analytical investigation of parametric excitation can be carried out by fitting to one another known solutions to the linear Eqs. (5.7)–(5.8) for consecutive adjacent time intervals.

The initial conditions for each subsequent time interval are chosen according to the physical model in the following way. Each initial value of the angular displacement $\varphi$ equals the value $\varphi(t)$ reached by the oscillator at the end of the preceding time interval. The initial value of the angular velocity $\dot\varphi$ is related to the angular velocity at the end of the preceding time interval by the law of conservation of the angular momentum:

$$(1 + m)\dot\varphi_1 = (1 - m)\dot\varphi_2. \tag{5.9}$$

In Eq. (5.9) $\dot\varphi_1$ is the angular velocity at the end of the preceding time interval, when the moment of inertia of the rotor has the value $J_1 = J_0(1 + m)$, and $\dot\varphi_2$ is the initial value for the following time interval, during which the moment of inertia is equal to $J_2 = J_0(1 - m)$. The change in the angular velocity at an abrupt variation of the inertia moment from the value $J_2$ to $J_1$ can be found in the same way.

That we may use the conservation of angular momentum, as expressed in Eq. (5.9), is allowed because, at sufficiently rapid displacement of the weights along the rotor, we can neglect the influence of the spring and consider the rotor as if it were freely rotating about its axis. This assumption is valid provided the duration of the displacement of the weights is a small portion of the natural period.

Considering conditions for which Eqs. (5.7)–(5.8) yield solutions with increasing amplitudes, we can determine the ranges of frequency $\omega$ near the values $\omega_n = 2\omega_0/n$, within which the state of rest is unstable for a given modulation depth $m$. In these ranges of instability an arbitrarily small deflection from equilibrium is sufficient for the progressive growth of small initial oscillations.

### 5.1.7   The Mean Natural Period at Large Depth of Modulation

The threshold for the parametric excitation of the torsion pendulum is determined above for the resonant situations in which two cycles of the parametric modulation occur during one natural period or during three natural periods of oscillation. The estimate obtained, Eq. (5.6), is valid for small values of the modulation depth $m$.

For large values of the modulation depth $m$, the notion of a natural period needs a more precise definition. Let $T_0 = 2\pi/\omega_0 = 2\pi\sqrt{J_0/D}$ be the period of oscillation of the rotor when the weights are fixed in some middle positions, for which the moment of inertia equals $J_0$. The period is somewhat longer when the weights are moved further apart: $T_1 = T_0\sqrt{1 + m} \approx T_0(1 + m/2)$. The period is shorter when the weights are moved closer to one another: $T_2 = T_0\sqrt{1 - m} \approx T_0(1 - m/2)$.

It is convenient to define the average period $T_{av}$ not as the arithmetic mean $\frac{1}{2}(T_1 + T_2)$, but rather as the period that corresponds to the arithmetic mean frequency $\omega_{av} = \frac{1}{2}(\omega_1 + \omega_2)$, where $\omega_1 = 2\pi/T_1$ and $\omega_2 = 2\pi/T_2$. So we define $T_{av}$ by the relation:

$$T_{av} = \frac{2\pi}{\omega_{av}} = \frac{2T_1 T_2}{(T_1 + T_2)}. \tag{5.10}$$

The period $T$ of the parametric modulation that is exactly tuned to any of the parametric resonances is determined not only by the order $n$ of the resonance, but also by the depth of modulation $m$. In order to satisfy the resonant conditions, the increment in the phase of natural oscillations during one cycle of modulation must be equal to $\pi$, $2\pi$, $3\pi$, ..., $n\pi$, ... During the first half-cycle the phase increases by $\omega_1 T/2$, and during the second half-cycle by $\omega_2 T/2$. Consequently, instead of the approximate condition expressed by Eq. (5.2), we obtain:

$$\frac{\omega_1 + \omega_2}{2}T = n\pi, \qquad \text{or} \qquad T = n\frac{\pi}{\omega_{av}} = n\frac{T_{av}}{2}. \tag{5.11}$$

Thus, for a parametric resonance of some definite order $n$, the condition for exact tuning can be expressed in terms of the two natural periods, $T_1$ and $T_2$. This condition is $T = nT_{av}/2$, where $T_{av}$ is defined by Eq. (5.10). For moderate values of $m$ it is possible to use approximate expressions for the average frequency and period:

$$\omega_{av} = \frac{\omega_0}{2}\left(\frac{1}{\sqrt{1+m}} + \frac{1}{\sqrt{1-m}}\right) \approx \omega_0(1 + \frac{3}{8}m^2), \; T_{av} \approx T_0(1 - \frac{3}{8}m^2). \tag{5.12}$$

The difference between $T_{av}$ and $T_0$ reveals itself in terms proportional to the square of the depth of modulation $m$.

## 5.2 Frequency Ranges of Parametric Excitation

An infinite growth of the amplitude during parametric excitation is possible not only at exact tuning to one of the resonances but in certain *intervals* of $T$-values. These intervals, or the *ranges of instability*, surround the resonant values $T = T_{av}/2$, $T = T_{av}$, $T = 3T_{av}/2$, ... The width of the intervals increases with the depth $m$ of the parameter modulation.[1] Outside the intervals, the equilibrium position of a torsion pendulum is stable, and the amplitude of oscillations does not grow.

In order to determine the boundaries of the frequency ranges of parametric instability surrounding the resonant values $T = T_{av}/2$, $T = T_{av}$, $T = 3T_{av}/2$, ..., we can consider *stationary oscillations* that occur when the period of modulation $T$ corresponds to one of the boundaries. These stationary oscillations can

---

[1] Strictly speaking, for high resonances ($n \geq 2$) this statement is true only for small and moderate values of the depth of modulation (see Figure 5.6 on p. 108 with the diagram of the ranges of parametric resonance).

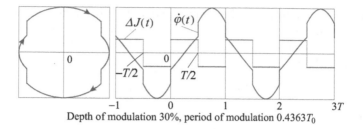

Depth of modulation 30%, period of modulation $0.4363T_0$

Figure 5.4: Stationary parametric oscillations at the lower boundary of the principal interval of instability (near $T = T_{av}/2$).

be represented as an alternation of natural oscillations with the periods $T_1$ and $T_2$. In the absence of friction the graphs of such oscillations are formed by segments of non-damped sine curves with the corresponding periods.

## 5.2.1   Main Interval of Parametric Excitation

We examine first the vicinity of the principal resonance occurring at $T = T_{av}/2$. Suppose that the period $T$ of the parametric variation is a little shorter than the resonant value $T = T_{av}/2$, so that $T$ corresponds to the left boundary of the interval of instability. In this case a little less than a quarter of the mean natural period $T_{av}$ elapses between consecutive abrupt increases and decreases of the moment of inertia. The graph of the angular velocity $\dot{\varphi}(t)$ for this periodic stationary process has the characteristic pattern shown in Figure 5.4. The segments of the graphs of free oscillations (which occur at time intervals during which the moment of inertia is constant) are alternating parts of sine or cosine curves with the periods $T_1$ and $T_2$. These segments are symmetrically truncated on both sides.

To find conditions at which such stationary oscillations take place, we can write the expressions for $\varphi(t)$ and $\dot{\varphi}(t)$ during the adjacent intervals in which the oscillator executes natural oscillations, and then fit these expressions to one another at the boundaries. Such fitting must provide a periodic stationary process.

We let the origin of time, $t = 0$, be the instant when the weights are shifted apart. The angular velocity is abruptly decreased in magnitude at this instant (see Figure 5.4). Then during the interval $(0,\ T/2)$ the graph describes a natural oscillation with the frequency $\omega_1 = \omega_0/\sqrt{1+m}$. It is convenient to represent this motion as a superposition of sine and cosine waves whose constant amplitudes are $A_1$ and $B_1$, respectively:

$$\begin{aligned}
\varphi_1(t) &= A_1 \sin \omega_1 t + B_1 \cos \omega_1 t, \\
\dot{\varphi}_1(t) &= A_1 \omega_1 \cos \omega_1 t - B_1 \omega_1 \sin \omega_1 t.
\end{aligned} \tag{5.13}$$

Similarly, during the interval $(-T/2,\ 0)$ the graph in Figure 5.4 is a segment of natural oscillation with the frequency $\omega_2 = \omega_0/\sqrt{1-m}$:

$$\begin{aligned}
\varphi_2(t) &= A_2 \sin \omega_2 t + B_2 \cos \omega_2 t, \\
\dot{\varphi}_2(t) &= A_2 \omega_2 \cos \omega_2 t - B_2 \omega_2 \sin \omega_2 t.
\end{aligned} \tag{5.14}$$

To determine the values of constants $A_1$, $B_1$ and $A_2$, $B_2$, we can use the conditions that must be satisfied when the segments of the graph are joined together, taking into account the periodicity of the stationary process.

At $t = 0$ the angle of deflection is the same for both $\varphi_1$ and $\varphi_2$: $\varphi_1(0) = \varphi_2(0)$. From this condition we find that $B_1 = B_2$. We later denote these equal constants simply by $B$. The angular velocity at $t = 0$ undergoes a sudden change:

$$(1+m)\dot{\varphi}_1(0) = (1-m)\dot{\varphi}_2(0).$$

This condition gives us the following relation between $A_2$ and $A_1$: $A_2 = kA_1 = kA$, where we have denoted $A_1$ simply by $A$ and introduced a dimensionless quantity $k$, which depends on the depth of modulation $m$:

$$k = \sqrt{\frac{1+m}{1-m}}. \tag{5.15}$$

Equations for the constants $A$ and $B$ are determined by the conditions at the instants $-T/2$ and $T/2$. For stationary periodic oscillations, corresponding to the principal resonance (and to all resonances of odd numbers $n = 1, 3, \ldots$ in Eq. (5.11)), these conditions are:

$$\varphi_1(T/2) = -\varphi_2(-T/2), \quad (1+m)\dot{\varphi}_1(T/2) = -(1-m)\dot{\varphi}_2(-T/2). \tag{5.16}$$

Substituting $\varphi$ and $\dot{\varphi}$ from Eq. (5.14) in Eq. (5.16), we obtain the system of homogeneous equations for the unknown quantities $A$ and $B$:

$$\begin{aligned}
(S_1 - kS_2)A + (C_1 + C_2)B &= 0, \\
k(C_1 + C_2)A - (kS_1 - S_2)B &= 0.
\end{aligned} \tag{5.17}$$

In Eqs. (5.17) the following notations are used:

$$\begin{aligned}
C_1 &= \cos(\omega_1 T/2), \quad C_2 = \cos(\omega_2 T/2), \\
S_1 &= \sin(\omega_1 T/2), \quad S_2 = \sin(\omega_2 T/2).
\end{aligned} \tag{5.18}$$

The homogeneous system of equations for $A$ and $B$, Eqs. (5.17), has a non-trivial (non-zero) solution only if its determinant is zero:

$$2kC_1C_2 - (1 + k^2)S_1S_2 + 2k = 0. \tag{5.19}$$

This condition for the existence of a non-zero solution to Eqs. (5.17) gives us an equation for the unknown variable $T$, which enters in Eq. (5.19) as the arguments of sine and cosine functions in $S_1$, $S_2$ and $C_1$, $C_2$. This equation determines the desired boundaries of the interval of instability for a given value of the modulation depth $m$, which enters in the parameter $k$ defined by Eq. (5.15). These boundaries $T_-$ and $T_+$ are given by the roots of the equation, Eq. (5.19).

To find approximate solutions $T$ to this transcendental equation, Eq. (5.19), we transform it into a more convenient form. We first represent in Eq. (5.19) the products $C_1 C_2$ and $S_1 S_2$ as follows:

$$C_1 C_2 = \frac{1}{2}\left(\cos \frac{\Delta \omega T}{2} + \cos \omega_{av} T\right), \qquad S_1 S_2 = \frac{1}{2}\left(\cos \frac{\Delta \omega T}{2} - \cos \omega_{av} T\right),$$

where $\Delta \omega = \omega_2 - \omega_1$. Then, using the identity $\cos \alpha = 2\cos^2(\alpha/2) - 1$, we reduce Eq. (5.19) to the following form:

$$(1 + k) \cos \frac{\omega_{av} T}{2} = \pm|1 - k| \cos \frac{\Delta \omega T}{4}. \tag{5.20}$$

For the boundaries of the instability interval that contains the principal resonance $n = 1$, we search for a solution $T$ of Eq. (5.20) in the vicinity of $T = T_0/2$. For a given value of the depth of modulation $m$, Eq. (5.20) in the neighborhood of $T_0/2 \approx T_{av}/2$ has two solutions that correspond to the boundaries $T_-$ and $T_+$ of the instability interval. The phase diagram and the graph of the angular velocity for the right boundary of the main interval ($n = 1$) are shown in Figure 5.5.

To find the boundaries $T_-$ and $T_+$ of the instability interval, we replace $T$ in the argument of the cosine on the left-hand side of Eq. (5.20) by $T_{av}/2 + \Delta T$, where $\Delta T \ll T_0$. Since $\omega_{av} T_{av} = 2\pi$, we can replace the cosine in Eq. (5.20) with $-\sin(\omega_{av} \Delta T/2)$. Then Eq. (5.20) becomes:

$$\sin \frac{\omega_{av} \Delta T}{2} = \mp \frac{|1 - k|}{1 + k} \sin \frac{\Delta \omega (T_{av}/2 + \Delta T)}{4}. \tag{5.21}$$

This equation for $\Delta T$ can be solved numerically by iteration. We start with $\Delta T = 0$ as an approximation of the zeroth order, substituting it into the right-hand side of Eq. (5.21), taken, say, with the upper sign. Then the left-hand side of Eq. (5.21) gives us the value of $\Delta T$ to the first order. We substitute this first-order value into the right-hand side of Eq. (5.21), and on the left-hand side we obtain $\Delta T$ to the second order. This procedure is iterated until a self-consistent value of $\Delta T$ for the left boundary is obtained. To determine $\Delta T$ for the right boundary, we use the same procedure, taking the lower sign on the right-hand side of Eq. (5.21).

After the substitution of one of the roots $T_-$ or $T_+$ of Eq. (5.21) into Eqs. (5.17) both equations for $A$ and $B$ become equivalent and permit us to find only the ratio $A/B$. This limitation means that the amplitude of stationary oscillations at the boundary of the instability interval can be arbitrarily large. This amplitude depends on the initial conditions. Nevertheless, these oscillations have a definite shape that is determined by the ratio of the amplitudes $A$ and $B$ of the sine and cosine functions whose segments form the pattern of the stationary parametric oscillation (see Figures 5.4 and 5.5).

The periods of modulation $T_-$ and $T_+$, corresponding respectively to the left and right boundaries of the instability interval that contains the principal parametric resonance $n = 1$, calculated numerically for different vales of the modulation

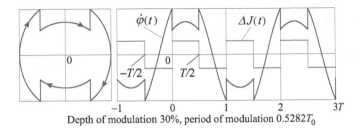

Depth of modulation 30%, period of modulation $0.5282T_0$

Figure 5.5: Stationary parametric oscillations at the upper boundary of the principal interval of instability (near $T = T_{av}/2$).

depth $m$ with the help of the above-described procedure, are shown by the border lines of the first "tongue" in Figure 5.6. The central curve of this "tongue" gives the period of modulation $T = T_{av}/2$ as a function of the modulation depth $m$, which corresponds to exact tuning to the principal parametric resonance. The other "tongues" in Figure 5.6 show the intervals of parametric instability of high orders, discussed further on in this section.

To obtain an approximate analytic solution to Eq. (5.21) that is valid for small values of the modulation depth $m$, we can simplify the expression on the right-hand side by assuming $k \approx 1 + m$, $|1 - k| \approx m$. We may also assume the value of the cosine to be 1. On the left-hand side of Eq. (5.21), the sine can be replaced by its small argument, in which $\omega_{av} = 2\pi/T_{av}$. Thus we obtain the following approximate expression that is valid up to terms to the second order in $m$:

$$T_{\mp} = \frac{1}{2}\left(1 \mp \frac{m}{\pi}\right) T_{av}. \tag{5.22}$$

Since the natural period $T_0 = 2\pi\sqrt{D/J_0}$ is used in the relevant simulation program as an appropriate time unit for the input of the period of modulation $T$, we express the values of $T_{\mp}$ given by Eq. (5.22) also in terms of $T_0$:

$$T_{\mp} = \frac{1}{2}\left(1 \mp \frac{m}{\pi} - \frac{3m^2}{8}\right) T_0. \tag{5.23}$$

## 5.2.2   Third-Order Interval of Parametric Instability

In a similar way we can determine the boundaries of the instability interval in the vicinity of a resonance of the higher order $n = 3$. At the third order resonance ($n = 3$) two cycles of variation of the inertia moment occur during approximately three natural periods of oscillation ($T \approx 3T_{av}/2$).

The phase trajectories and the time-dependent graphs of stationary oscillations at the left and right boundaries of the third interval are shown in Figure 5.7.

The phase orbit of the periodic oscillation closes after two cycles of modulation. This orbit is formed by two concentric ellipses that correspond to natural

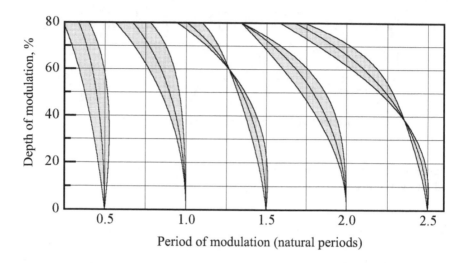

Figure 5.6: Intervals of parametric instability at square-wave modulation of the inertia moment in the absence of friction.

oscillations of the flywheel with frequencies $\omega_1$ and $\omega_2$. The representative point moves clockwise along this orbit, jumping from one ellipse to the other each time the weights are shifted along the rod of the flywheel. The numbers in Figure 5.7 make it easier to follow how the representative point describes this orbit: Equivalent points of the phase orbit and the graph of the angular velocity are marked by equal numbers.

Considering conditions at which the graphs of natural oscillations with frequencies $\omega_1$ and $\omega_2$ on the left boundary fit one another for adjacent time intervals and produce the periodic process shown in Figure 5.7, we get the same Eqs. (5.16)–(5.17) for $A_1$ and $A_2$, as well as Eq. (5.22) for the period of modulation. Actually, this is true for all intervals of parametric instability of odd orders. Similarly, for the right boundary we get the same equations for $B_1$ and $B_2$ as in case $n = 1$, and also Eq. (5.22) with the opposite sign for determination of the corresponding period of modulation $T$. However, if we are interested in the third interval, we should search for a solution to these equations in the vicinity of $T = 3T_{\mathrm{av}}/2$, as well as for any other interval of odd order $n$ — in the vicinity of $T = nT_{\mathrm{av}}/2$. The boundaries of intervals of the third and fifth orders, obtained by a numerical solution, are also shown in Figure 5.6.

For small values of the depth of modulation $m$, we can find approximate analytic expressions for the lower and the upper boundaries of the interval that are valid up to quadratic terms in $m$:

$$T_{\mp} = \left( \frac{3}{2} \mp \frac{m}{2\pi} \right) T_{\mathrm{av}}. \tag{5.24}$$

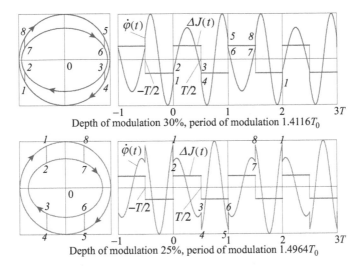

Depth of modulation 30%, period of modulation $1.4116T_0$

Depth of modulation 25%, period of modulation $1.4964T_0$

Figure 5.7: The phase trajectories (left) and the time-dependent graphs of the angular velocity (right) of stationary parametric oscillations at the left (upper panel) and right (lower panel) boundaries of the interval of instability near $T = 3T_{av}/2$.

In terms of the mean natural period $T_0$ these boundaries are expressed as follows:

$$T_{\mp} = \left( \frac{3}{2} \mp \frac{m}{2\pi} - \frac{9m^2}{16} \right) T_0. \qquad (5.25)$$

In this approximation, the third interval has the same width $(m/\pi)T_0$ as does the interval of instability in the vicinity of the principal resonance. However, this interval is distinguished by a greater asymmetry: Its central point is displaced to the left of the value $T = 3T_0/2$ by $(9/16)m^2T_0$.

## 5.2.3 Frequency Ranges for Resonances of Even Orders

For small and moderate square-wave modulation of the moment of inertia, parametric resonance of the order $n = 2$ (one cycle of the parametric variation during one natural period of oscillation) is relatively weak compared to the resonances $n = 1$ and $n = 3$ considered above. In the case in which $n = 2$ the abrupt changes of the moment of inertia induce both an increase and a decrease of the energy only once during each natural period. The growth of oscillations occurs only if the increase in energy at the instant when the weights are drawn closer is greater than the decrease in energy when the weights are drawn apart. This is possible only if the weights are shifted toward the axis when the angular velocity of the rotor is greater in magnitude than it is when they are shifted apart. For $T \approx T_{av}$, these conditions can be fulfilled only because there is a small difference between the natural periods $T_1$ and $T_2$ of the rotor, where $T_1$ is the period with the weights

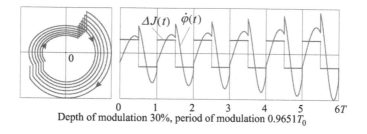

Depth of modulation 30%, period of modulation $0.9651T_0$

Figure 5.8: The phase trajectory (left) and the graph of the angular velocity (right) of oscillations at parametric resonance of the second order ($T = T_{av}$).

shifted apart and $T_2$ is the period with them shifted together. This difference is proportional to $m$.

The growth of oscillations at parametric resonance of the second order is shown in Figure 5.8. In this case, the investment of energy during a period is proportional to the *square* of the depth of modulation $m$, while in the cases of resonances with $n = 1$ and $n = 3$ the investment of energy is proportional to the first power of $m$. Therefore, for the same value of the damping constant $\gamma$ (the same quality factor $Q$), a considerably greater depth of modulation is required here to exceed the threshold of parametric excitation.

The interval of instability in the vicinity of resonance with $n = 2$ is considerably narrower compared to the corresponding intervals of the resonances with $n = 1$ and $n = 3$. Its width is also proportional only to the square of $m$ (for small values of $m$).

To determine the boundaries of this interval of instability, we can consider, as is done above for other resonances, stationary oscillations for $T \approx T_0$ formed by alternating segments of free sinusoidal oscillations with the periods $T_1$ and $T_2$. The graph of the angular velocity and the phase trajectory of such stationary periodic oscillations for one of the boundaries are shown in Figure 5.9. During oscillations occurring at the boundary of the instability interval, the abrupt increment and decrement in the angular velocity exactly compensate each other.

To describe these stationary oscillations, we can use the same expressions for $\varphi(t)$ and $\dot{\varphi}(t)$ as we use in Eqs. (5.13)–(5.14). The conditions for joining the graphs at $t = 0$ are also the same. However, differences begin with the equations for the constants $A$ and $B$. They are determined by the conditions of periodicity at the instants $-T/2$ and $T/2$. For stationary periodic oscillations, corresponding to resonance with $n = 2$ (and for all resonances of even orders $n = 2, 4, \ldots$ in Eq. (5.11)), these conditions are:

$$\varphi_1(T/2) = \varphi_2(-T/2), \qquad (1 + m)\dot{\varphi}_1(T/2) = (1 - m)\dot{\varphi}_2(-T/2), \quad (5.26)$$

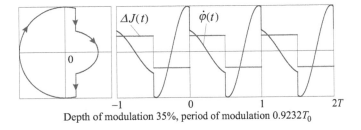

Depth of modulation 35%, period of modulation $0.9232T_0$

Figure 5.9: Stationary parametric oscillations at one of the boundaries of the interval of instability of the second order (near $T = T_{av} \approx T_0$).

and we obtain the system of equations for the amplitudes $A$ and $B$:

$$(S_1 + kS_2)A + (C_1 - C_2)B = 0,$$
$$k(C_1 - C_2)A - (kS_1 + S_2)B = 0, \tag{5.27}$$

where $S_1$, $C_1$ and $S_2$, $C_2$ are defined by the same Eqs. (5.18). The homogeneous system of equations for $A$ and $B$, Eqs. (5.27), has a non-trivial solution if its determinant is zero:

$$2kC_1C_2 - (1 + k^2)S_1S_2 - 2k = 0. \tag{5.28}$$

In order to find the values $T_{\mp} = T_{av} + \Delta T$ for the instability interval with $n = 2$ from Eq. (5.28), we transform the products $C_1C_2$ and $S_1S_2$ in Eq. (5.28) by using the identity $\cos \alpha = 1 - 2\sin^2(\alpha/2)$:

$$(1 + k) \sin \frac{\omega_{av}T}{2} = \pm|1 - k| \sin \frac{\Delta\omega T}{4}. \tag{5.29}$$

We next replace $T$ in the argument of the sine on the left-hand side of Eq. (5.29) by $T_{av} + \Delta T$, where $\Delta T \ll T_0$. Since $\omega_{av}T_{av} = 2\pi$, we can write this sine as $-\sin(\omega_{av}\Delta T/2)$. Then Eq. (5.20) becomes:

$$\sin \frac{\omega_{av}\Delta T}{2} = \mp \frac{|1 - k|}{1 + k} \cos \frac{\Delta\omega(T_{av} + \Delta T)}{4}. \tag{5.30}$$

This equation gives the left boundary $T_-$ of the instability interval when we take the upper sign in its right-hand side, and the right boundary $T_+$ when we take the lower sign. Stationary oscillations, which correspond to the right boundary, are shown in Figure 5.10.

Equation (5.30) for $\Delta T$ can be also solved numerically by iteration. Substituting $T_-$ or $T_+$ obtained from (5.30) into one of the equations in (5.17), we get the ratio of the amplitudes $A$ and $B$ that determines the pattern of stationary oscillations at the corresponding boundary of the instability interval. We note how narrow the intervals of even resonances ($n = 2, 4$) are for small values of $m$. With

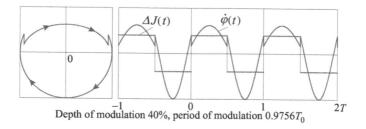

Depth of modulation 40%, period of modulation $0.9756T_0$

Figure 5.10: Stationary parametric oscillations at the other boundary of the interval of instability of the second order (near $T = T_{av} \approx T_0$).

the growth of $m$ the intervals expand and become comparable with the intervals of odd orders.

For moderate values of the depth of modulation, it is possible to find an approximate analytical solution of Eq. (5.30):

$$T_{\mp} = \left(1 \mp \frac{1}{4}m^2\right) T_{av}. \qquad (5.31)$$

In terms of $T_0$ the boundaries of the second interval are:

$$T_{\mp} = T_0 + \left(\mp\frac{1}{4} - \frac{3}{8}\right) m^2 T_0, \qquad (5.32)$$

i.e., $T_- = T_0 - (5/8)m^2 T_0$, $T_+ = T_0 - (1/8)m^2 T_0$. As mentioned above, the width $T_+ - T_- = (m^2/2)T_0$ of this interval of instability is proportional to the square of the modulation depth.

The intervals of instability for the first five parametric resonances at square-wave modulation of the inertia moment in the absence of friction are shown by the shaded "tongues" in Figure 5.6) for various values of the modulation depth $m$. The diagram is obtained by numerical solution of the equations that are discussed above. We note how narrow the intervals of even resonances ($n = 2, 4$) are for small values of $m$. With the growth of $m$ the intervals expand and become comparable with the intervals of odd orders.

## 5.2.4   Intersections of the Boundaries at Large Modulation

Figure 5.6 shows that at some definite values of $m$ both boundaries of intervals with $n > 2$ coincide (we may consider that they *intersect*). Thus at these values of $m$ the corresponding intervals of parametric resonance disappear. These values of $m$ correspond to the natural periods $T_1$ and $T_2$ of oscillation (associated with the weights far apart and close to each other), whose ratio is $2 : 1$, $3 : 1$, and $3 : 2$. For the corresponding values of the modulation depth $m$ and the period of modulation $T$, steady oscillations occur for arbitrary initial conditions.

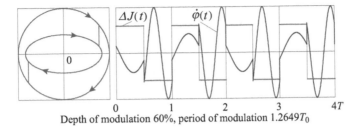

Depth of modulation 60%, period of modulation $1.2649T_0$

Figure 5.11: The phase trajectory and the time-dependent graph of the angular velocity for stationary oscillations at the intersection of both boundaries of the third interval.

For the first intersection (ratio 2 : 1) exactly one half of the natural oscillation with period $T_1$ is completed during the first half of the modulation cycle (see Figure 5.11). On the phase diagram, the representing point traces a half of the smaller ellipse, and then abruptly jumps down to the larger ellipse. During the second half of the modulation cycle the oscillator executes exactly a whole natural oscillation with period $T_2 = T_1/2$, so that the representing point passes in the phase plane along the whole larger ellipse, and then jumps up to the smaller ellipse along the same vertical segment.

During the next modulation cycle the representing point first generates the other half of the smaller ellipse, and then again the whole larger ellipse. Therefore during any two adjacent cycles of modulation the representing point passes once along the closed smaller ellipse and twice along the larger one, finally returning to the initial point of the phase plane. We see that such an oscillation is periodic for arbitrary initial conditions. This means that for the corresponding values of the modulation depth $m$ and the period of modulation $T$ the growth of amplitude is impossible even in the absence of friction (the instability interval vanishes).

Similar explanations can be suggested for other cases in Figure 5.6 in which the boundaries intersect.

### 5.2.5 Intervals of Excitation in the Presence of Friction

When there is friction in the system, the intervals of the period of modulation that correspond to the parametric instability become narrower, and for strong enough friction (below the threshold) the intervals disappear.

Next we show that above the threshold approximate values for the boundaries of the first interval are given by Eq. (5.22) provided we substitute for $m$ the expression $\sqrt{m^2 - m_{\min}^2}$ with the threshold value $m_{\min} = \pi/(2Q)$ defined by Eq. (5.6).

For the third interval, we can use Eq. (5.24), substituting $\sqrt{m^2 - m_{\min}^2}$ for $m$, with $m_{\min} = 3\pi/(2Q)$. When $m$ is equal to the threshold value $m_{\min}$, the corresponding interval of parametric resonance disappears.

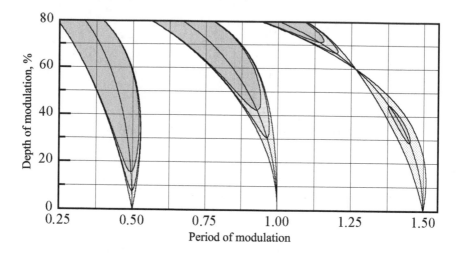

Figure 5.12: Intervals of parametric excitation at square-wave modulation of the moment of inertia without friction, for $Q = 20$, and for $Q = 10$.

The boundaries of the second interval of parametric resonance in the presence of friction are approximately given by Eq. (5.31) provided we substitute for $m^2$ the expression $\sqrt{m^4 - m_{min}^4}$ with the threshold value $m_{min} = \sqrt{2/Q}$, which corresponds to the second parametric resonance.

The diagram in Figure 5.12 shows the boundaries of the first three intervals of parametric resonance for $Q = 20$ and $Q = 10$ (and also in the absence of friction). Note the "island" of parametric resonance for $n = 3$ and $Q = 20$. This resonance disappears when the depth of modulation exceeds 45% and reappears when $m$ exceeds approximately 66%.

### 5.2.5.1 Main Interval of Instability in the Presence of Friction

Stationary oscillations occurring at the left boundary of the instability interval in the vicinity of the principal parametric resonance in the presence of friction are shown in Figure 5.13 (compare with Figure 5.4). Twice during the full cycle the angular velocity abruptly increases, and twice it decreases. The increments are greater than the decrements, so that as a whole the energy received by the rotor exceeds the energy given away. This surplus compensates for the dissipation of the energy that occurs at natural oscillation during the intervals between the abrupt displacements of the weights along the rod of the flywheel.

To find conditions at which such stationary oscillations take place, we can write the expressions for $\varphi(t)$ and $\dot{\varphi}(t)$ during the adjacent intervals when the oscillator executes damped natural oscillations, and then fit these expressions to one another at the boundaries. We choose as the time origin $t = 0$ the instant when the weights are shifted apart, and the angular velocity is decreased in magnitude. Then during the interval $(0, T/2)$ the graph describes a damped natural oscilla-

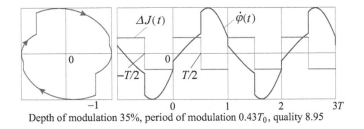

Depth of modulation 35%, period of modulation $0.43T_0$, quality 8.95

Figure 5.13: Stationary oscillations in the presence of friction at the left boundary of the principal instability interval.

tion with the frequency $\omega_1 = \omega_0/\sqrt{1+m}$. Similarly to Eq. (5.13), p. 104, it is convenient to represent this motion as a superposition of damped oscillations of sine and cosine type with some constants $A_1$ and $B_1$:

$$\varphi_1(t) = (A_1 \sin \omega_1 t + B_1 \cos \omega_1 t) e^{-\gamma t},$$
$$\dot{\varphi}_1(t) \approx (A_1 \omega_1 \cos \omega_1 t - B_1 \omega_1 \sin \omega_1 t) e^{-\gamma t}. \tag{5.33}$$

The latter expression for $\dot{\varphi}(t)$ is valid for relatively weak friction ($\gamma \ll \omega_0$). To obtain it, we differentiate $\varphi(t)$ with respect to the time, considering the exponential factor $e^{-\gamma t}$ to be approximately constant. Indeed, at weak damping the main contribution to the time derivative originates from the oscillating factors $\sin \omega_1 t$ and $\cos \omega_1 t$ in the expression for $\varphi(t)$. Similarly, during the interval $(-T/2, 0)$ the graph in Figure 5.13 is a segment of damped natural oscillation with the frequency $\omega_2$:

$$\varphi_2(t) = (A_2 \sin \omega_2 t + B_2 \cos \omega_2 t) e^{-\gamma t},$$
$$\dot{\varphi}_2(t) \approx (A_2 \omega_2 \cos \omega_2 t - B_2 \omega_2 \sin \omega_2 t) e^{-\gamma t}. \tag{5.34}$$

Further calculations are similar to those leading from Eqs. (5.13) and (5.14) to (5.19), but instead of Eq. (5.19) we get the following equation:

$$2kC_1C_2 - (1 + k^2)S_1S_2 + k(p + 1/p) = 0, \tag{5.35}$$

where the notation $p = e^{-\gamma T}$ is used. This condition of existence of a non-zero solution for constants $A_1$ and $B_1$ gives us an equation that determines the desirable boundaries of the interval of instability. These boundaries are given by the values of the unknown variable $T$ (the roots of the equation), which enters in Eq. (5.35) as the arguments of sine and cosine functions in $S_1$, $S_2$ and $C_1$, $C_2$, and also as the argument of the exponent in $p = e^{-\gamma T}$. To find approximate solutions $T$ to this transcendental equation, we transform it into a more convenient form in the same way as in Section 5.2.1, p. 104:

$$(1 + k) \cos \frac{\omega_{\text{av}} T}{2} = \pm \sqrt{(1 - k)^2 \cos^2 \frac{\Delta \omega T}{4} - k (p + 1/p - 2)}. \tag{5.36}$$

To find the boundaries of the interval that contains the principal resonance, we should search for a solution $T$ of Eq. (5.36) in the vicinity of $T = T_0/2 \approx T_{av}/2$. If for a given value of the quality factor $Q$ ($Q$ enters in $p = e^{-\gamma T}$) the depth of modulation $m$ exceeds the threshold value, Eq. (5.36) has two solutions that correspond to the desirable boundaries $T_-$ and $T_+$ of the instability interval. These solutions exist if the expression under the radical sign in Eq. (5.36) is positive. Its zero value corresponds to the threshold conditions:

$$\frac{(1-k)^2}{k} \cos^2 \frac{\Delta\omega T}{4} = p + 1/p - 2. \tag{5.37}$$

To evaluate the threshold value of $Q$ for small values of the modulation depth $m$, we may assume here $k \approx 1 + m$, and $\cos(\Delta\omega T/4) \approx 1$. In the right-hand side of Eq. (5.37), in $p = e^{-\gamma T}$, we can consider $\gamma T \approx \gamma T_0/2 = \pi/(2Q) \ll 1$, so that $p + 1/p - 2 \approx (\gamma T)^2 = (\pi/2Q)^2$. Thus, for the threshold of the principal parametric resonance we obtain:

$$Q_{min} \approx \frac{\pi}{2m}\left(1 + \frac{m}{2}\right) \approx \frac{\pi}{2m}, \quad m_{min} \approx \frac{\pi}{2Q}\left(1 + \frac{\pi}{4Q}\right) \approx \frac{\pi}{2Q}. \tag{5.38}$$

At the threshold the expression under the radical sign in Eq. (5.37) is zero. Both its roots (the boundaries of the instability interval) merge. This occurs when the cosine in the left-hand side of Eq. (5.37) is zero, that is, when its argument equals $\pi/2$:

$$\omega_{av}\frac{T}{2} = \frac{\pi}{2}, \quad \text{or} \quad T = \frac{\pi}{\omega_{av}} = \frac{1}{2}T_{av},$$

so that the threshold conditions, Eqs. (5.38), correspond to exact tuning to resonance, when $T = T_{av}/2$.

To find the boundaries $T_-$ and $T_+$ of the instability interval, we represent $T$ in the argument of the cosine function in the left-hand side of Eq. (5.36) as $T_{av}/2 + \Delta T$, where $\Delta T \ll T_0$. Since $\omega_{av}T_{av} = 2\pi$, we can write this cosine as $-\sin(\omega_{av}\Delta T/2)$. Then Eq. (5.36) becomes:

$$\sin\frac{\omega_{av}\Delta T}{2} = \mp\frac{1}{1+k}\sqrt{(k-1)^2\cos^2\frac{\Delta\omega(\frac{1}{2}T_{av} + \Delta T)}{4} - k\frac{(p-1)^2}{p}}. \tag{5.39}$$

For zero friction $p = 1$, and Eq. (5.39) coincides with (5.21), p. 106. The diagram in Figure 5.12 (p. 114) is obtained by numerically solving this equation for $\Delta T$ by iteration.

To obtain an approximate solution of Eq. (5.39) that is valid for small values of the modulation depth $m$ up to terms to the second order of $m$, we can simplify the expression under the radical sign in the right-hand side of Eq. (5.36), assuming $k \approx 1 + m$, $(1-k)^2 \approx m^2$, and the value of the cosine function to be 1. The last term of the radicand can be represented as $(\pi/2Q)^2 \approx m_{min}^2$. In the left-hand side the sine can be replaced with its small argument, where $\omega_{av} = 2\pi/T_{av}$. Thus we obtain:

$$\frac{\Delta T}{T_{av}} \approx \mp\frac{1}{2\pi}\sqrt{m^2 - m_{min}^2}, \quad \text{or} \quad T_\mp = \frac{T_{av}}{2}\left(1 \mp \frac{1}{\pi}\sqrt{m^2 - m_{min}^2}\right). \tag{5.40}$$

For the case of zero friction $m_{\min} = 0$, and these approximate expressions for the boundaries of the instability interval reduce to Eq. (5.22), p. 107. For the threshold conditions $m = m_{\min}$, and both boundaries of the interval merge, that is, the interval disappears.

### 5.2.5.2. The Second Interval of Instability in the Presence of Friction

When friction is taken into account, we arrive at, instead of Eq. (5.29), p. 111, the following equation for the boundaries of the second interval of parametric instability:

$$(1 + k)\sin\frac{\omega_{\text{av}}T}{2} = \pm\sqrt{(1 - k)^2 \sin^2(\Delta\omega T/4) - k(p + 1/p - 2)}. \quad (5.41)$$

We should search for its solution $T$ in the vicinity of $T = T_0 \approx T_{\text{av}}$. If for a given value of the quality factor $Q$ ($Q$ enters in $p = e^{-\gamma T}$) the depth of modulation $m$ exceeds the threshold value, Eq. (5.41) has two solutions that correspond to the boundaries $T_-$ and $T_+$ of the instability interval. These solutions exist if the expression under the radical sign in Eq. (5.41) is positive. Its zero value corresponds to the threshold conditions:

$$\frac{(k - 1)^2}{k}\sin^2(\Delta\omega T_{\text{av}}/4) = \frac{(p - 1)^2}{p}. \quad (5.42)$$

To estimate the threshold value of $Q$ for small values of the modulation depth $m$, we may assume here $k \approx 1 + m$, and $\sin(\Delta\omega T/4) \approx \Delta\omega T/4$. In the right-hand side of Eq. (5.42), in $p = e^{-\gamma T}$, we can consider $\gamma T \approx \gamma T_0 = \pi/Q \ll 1$, so that $p + 1/p - 2 = (p - 1)^2/p \approx (\gamma T)^2 = (\pi/Q)^2$. Thus, for the threshold of the second parametric resonance we obtain:

$$Q_{\min} \approx \frac{2}{m^2}, \qquad m_{\min} \approx \sqrt{\frac{2}{Q}}. \quad (5.43)$$

The threshold conditions correspond to exact tuning to resonance, when $T = T_{\text{av}}$.

To find the boundaries $T_-$ and $T_+$ of the instability interval, we represent $T$ in the argument of the sine function in the left-hand side of Eq. (5.41) as $T_{\text{av}} + \Delta T$, where $\Delta T \ll T_{\text{av}} \approx T_0$. Since $\omega_{\text{av}}T_{\text{av}} = 2\pi$, we can write this sine as $-\sin(\omega_{\text{av}}\Delta T/2)$. Then Eq. (5.41) becomes:

$$\sin\frac{\omega_{\text{av}}\Delta T}{2} = \mp\frac{1}{1 + k}\sqrt{(k - 1)^2 \sin^2\frac{\Delta\omega(T_{\text{av}} + \Delta T)}{4} - k\frac{(p - 1)^2}{p}}. \quad (5.44)$$

This form of the equation is convenient for numerical solution by iteration. For the zero friction, $p = 1$, and Eq. (5.44) coincides with Eq. (5.30), p. 111. To obtain an approximate solution of Eq. (5.44), valid for small values of the modulation depth $m$ up to the terms of the second order of $m$, we can simplify the expression under the radical sign in the right-hand side of Eq. (5.44), assuming $k \approx 1 + m$, $(1 - k)^2 \approx m^2$, and $\sin\Delta\omega(T_{\text{av}} + \Delta T)/4 \approx \Delta\omega T_{\text{av}}$. The last term of the radicand can be represented as $(2/Q)^2 \approx m_{\min}^4$. In the left-hand side the sine can be replaced by its small argument, in which $\omega_{av} = 2\pi/T_{av}$. Thus for the boundaries of the second instability interval we obtain the following approximate expressions:

$$\frac{\Delta T}{T_{av}} \approx \mp\frac{1}{4}\sqrt{m^4 - m_{\min}^4}, \quad \text{or} \quad T_{\mp} = \left(1 \mp \frac{1}{4}\sqrt{m^4 - m_{\min}^4}\right)T_{av}. \quad (5.45)$$

In the presence of friction, for a given value $m$ of the depth of square-wave modulation, only several first intervals of parametric resonance can exist (the corresponding "tongues" of instability appear only if $m$ exceeds the threshold value), in contrast to the idealized case of zero friction (see the diagram in Figure 5.12, p. 114).

## 5.3   Concluding Remarks

We have shown above that a linear torsion oscillator whose moment of inertia is subjected to square-wave modulation by mass reconfiguration gives a very convenient example in which the phenomenon of parametric resonance can be clearly explained physically with all its peculiarities and even investigated quantitatively by rather modest mathematical means.

In a linear system, if the threshold of parametric excitation is exceeded, the amplitude of oscillations increases exponentially with time. In contrast to forced oscillations, linear viscous friction is unable to restrict the growth of the amplitude at parametric resonance. In real systems the growth of the amplitude is restricted by nonlinear effects that cause the period to depend on the amplitude. During parametric excitation the growth of the amplitude causes variation of the natural period and thereby violates the conditions of resonance.

We note that even if the equilibrium of the system is unstable due to modulation of the parameter (that is, if the conditions of parametric excitation are fulfilled), when the initial values of $\varphi$ and $\dot{\varphi}$ are zero, they remain zero over the course of time. This behavior is in contrast to that of resonance arising from forced oscillations. In the latter instance, the amplitude increases with time even if the initial conditions are zero. In other words, if parametric resonance is to be excited, the system must already be oscillating, at least slightly, when the parametric variation first occurs.

## 5.4   Questions, Problems, Suggestions

### 5.4.1   Principal Parametric Resonance

#### 5.4.1.1*   Principal Resonance ($n = 1$) in the Absence of Friction.

(a) Input a moderate value of the depth $m$ of modulation of the moment of inertia (about 10–15 percent). Choose the period of modulation $T$ to be equal to one half the mean natural period of the oscillator. What kind of initial conditions ought you to enter in order to generate from the very beginning the fastest growth of the amplitude? Remember that at the initial moment, $t = 0$, the weights are suddenly moved away from one another, further from the axis of rotation.

(b) What initial conditions would lead at first to a fading away of oscillations that are already present? Using the plots of the oscillations, explain the physical reason for the increase or decrease in amplitude. Take into account the phase relationship between the natural oscillation of the rotor and the periodic changes in its moment of inertia. Why is it that some time later a phase relation is established that generates a growth in the amplitude?

(c) Try to understand the reasons that determine the lapse of time between the initial fading of the amplitude and its subsequent infinite growth.

### 5.4.1.2* The Growth of the Amplitude at the Principal Resonance without Friction.

(a) If the modulation is to generate the principal parametric resonance, what rule governs the growth of the amplitude when there is an initial deflecton and an initial angular velocity of zero? Calculate the depth of modulation $m$ that, in the absence of friction, generates a doubling of the amplitude after 10 cycles of the parametric modulation. Verify your result with a simulation experiment on the computer.

(b) What difference do you find in your observations of part (a) if you set the initial deflection to be opposite the deflection in part (a)?

### 5.4.1.3* The Threshold for the Principal Resonance.

(a) Choosing a moderate value for the modulation depth (say, $m = 0.15$), estimate the threshold (minimal) value of the quality factor $Q_{min}$ that corresponds to stationary oscillations (i.e., to parametric regeneration) when the modulation is tuned to the principal resonance $(T = T_0/2)$.

(b) Make your calculated estimation of the threshold value $Q_{min}$ more exact by using an experiment on the computer. Describe the character of the plots and of the phase trajectory under conditions of parametric regeneration and explain their features.

(c) Is the mode of stationary oscillations at the threshold (for $Q = Q_{min}$) stable with respect to small deviations in the properties of the system? Is the mode stable with respect to small deviations in the initial conditions?

(d)** The threshold value of the quality factor for any given modulation depth $m$ is absolutely minimal when the modulation is *exactly* tuned to resonance. For small values of $m$ the principal resonance occurs when $T = T_0/2$. However, when $m$ increases, the resonant value of the modulation period $T$ departs from $T_0/2$. Find this resonant value of $T$ for an arbitrarily large modulation depth $m$ and estimate values of $T$ for $m = 15\%$ and $m = 40\%$.

### 5.4.1.4* The Amplitude Growth over the Threshold.

(a) For the case in which $T = T_0/2$ and $m = 15\%$, by what factor does the amplitude of oscillation increase during 10 cycles of parametric oscillation if $Q = 2Q_{min}$? Does the answer depend on the initial conditions? Verify your answer with a simulation experiment.

(b) What is the amplitude of oscillation after the next 10 cycles of modulation? Why does friction not restrict the growth of the amplitude of parametrically excited oscillations?

### 5.4.1.5** The Principal Interval of Parametric Resonance in the Absence of Friction.

(a) Calculate the values of the period of modulation $T$ corresponding to the boundaries of the instability interval at a given modulation depth $m$ (in the approximation $m \ll 1$) for the case when friction is absent.

(b) How does the width of the interval depend on the depth of modulation? Do the terms of second order influence the width of the interval?

### 5.4.1.6* The Initial Conditions for Steady Oscillations.

(a) Enter the value of the period of modulation corresponding to the left boundary of the instability interval at a given value $m$ of the modulation depth. Choose the absence of friction. Input some initial deflection. What value of the initial angular velocity ought you to enter for a given angular deflection in order that stationary oscillations of a constant amplitude occur from the beginning of the modulation?

(b) Verify your calculated approximate values of $T$ for either boundary by simulating an experiment, and find more precise values. Explain the appearance of characteristic features of the plots and the phase trajectories of stationary oscillations corresponding to each boundary of the instability interval.

(c) For a given value of the initial displacement $\varphi_0$, and for the calculated value $\dot{\varphi}(0)$ of the initial angular velocity that provides stationary oscillations (at each of the boundaries of the interval of instability), calculate the amplitude of these oscillations. Verify the theoretical value by the experiment.

### 5.4.1.7** The Threshold of Excitation within the Instability Interval.

(a) Choose a value $T$ of the period of modulation somewhere between the limits of the interval of instability, e.g., approximately halfway between the resonant value and one of the boundaries. Evaluate experimentally the growth of the amplitude in the absence of friction, and from your observations, calculate the threshold value of the quality factor $Q = Q_{min}$ for parametric excitation at the given value $T$ of the modulation period.

(b) Verify your result experimentally and use the experiment to find a more exact value of $Q_{min}$. Compare the observed plots of these stationary oscillations with the plots of stationary (threshold) oscillations at exact tuning to resonance. What are the differences between the plots (and the phase trajectories) of stationary oscillations at the threshold within the interval of parametric excitation with friction, and the plots (and the phase trajectories) of stationary oscillations at the boundaries of the instability interval without friction?

(c) If the threshold is exceeded, why does the amplitude continue to increase indefinitely? In other words, why is friction unable to restrict the growth of the amplitude of parametrically excited oscillations?

(d) For small values of the modulation depth $m \ll 1$, calculate up to terms of second order in $m$ the threshold value $Q = Q_{min}$ of the quality factor for the period of modulation $T$ lying somewhere within the interval of instability. Compare your theoretical result with the value that you have obtained experimentally in parts (a) and (b).

### 5.4.1.8** The Interval of Instability in the Presence of Friction.

(a) For some depth of modulation $m$, the frequency interval of parametric excitation shrinks because of friction and disappears as the quality factor reaches the threshold value. Let the quality factor $Q$ be greater than the threshold value $Q_{min}$. Find the values $T_-$ and $T_+$ of the modulation period $T$ that correspond to the boundaries of the instability interval for a given $m$ and $Q$ (in the approximation $m \ll 1$). Express these values in terms of $m$ and $m_{min}$, where $m_{min} = \pi/(2Q)$ is the approximate threshold value of the modulation depth $m$ for a given quality factor $Q$.

(b) In order to observe steady oscillations corresponding to these boundaries as soon as the simulation begins, you need to set the initial conditions properly. For a given value $\varphi_0$ of the initial deflection, and for each of the boundaries of the interval, what initial velocity produces steady oscillations from the very beginning? Verify your answer by simulating the experiment.

### 5.4.1.9 Oscillations Outside the Interval of Parametric Resonance.
For a given value of $m$, enter a value $T$ of the modulation period lying somewhere outside the limits of the instability interval. Convince yourself that for any set of initial conditions the oscillations eventually fade away, even if the friction is very weak, and that the rotor comes to rest at the equilibrium position in spite of the forced periodic changes in its moment of inertia.

## 5.4.2 Manual Control of the Parameter

There is an option in the relevant simulation program in which the abrupt changes of the moment of inertia do not occur with a periodicity chosen beforehand, but respond instead to signals that you send by the mouse or from the keyboard. First you choose the depth $m$ of modulation and the quality factor $Q$, input initial conditions and start natural free oscillations. Then you can control the oscillator behavior by changing manually its moment of inertia at moments you choose yourself. In this way you can acquire both an understanding of and a feel for the physical causes of parametric excitation.

In order to suppress the automatic modulation of the moment of inertia, mark the check-box "Manual Control." Then the button on the screen marked by arrows becomes enabled, and you can produce abrupt shifts of the weights toward and away from the axis by repeatedly clicking on this button or by toggling the Spacebar.

### 5.4.2.1 Exciting Oscillations by Manual Control of Modulation.
(a) Enter some value $m$ of the modulation depth (about 15–30 percent) and choose a value of the quality factor exceeding the threshold (for the given $m$) By

setting suitable initial conditions, you may excite natural oscillations of moderate swing (10–20 degrees). Try to increase the swing by toggling the Spacebar (or by repeatedly clicking on the button with arrows) regularly at appropriate moments. At what instants ought you to shift the weights towards the axis, and at what instants ought you to return them to their previous positions?

(b*) Is it possible to increase the amplitude to a given value (say, 180°) more quickly by toggling the manual control than by the strictly periodic programmed changes of the moment of inertia for the case of exact tuning to resonance (and with the same values of the depth of modulation and the quality factor)? Give convincing reasons for your answer.

**5.4.2.2 Damping of Oscillations by Manual Control.** After you have produced a large oscillation by appropriately changing the moment of inertia, try to make the oscillations dampen out as fast as possible by changing the moment of inertia. (You should choose a different phase for the changes.) At what time instants ought you to shift the weights toward and away from the axis in order to cause the quickest fading of oscillations?

## 5.4.3    Parametric Resonances of High Orders

### 5.4.3.1* The Third Parametric Resonance (n = 3) without Friction.

(a) Examine the parametric excitation of the rotor for abrupt changes of its moment of inertia with the period $T \approx 3T_0/2$ (approximately one and a half times the natural period, or about three cycles of the parameter modulation during two cycles of natural oscillations). What initial conditions ensure the growth of the amplitude from the beginning of the modulation?

(b) What value $m$ of the modulation depth in the absence of friction is necessary in order to double the initial oscillation during 15 cycles of the parameter modulation? After how many cycles does the amplitude double once more?

(c) For what initial conditions does the oscillation at first decrease? Why does this fading inevitably change after a while into an increase in the amplitude?

### 5.4.3.2* The Threshold for the Third Resonance.

(a) For small values $m$ of the modulation depth, calculate the threshold value $Q_{min}$ of the quality factor up to terms in the first order of $m$. How does this value depend on $m$? Compare your answer with the principal resonance, $n = 1$ (see Problem 5.4.1.3), and with the second resonance, $n = 2$ (see Problem 5.4.3.4). What might be a qualitative explanation for the difference?

(b) For $m \approx 20\%$ evaluate the minimal value $Q_{min}$ of the quality factor for which parametric resonance of the order $n = 3$ is possible. Improve your theoretical estimate by the simulation experiment. Explain the observed shape of the angular velocity plot and the form of the phase trajectory of stationary oscillations at $Q = Q_{min}$. The plots of oscillations occurring at the threshold of the third resonance are shown in Figure 5.14. What factor determines the amplitude of such oscillations?

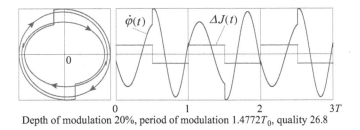

Depth of modulation 20%, period of modulation $1.4772T_0$, quality 26.8

Figure 5.14: The phase trajectory and the time-dependent graph of the angular velocity of stationary oscillations at the threshold of the third parametric resonance.

### 5.4.3.3** The Third Interval of Parametric Excitation.

(a) Calculate the values of the modulation period $T$ which, in the absence of friction, correspond to the boundaries of the third instability interval for a given modulation depth $m$ (in the approximation $m \ll 1$). How does the width of the interval depend on the depth of modulation? Do the terms of the second order influence the width of the interval?

(b) What value of the initial angular velocity ought you to enter for a given initial deflection $\varphi_0$ in order to get stationary oscillations of a constant amplitude from the very beginning of the modulation on each boundary of the instability interval? Verify your calculated values experimentally.

What are the shapes of the phase trajectories that correspond to the left and right boundaries of this interval?

(c) Explore the width of the third interval of parametric excitation without friction at arbitrarily large values of the modulation depth $m$. Note how the interval moves to the left and gradually shrinks as $m$ becomes greater (see the diagram in Figure 5.6, p. 108)).

At $m = 60\%$ both boundaries of the interval coincide. (You may also say that they intersect at this value of $m$.) This coincidence means that for the corresponding value of the modulation period $T$ you get steady oscillations for arbitrary initial conditions. What might be a physical explanation for this behavior?

**Hint**: What is the ratio of the natural periods for the maximum and minimum values of the moment of inertia for this value of the modulation depth?

### 5.4.3.4** The Third Instability Interval with Friction.

(a) At small values of $m$ the third parametric resonance occurs at $T = 3T_0/2$. However, with the growth of $m$ the resonant value of the modulation period $T$ departs from $3T_0/2$. Find an analytical expression for this resonant value of $T$ (for an arbitrarily large modulation depth $m$) and make a numerical estimate for $m = 15\%$ and $m = 40\%$.

(b) How does friction reduce the width of the third interval of parametric excitation? For a small depth of modulation $m \ll 1$, calculate approximate values of the period of modulation $T$ that correspond to the boundaries of the interval for a given value of the quality factor $Q$. Express the results in terms of $m$ and the threshold value $m_{\min} = 3\pi/(2Q)$ (see Problem 5.4.3.2) for the given $Q$-value.

### 5.4.3.5** Parametric Resonance of the Second Order (n = 2).

(a) Choosing a moderate value of the modulation depth ($m < 20\%$), excite parametric resonance of the order $n = 2$ (for which the period of modulation is approximately equal to the average natural period). Why does the growth in amplitude occur much more slowly in this case than it does for the principal resonance, and even more slowly than it does for the resonance of the order $n = 3$ (for the same value $m$ of the modulation depth)? Explain the observed shape of the phase diagram for $n = 2$. Try to determine experimentally the threshold value of the modulation depth for a given value of the quality factor (say, $Q = 15$).

(b) For small values of the modulation depth $m \ll 1$, try to calculate the threshold value of the quality factor $Q_{\min}$. (You need to keep the terms of the second order in $m$). How does the threshold value of $Q$ depend on $m$? Compare your calculated value with the threshold of principal resonance and of the third resonance. Explain the difference qualitatively. Also compare the theoretical threshold value with your experimental result of part (a).

### 5.4.3.6** The Second Interval of Parametric Excitation.

(a) For small values of the modulation depth $m \ll 1$, calculate the width of the interval (you need to keep the terms to the second order of $m$). How does the width depend on $m$?

(b) Excite and experimentally examine stationary oscillations without friction that correspond to the boundaries of the second instability interval (near the resonance for $n = 2$). For small values of the modulation depth $m \ll 1$, why is this interval considerably narrower than the interval for resonance of a higher order $n = 3$?

(c) Why do two different phase trajectories correspond to each boundary of the interval? What is the difference between the two stationary oscillations that correspond to the same boundary? How can each one of them be excited? What initial conditions ensure steady oscillations from the beginning of modulation?

### 5.4.3.7*** The Second Interval of Parametric Excitation with Friction.

How does friction influence the width of the second interval of parametric excitation? For a small depth of modulation $m \ll 1$, calculate approximate values of the period of modulation $T$ that correspond to the boundaries of the interval for a given value of the quality factor. Write down the results in terms of $m$ and the threshold value $Q_{\min}$ (see Problem 3.5) for the given value of $Q$.

# Chapter 6

# Sinusoidal Modulation of the Parameter

**Annotation.** The general concepts of the parametric excitation of a linear system and the conditions necessary for parametric resonance are considered in Chapter 5, in which we examine the square-wave modulation of a parameter. The present chapter deals with another case of parametric excitation of a linear torsion spring oscillator, namely that of excitation arising from a smooth, harmonic motion of the weights along the rod of the flywheel.

To provide a growth of energy during a smooth periodic modulation of the moment of inertia, the motion of the weights toward the axis of rotation must occur while the angular velocity of the flywheel is greater in magnitude than its angular velocity when the the weights are returned to the ends of the rod. To be sure, a smooth parametric modulation is less efficient for the delivery of energy to the flywheel than is a square-wave modulation. During a square-wave modulation, the shifts of the weights toward and away from the axis occur instantly and at the most favorable moments. On the other hand, during smooth modulations these processes are protracted in time. Hence the threshold value of the depth of modulation for smooth modulations is somewhat greater than it is for square-wave modulation. Nevertheless, if the threshold of excitation is exceeded, the amplitude during smooth modulation also grows indefinitely.

## 6.1 Summary of the Theory: Basic Concepts

### 6.1.1 The Physical System

The physical system investigated in the present chapter and simulated in the relevant computer program is the same linear torsion spring oscillator that is used for the discussion of square-wave modulation (Chapter 5). A flywheel (rotor), which is a balanced rod with two identical weights, can rotate about an axis that passes through its center (see Figure 5.1, p. 95). When in its equilibrium position, one end of the rod points toward the zero of the dial. When the rod is displaced from the equilibrium position, the spiral spring is twisted and produces a restoring torque that is proportional to the angle of deflection.

To provide modulation of a system parameter, we assume that the weights can be shifted simultaneously along the rod in opposite directions into other symmetrical positions so that the rotor as a whole remains balanced. However, its moment of inertia $J$ is changed by such displacements of the weights. When the weights are shifted toward or away from the axis, the moment of inertia decreases or increases respectively. Thus the moment of inertia of the rotor is the parameter to be modulated in the investigated physical system. As the moment of inertia $J$ is changed, so also is the natural frequency $\omega_0 = \sqrt{D/J}$ of the torsional oscillations of the rotor. Periodic modulation of the moment of inertia can cause, under certain conditions, a growth of (initially small) natural rotary oscillations of the rod.

### 6.1.2 Physical Reasons for Parametric Excitation at Smooth Modulation

To understand how a change in the moment of inertia can increase or decrease the angular velocity of the rotor, let us imagine for a while that the spiral spring is absent. Then the angular momentum of the system would remain constant as the weights are being moved along the rod. Thus the resulting reduction in the moment of inertia is accompanied by an increment in the angular velocity, and the rotor acquires additional energy. The system is similar in some sense to a spinning figure skater, whose rotation accelerates as she moves her initially stretched arms closer to her body.

The greater the initial angular velocity, the greater the increment in the velocity and the energy. This additional energy is supplied to the rotor by the source that moves the weights along the rod. On the other hand, if the weights are moved apart along the rotating rod, the angular velocity and the energy of the rotor diminish. The decrease in energy is transmitted back to the source.

In order that increments in energy occur regularly and exceed the amounts of energy returned, i.e., in order that, as a whole, the modulation of the moment of inertia regularly feeds the oscillator with energy, the period and the phase of modulation must satisfy certain conditions.

An example of parametric excitation of the torsion oscillator during the sinusoidal motion of the weights along the rod with the period that equals one half of the natural period is shown in Figure 6.1.

### 6.1.3 Conditions of Parametric Resonance

To provide a growth of energy by modulation of the moment of inertia, the motion of the weights toward the axis of rotation must occur while the angular velocity of the rotor is on the average greater in magnitude than it is when the weights are moved apart to the ends of the rod. The graphs in Figure 6.1 correspond to this

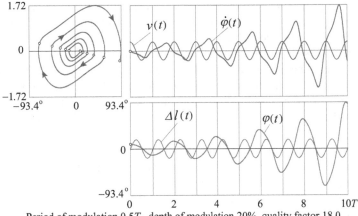

Period of modulation $0.5T_0$, depth of modulation 20%, quality factor 18.0

Figure 6.1: The phase trajectory, the graphs of the angular velocity and displacement of the rotor, and of the weights motion in conditions of the principal parametric resonance.

case: We see clearly that during the intervals of negative values of $v$ the angular velocity $\dot{\varphi}$ is greater in magnitude than during the intervals of positive $v$. Otherwise the modulation of the moment of inertia aids the damping of the natural oscillations.

The strongest parametric oscillations are excited when the cycle of modulation is repeated twice during one period $T_0$ of natural oscillations in the system, i.e., when the frequency $\omega$ of parametric modulation is twice the natural frequency $\omega_0$ of the system. But the delivery of energy is also possible when the parameter changes once during one period, twice during three periods, and so on. That is, parametric resonance is possible when one of the following conditions for the frequency $\omega$ of modulation (or for the period of modulation $T = 2\pi/\omega$) is fulfilled:

$$\omega = 2\omega_0/n, \quad T = nT_0/2, \tag{6.1}$$

where $n = 1, 2, \ldots$. For a given amplitude of modulation of the parameter, the higher the order $n$ of parametric resonance, the less (in general) the amount of energy delivered to the oscillating system during one period.

One of the most interesting characteristics of parametric resonance is the possibility of exciting increasing oscillations not only at the frequencies $\omega_n$ given in Eq. (6.1), but also in intervals of frequencies lying on either side of the values $\omega_n$ (in the *ranges of instability*). These intervals become wider as the range of parametric variation is extended, that is, as the depth of modulation is increased.

An important distinction between parametric excitation and forced oscillations is related to the dependence of the growth of energy on the energy already stored in the system. While for a direct forced excitation the increment of energy during one period is proportional to the *amplitude* of oscillations, i.e., to the square root of the energy, at parametric resonance the increment of energy is proportional to the *energy* stored in the system.

Energy losses caused by friction (unavoidable in any real system) are also proportional to the energy already stored. In the case of direct forced excitation, an arbitrarily small external force gives rise to resonance. However, energy losses restrict the growth of the amplitude because these losses grow with the energy faster than does the investment of energy arising from the work done by the external force.

In the case of parametric resonance, both the investment of energy caused by the modulation of a parameter and the frictional losses are proportional to the energy stored (to the square of the amplitude), and so their ratio does not depend on the amplitude. Therefore, parametric resonance is possible only when a threshold is exceeded, that is, when the increment of energy during a period (caused by the parametric variation) is larger than the amount of energy dissipated during the same time. To satisfy this requirement, the range of the parametric variation (the depth of modulation) must exceed some critical value. This threshold value of the depth of modulation depends on friction (see Section 6.1.5). However, if the threshold is exceeded, the frictional losses of energy cannot restrict the growth of the amplitude. In a linear system the amplitude of parametrically excited oscillations must grow indefinitely.

In a nonlinear system the natural period depends on the amplitude of oscillations. If conditions for parametric resonance are fulfilled at small oscillations and the amplitude begins to grow, the conditions of resonance become violated at large amplitudes. In a real system the growth of the amplitude over the threshold is restricted by nonlinear effects.

## 6.1.4   Energy Transformations at Parametric Excitation

To understand why it is possible to excite torsional oscillations of the rotor by a smooth (e.g., sinusoidal) motion of the weights along the rod, and what conditions are necessary for parametric resonance, we make use of the conservation of energy.

In the case of the parametric excitation of oscillations, additional energy must be transmitted to the rotor by the source that makes the weights move periodically along the rod. Therefore, we calculate the work done by the source during one period of oscillation and find those conditions under which this work is positive.

To produce a parametric excitation of the oscillator, certain phase relationships between the variations of its angular velocity and the changes in its moment of inertia must hold. Namely, the weights must be drawn toward the axis when the angular velocity of the flywheel is on the average greater in magnitude than when the weights are moved apart. Otherwise the modulation of the moment of inertia aids the damping of the natural oscillation.

In the model we let the forced motion of the weights along the rod be exactly sinusoidal, and so their distance $l$ from the axis of rotation varies with time according to the following expression:

$$l(t) = l_0(1 + m \sin \omega t). \tag{6.2}$$

Here $l_0$ is the mean distance of the weights from the axis of rotation, and $m$ is the dimensionless amplitude of their harmonic motion along the rod ($m < 1$). From Eq. (6.2) we find that a weight moves along the rod with a velocity (relative to the rod) that changes with time as $\cos \omega t$:

$$v(t) = dl/dt = \omega l_0 m \cos \omega t. \tag{6.3}$$

The relative acceleration of the weight in its motion along the rod is:

$$a_r(t) = dv/dt = -\omega^2 l_0 m \sin \omega t. \tag{6.4}$$

In order to find the force $F$ exerted on the weight by the device that makes it move along the rod, we use a non-inertial reference frame rotating with the rod. Using Newton's second law applied to the motion of the weight in this rotating frame of reference, we must take into account the pseudo centrifugal force of inertia acting on the weight, $M\dot{\varphi}^2(t)l(t)$, where $M$ is the mass of the weight and $\dot{\varphi}(t)$ is the angular velocity of the rod:

$$M a_r(t) = F(t) + M\dot{\varphi}^2(t)l(t). \tag{6.5}$$

We are interested in the work of this force $F(t)$ done during one period of oscillation. The amount of this work (for both weights) equals the change in the energy of oscillations during one period. For the infinitesimal element of work $dW$ performed during a time interval $dt$, during which the weight is displaced along the rod a distance $dl = v(t)dt$, we can write the following expression:

$$dW = F(t)dl = F(t)v(t)dt = [\, Ma_r(t) - M\dot{\varphi}^2(t)l(t) \,]v(t)dt. \tag{6.6}$$

As we see from Eq. (6.3), the radial velocity $v(t)$ of the weight in Eq. (6.6) is proportional to the dimensionless amplitude $m$ of its forced motion along the rod. If we restrict our calculations to the first order of the small parameter $m$, we need keep only the second term in square brackets in Eq. (6.6), and we can substitute for $l(t)$ its mean value $l_0$ in the equation:

$$dW \approx -M\dot{\varphi}^2(t)l_0 v(t)dt = -M\dot{\varphi}^2(t)l_0^2 \omega m \cos \omega t. \tag{6.7}$$

As we noted above, the most favorable condition for the parametric excitation of the rotor occurs if the weights execute two full cycles of the forced motion during one mean period of natural oscillation. In other words, the frequency $\omega$ in Eq. (6.2) and Eq. (6.7) must be approximately twice the mean natural frequency $\omega_0 = 2\pi/T_0$ of oscillation of the rotor. (Here $\omega_0$ is the frequency of free oscillations of the rotor with the weights fixed at their average distance $l_0$ from the axis). The frequency of modulation $\omega$ in Eq. (6.2) that is equal to the doubled value of the mean natural frequency ($\omega = 2\omega_0$) is exactly tuned to the principal resonance ($n = 1$) for small values of the dimensionless amplitude $m$.

In addition, it is necessary that a certain phase relation between the forced motion of the weights and the torsional oscillations of the rotor be satisfied: Namely, the weights must move with maximal relative velocity toward the axis of rotation at moments when the oscillating rod moves with its greatest angular velocity (which it does when it is near its equilibrium position). This phase relation is satisfied for the motion of the weights described by Eqs. (6.2)–(6.3) provided we assume the following time dependence for the torsional oscillations of the rotor:

$$\varphi(t) = \varphi_m \cos\omega_0 t; \quad \dot{\varphi}(t) = -\varphi_m\omega_0 \sin\omega_0 t. \tag{6.8}$$

These are only approximate expressions because, strictly speaking, the torsional oscillations of the rotor are not harmonic. Deviations from a sinusoidal oscillation are caused by the motion of the weights since this motion influences the angular velocity and the moment of inertia of the rotor.

After the substitution of $\omega = 2\omega_0$ and $\dot{\varphi}(t)$ from Eq. (6.8) into Eq. (6.7) we can integrate $dW$ given by (6.7) over a period $T_0 = 2\pi/\omega_0$, taking into account that

$$\int_0^{T_0} \cos^2 \omega_0 t dt = T_0/2.$$

Finally we find that (up to terms of the first order in the small value $m$), the work $W$ of the force $F(t)$ done during a period $T_0$ is given by the following expression:

$$W = \frac{1}{2}M\varphi_m^2\omega_0^2 l_0^2 \cdot 2\pi m. \tag{6.9}$$

The same relationship is valid for the second weight. And so as a whole the forces exerted on the weights by the device that makes them move along the rod perform positive work $W > 0$ during a period and increase the energy of the oscillator by the amount:

$$\Delta E = 2W = M\varphi_m^2\omega_0^2 l_0^2 \cdot 2\pi m. \tag{6.10}$$

For simplicity, we let the rod be very light compared to the weights so that we can consider all kinetic energy of the rotor to be the kinetic energy of these massive weights. The total energy $E$ of the oscillator is equal to the maximal value of its kinetic energy, which is attained at the instants when the oscillating rotor moves near its equilibrium position and has its greatest angular velocity $\omega_0\varphi_m$:

$$E = M\varphi_m^2\omega_0^2 l_0^2.$$

Comparing this expression with the right side of Eq. (6.10), we see the most essential feature of parametric resonance, namely that the investment of energy $\Delta E$ due to modulation of a parameter is proportional to the energy $E$ already stored in the oscillator:

$$\Delta E = 2\pi m E. \tag{6.11}$$

Eq. (6.11) means that at parametric resonance the total energy $E$ of oscillations, averaged over a period $T_0 = 2\pi/\omega_0$ of oscillation, grows exponentially with time:

$$\frac{dE}{dt} = m\omega_0 E, \quad E(t) = E_0 \exp(2st), \quad \text{where} \quad 2s = m\omega_0. \tag{6.12}$$

## 6.1.5 The Threshold of Parametric Excitation

The exponential growth of energy in conditions of the principal parametric resonance described by Eq. (6.12) occurs in the absence of friction. Dissipation of the mean energy $E$ due to viscous friction is also described by an exponential function:

$$\frac{dE}{dt} = -2\gamma E, \quad E(t) = E_0 \exp(-2\gamma t). \tag{6.13}$$

The threshold of parametric resonance corresponds to the case for which these energy losses are just compensated for by the delivery of energy arising from the forced periodic motion of the weights. In this instance, $\gamma = s$. Thus we can find the minimal value of $m$ (for a given value of $\gamma$ or of the quality factor $Q$) that makes parametric excitation possible:

$$m_{\min} = \frac{2\gamma}{\omega_0} = \frac{1}{Q}. \tag{6.14}$$

Equivalently, the threshold condition can be expressed in terms of the maximal value of the damping constant $\gamma$ (or the minimal quality factor $Q$) for a given value $m$ of the amplitude in (6.2):

$$\gamma_{\max} = \frac{1}{2}m\omega_0, \quad Q_{\min} = \frac{\omega_0}{2\gamma_{\max}} = \frac{1}{m}. \tag{6.15}$$

These results concerning the threshold of parametric excitation are approximate and are valid only for small values of the dimensionless amplitude $m$ of the forced motion of the weights along the rod. The simulation program performs numerical integration of the differential equation of motion. This integration is not restricted to small values of $m$. Thus the simulation allows us to determine the threshold conditions experimentally with greater accuracy.

Steady oscillations occurring at the threshold are called *parametric regeneration*. They are shown in Figure 6.2. These graphs should be compared with those shown in Figure 6.1, which displays plots of resonant oscillations occurring above the threshold, where the amplitude grows exponentially in spite of the friction.

When the depth of modulation exceeds the threshold value, the (averaged over the period) energy of oscillations increases exponentially with time. The growth of the energy is again described by Eq. (6.12), p. 131. However, now the index of growth $2\alpha$ is determined by the amount by which the energy delivered through parametric modulation exceeds the simultaneous losses of energy caused by friction: $2\alpha = m\omega_0 - 2\gamma$. The energy of oscillations is proportional to the square of

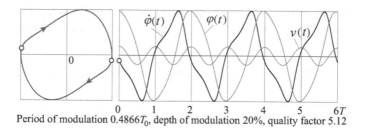

Period of modulation $0.4866 T_0$, depth of modulation 20%, quality factor 5.12

Figure 6.2: The phase trajectory (left), the graphs of the angular velocity and of radial velocity of the weights (right) for oscillations at the threshold condition $m \approx 1/Q$.

the amplitude. Therefore the amplitude of parametrically excited oscillations also increases exponentially with time (see Figure 6.1): $a(t) = a_0 \exp(\alpha t)$ with the index $\alpha$ (one half the index $2\alpha$ of the growth in energy). For the principal resonance we have $\alpha = m\omega_0/2 - \gamma$.

We can now compare the threshold values of the depth of modulation for the case of square-wave modulation with that of harmonic modulation. For $m \ll 1$, as will be seen below from Eq. (6.18), the moment of inertia is modulated nearly harmonically with the depth $m_J \approx 2m$. (The depth of modulation of the moment of inertia is approximately twice the depth of modulation of the distance $l$ between the axis and the weights because the moment of inertia is proportional to the square of the distance $l$.) So for harmonic modulation, Eq. (6.14) gives the threshold depth of the modulation of the moment of inertia: $m_J = 2/Q$. This value is somewhat greater than $m_J = \pi/(2Q)$ given by Eq. (5.6), p. 99 for the case of square-wave modulation (in Eq. (5.6) the depth of modulation of the moment of inertia $m_J$ is denoted simply as $m$). As already mentioned, square-wave modulation provides more favorable conditions for the transfer of energy to the oscillator from the source that moves the weights along the rod of the flywheel.

## 6.1.6 Differential Equation for Sinusoidal Motion of the Weights along the Rod

For simplicity we consider the rod itself to be very light, so that the moment of inertia $J$ of the rotor is due principally to the weights: $J = 2Ml^2(t)$. The angular momentum $J\dot{\varphi}(t)$ changes with time according to the equation:

$$\frac{d}{dt}(J\dot{\varphi}) = -D\varphi, \tag{6.16}$$

where $-D\varphi$ is the restoring torque of the spring. Substituting into Eq. (6.16) $l(t)$ from Eq. (6.2) and taking into account the expression $\omega_0^2 = D/J_0$ ($J_0 = 2Ml_0^2$ is the moment of inertia with the weights in their mean positions), we finally obtain:

$$\frac{d}{dt}\left[(1 + m\sin\omega t)^2\dot{\varphi}\right] = -\omega_0^2\varphi - 2\gamma\dot{\varphi}. \tag{6.17}$$

We have also added the drag torque of viscous friction to the right-hand side of Eq. (6.17). This equation is solved numerically in the relevant computer program in real time during the simulation of oscillations at sinusoidal motion of the weights.

We note that the harmonic motion of the weights along the rod described by Eq. (6.2) does not mean that the moment of inertia is harmonically modulated. Indeed, $J$ is proportional to the *square* of the distance $l(t)$ rather than to its first power. The time dependence of $J(t)$ includes the second harmonic of the frequency $\omega$. Only for small values of the amplitude $m$ (when $m \ll 1$) can we consider the modulation of the moment of inertia to be approximately sinusoidal:

$$J(t) = 2Ml^2(t) = 2Ml_0^2(1 + m\sin\omega t)^2 \approx$$
$$2Ml_0^2(1 + 2m\sin\omega t) = J_0(1 + m_J\sin\omega t), \tag{6.18}$$

where $J_0 = 2Ml_0^2$ is the mean value of the moment of inertia, and $m_J = 2m$ is the depth of its modulation. (We note that the value of $m_J$ is twice the value of $m$.) If we are interested only in an approximate solution valid up to terms of the first order in the small parameter $m$, then instead of the exact differential equation of motion, Eq. (6.17), we can solve the following approximate equation:

$$\ddot{\varphi} + 2\gamma\dot{\varphi} + \omega_0^2(1 - 2m\sin\omega t)\varphi = 0. \tag{6.19}$$

We ignore here the modulation of the coefficient of $\dot{\varphi}$ because for parametric resonance the variation of only those parameters that store energy (the moment of inertia and the torsion spring constant) is essential. Modulation of the damping constant $\gamma$ cannot excite oscillations.

Equation (6.19) is a special case of Hill's equation, Eq. (5.1), p. 93, with sinusoidal time dependence of the parameter. It is called *Mathieu's equation*. The theory of Mathieu's equation has been fully developed, and all significant properties of its solutions are well known. A complete mathematical analysis of its solutions is rather complicated and is usually restricted to the determination of the frequency intervals in which the state of rest in the equilibrium position becomes unstable: At arbitrarily small deviations from the state of rest, the amplitude of incipient small oscillations begins to increase progressively with time. The boundaries of these *intervals of instability* depend on the depth of modulation $2m$. It is worth mentioning that even inside the intervals (when conditions for parametric resonance are satisfied), if $\varphi(0)$ and $\dot{\varphi}(0)$ are exactly zero simultaneously, they remain zero. This property contrasts with the usual case of resonance in which the system is acted upon by a periodic external force. In this case, the amplitude of oscillations begins to grow from a state of rest in the equilibrium position.

We emphasize that the application of the theory of Mathieu's equation to the simulated system under consideration is restricted to the linear order in $m$. For finite values of the depth of modulation $m$, the resonant frequencies and the boundaries of the intervals of instability for the simulated system differ from those predicted by Mathieu's equation. We shall see this point below, where we develop the theory of our system up to the terms of the second order of $m$.

## 6.2    The Intervals of Parametric Instability

In this section we shall show how to use the differential equation of motion, Eq. (6.17), in order to determine the intervals of the frequency $\omega$ of the forced oscillations of the weights along the rod, in which the state of rest of the rotor in the equilibrium position becomes unstable and parametric excitation of its oscillations occurs at deviations from equilibrium no matter how small they are.

### 6.2.1    The Principal Interval of Instability

In the vicinity of the principal resonance the frequency of modulation is approximately twice the natural frequency ($\omega \approx 2\omega_0$), and we can express $\omega$ in the form $\omega = 2\omega_0 + \varepsilon$, where $\varepsilon$ is a small detuning from resonance ($|\varepsilon| \ll \omega_0$). We then propose that an approximate solution to Eq. (6.17) represents a nearly harmonic motion with the frequency $\tilde{\omega} = \omega/2 = \omega_0 + \varepsilon/2$. We let the amplitude and phase of $\varphi(t)$ slowly vary with time:

$$\varphi(t) = p(t) \cos \tilde{\omega}t + q(t) \sin \tilde{\omega}t. \tag{6.20}$$

Here $p(t)$ and $q(t)$ are functions of time that vary slowly relative to the oscillating sine and cosine functions. In the exact solution to Eq. (6.17) there are also higher harmonics with the frequencies $3\tilde{\omega}$, $5\tilde{\omega}$, ... , but their contribution is proportional to higher powers of the small parameter $m \ll 1$. We do not include these higher harmonics in the approximate solution expressed by Eq. (6.20).

The time variation of the amplitudes $p(t)$ and $q(t)$ is caused by the modulation of the square of the natural frequency, and so the derivatives of the functions $p(t)$ and $q(t)$ are also proportional to the small quantity $m$. Substituting $\varphi$ from Eq. (6.20) into the differential equation, Eq. (6.17), we can express the products of the sine and cosine functions in the following way:

$$\sin 2\tilde{\omega}t \cos \tilde{\omega}t = (\sin \tilde{\omega}t + \sin 3\tilde{\omega}t)/2,$$
$$\sin 2\tilde{\omega}t \sin \tilde{\omega}t = (\cos \tilde{\omega}t - \cos 3\tilde{\omega}t)/2,$$

and omit in the equation the higher harmonics with the frequency $3\tilde{\omega}$. Thus for the functions $p(t)$ and $q(t)$ we obtain the following system of differential equations of the first order:

$$2\tilde{\omega}\dot{q} - (\tilde{\omega}^2 - \omega_0^2)p + (2\gamma\tilde{\omega} - m\omega_0^2)q = 0,$$
$$-2\tilde{\omega}\dot{p} - (2\gamma\tilde{\omega} + m\omega_0^2)p - (\tilde{\omega}^2 - \omega_0^2)q = 0. \tag{6.21}$$

We have omitted here the terms $2\gamma\dot{p}$ and $2\gamma\dot{q}$ since parametric excitation is possible only if friction is small enough (from Eq. (6.15) we see that $2\gamma < m\omega_0$). The contribution of these omitted terms to Eq. (6.21) is of the order $m^2$.

According to general rules, the solution to these equations can be searched for in the form $\exp \alpha t$. The condition for the existence of a nontrivial (nonzero) solution of this system of homogeneous equations gives the following expression for $\alpha$:

$$\alpha \approx \frac{1}{2}\sqrt{(m\omega_0)^2 - \varepsilon^2} - \gamma. \tag{6.22}$$

Here we have taken into account that $\tilde{\omega}^2 \approx \omega_0^2 + \omega_0\varepsilon$. If there is an exact tuning to resonance, the deviation in frequency $\varepsilon$ vanishes ($\varepsilon = 0$), and Eq. (6.22) gives the following value for the index $\alpha$ that determines the exponential growth in the amplitude of parametrically excited oscillations:

$$\alpha \approx m\omega_0/2 - \gamma. \tag{6.23}$$

The amplitude of oscillation grows if $\alpha > 0$. Therefore, for the threshold of parametric resonance we obtain $m = 2\gamma/\omega_0 = 1/Q$. The same value for the threshold of parametric excitation under conditions of exact tuning to resonance is obtained above using the conservation of energy; see Eq. (6.14), p. 131. For zero friction, the index of the exponential resonant growth in the amplitude is proportional to the depth of modulation: $\alpha = m\omega_0/2$.

For the case in which friction is absent ($\gamma = 0$), and for a given value $m$ of the depth of modulation, we find from Eq. (6.22) that increasing with time solutions of the linearized differential equation, Eq. (6.17), exist in some interval of frequencies that extends by $\Delta\omega = m\omega_0$ on either side of the resonant value $\omega_{\text{res}} = 2\omega_0$. Thus, the half-width $\Delta\omega$ of the interval of instability is given by:

$$\Delta\omega = m\omega_0. \tag{6.24}$$

For zero friction, the width $2\Delta\omega$ of the interval within which parametric resonance occurs is proportional to the amplitude $m$ of the forced periodic motion of the weights. For a value $\omega$ of the frequency of modulation lying somewhere within the interval, the amplitude of parametrically excited oscillations grows exponentially with time as $\exp(\alpha t)$, where the index $\alpha$ of the growth is given by Eq. (6.22) with $\gamma = 0$:

$$\alpha = \frac{1}{2}\sqrt{(m\omega_0)^2 - (\omega - \omega_{\text{res}})^2} \tag{6.25}$$

(for $|\omega - \omega_{\text{res}}| \leq m\omega_0$). The value of $\alpha$ is zero at the boundaries $\omega_\pm$ of the interval of instability: $\omega_\pm = \omega_{\text{res}} \pm m\omega_0$. At these boundaries, stationary oscillations of constant amplitude are possible. An example of such oscillations is shown Figure 6.3.

The symmetric shape of these graphs shows that on average there is no energy transfer to the frictionless oscillator: The energy gained during one half-cycle of modulation is returned back during the next half-cycle of these oscillations on the boundary of the instability interval.

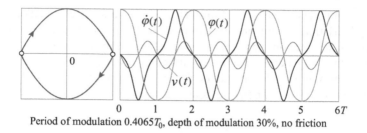

Period of modulation $0.4065T_0$, depth of modulation 30%, no friction

Figure 6.3: The phase trajectory of stationary oscillations occurring at the left boundary of the principal instability interval (left), the time dependent graphs of the angular velocity and displacement of the rotor, and of the radial velocity of the weights (right).

The non-sinusoidal shape of the graphs (see Figure 6.3) shows clearly that the spectrum of steady-state periodic oscillations on the boundaries of the instability, besides the principal harmonic whose frequency $\tilde{\omega}$ equals one half the frequency of modulation $\omega = 2\pi/T$, also contains harmonics of high orders.

Therefore, in order to obtain more precise values for the frequencies of modulation $\omega_\pm$ that correspond to the boundaries of the instability interval, we need to include these high harmonics in the approximate solution of Eq. (6.17). Their frequency $3\tilde{\omega}$, $5\tilde{\omega}$, ... is an odd-number multiple of the fundamental frequency $\tilde{\omega} \approx \omega_0/2$. We look for a solution containing terms up to the second order in $m$. We assume that this solution has the following form:

$$\varphi(t) = p_0 \cos \tilde{\omega}t + q_0 \sin \tilde{\omega}t + p_1 \cos 3\tilde{\omega}t + q_1 \sin 3\tilde{\omega}t. \qquad (6.26)$$

If we are interested only in the boundaries $\omega_\pm$ of the interval of instability, where oscillations are stationary and their amplitude does not vary with time, we can assume the coefficients $p_0$, $q_0$, $p_1$, and $q_1$ to be constant.

Substituting Eq. (6.26) in Eq. (6.17), we can omit the terms with the frequency $5\tilde{\omega}$. In the terms with the frequency $\tilde{\omega}$ we need to keep quantities up to the first and second order in $m$, while in the terms with the frequency $3\tilde{\omega}$ we need to keep only the terms of the first order.

Finally we arrive at a system of homogeneous equations for $p_0$, $q_0$, and $p_1$, $q_1$. The condition for the existence of a nontrivial solution of the system gives us the desired boundaries. These boundaries (for $\gamma = 0$) are given by the following frequencies $\omega_\pm$:

$$\omega_\pm = 2\omega_0 \pm m\omega_0 + \frac{11}{8}m^2\omega_0. \qquad (6.27)$$

In the simulation computer program we are to enter not the frequency $\omega$ of the parametric modulation, but rather the period $T = 2\pi/\omega$. And so, as a convenient reference here, we also indicate the boundaries of the instability interval near the

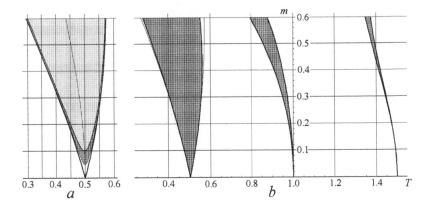

Figure 6.4: Principal instability interval ($a$) and the diagram showing the boundaries of the first three intervals ($b$). Thin curves that deviate slightly at large $m$ values from the boundaries of the principal interval are plotted according to the approximate expression (6.28).

principal resonance in units of $T_0$:

$$T = \frac{T_0}{2}\left(1 \mp \frac{1}{2}m - \frac{7}{16}m^2\right). \qquad (6.28)$$

The term of the second order in $m$ has the same value for both boundaries of the interval. It does not influence the width of the interval, shifting it as a whole by a value proportional to $m^2$.

The structure of the principal interval of parametric instability is shown in Figure 6.4$a$ for the absence of friction (thick bounding curves), for $Q = 20$, and $Q = 10$ (thin inner curves). It is more convenient to express the boundaries using not the frequency $\omega$ of the parametric modulation, but rather the period $T = 2\pi/\omega$. This convention is usually used in presenting the stability map for Mathieu-type systems by the so-called Incze–Strutt diagrams. We also use it in the simulation program and for all figures in this chapter. The dashed regions in Figure 6.4$b$ show the first three intervals of parametric instability in one $T - m$ diagram. [1]

In the presence of viscous friction the principal interval shrinks. The interval disappears if $Q < 1/m$: Its boundaries merge at the threshold. Equation (6.27) gives for the threshold the value $m_{min} = 1/Q$, which has been found above, in Eq. (6.14), from considerations based on the energy conservation.

An example of steady oscillations occurring at the left boundary of the principal instability interval is shown in Figure 6.3, p. 136. The upper panel of Figure 6.5 shows the phase diagram and the graphs of steady oscillations at the right boundary of the interval in the absence of friction. We note the departure of the shape of

---

[1] Actually the curves in Figure 6.4 are plotted with the help of somewhat more complicated formulas than Eq. (6.28) (not cited here), which are obtained by holding several more harmonic components in the trial function $\varphi(t)$.

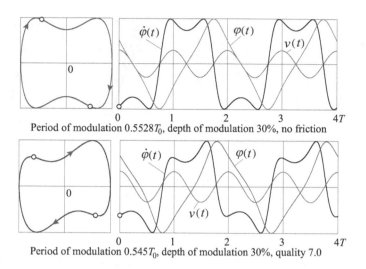

Figure 6.5: The phase trajectory of stationary oscillations occurring at the right boundary of the principal instability interval (left), the time-dependent graphs of the angular velocity and angular deflection of the rotor, and of radial velocity of the weights (right).

these graphs from a sine curve, which is caused by the contribution of higher harmonics (mainly of the third harmonic with the frequency $3\tilde{\omega} = \frac{3}{2}\omega$). The ratio of the amplitude of the third harmonic to the amplitude of the fundamental harmonic is approximately the same for both boundaries ($|C_3/C_1| \approx \frac{3}{8}m$). The difference in the patterns of oscillations at the left and right boundaries (compare the graphs in Figures 6.3 and 6.5) is explained by a different phase shift of the third harmonic with respect to the fundamental one.

The lower panel of Figure 6.5 corresponds to the right boundary in the presence of friction. From the asymmetry of the graph it is clear that in this case the energy received by the oscillator is greater than the energy returned back: During the intervals of negative values of $v$ (while the weights are moving toward the axis) the angular velocity $\dot{\varphi}$ is greater in magnitude. The energy excess compensates for the frictional losses, providing the stationary oscillations. Outside the instability interval, the modulation of the moment of inertia causes only a few changes in the shape of those decaying natural oscillations that may have been excited.

The simulations show that stationary oscillations at the boundaries of the principal resonance also include the fifth and even seventh harmonic components with frequencies $\frac{5}{2}\omega$ and $\frac{7}{2}\omega$ respectively. To find the boundaries with greater precision, we should include these high harmonics into the trial function $\varphi(t)$, Eq. (6.26). For the frictionless oscillator it is more convenient to choose the time origin in such a way that the motion of the weights along the rod is described in Eq. (6.2) by $l(t) = l_0(1 + m\cos\omega t)$ instead of the sine function. In this case the sine and cosine harmonics do not mix, that is, the stationary oscillations at the left boundary of the interval include only harmonics of the cosine type, and at the right boundary—of the sine type.

The final analytical expressions for the frequencies (and periods) of modulation and for the relative contributions of high harmonics (as functions of $m$) at the boundaries of the instability interval are complicated and hence not cited here. However, they show a very good agreement with the simulations. We cite here the calculated values for a certain modulation depth $m = 0.3$ (30%). The corresponding experimental values (obtained in the simulation) are shown in the parenthesis:

Left (cosine-type) boundary: Period $T/T_0 = 0.4066$ (0.4066);
$C_3/C_1 = -0.103$ ($-0.101$); $C_5/C_1 = 0.015$ (0.016); $C_7/C_1 = 0.002$ (0.001).

Right (sine-type) boundary: Period $T/T_0 = 0.5528$ (0.5528);
$S_3/S_1 = -0.129$ ($-0.129$); $S_5/S_1 = 0.020$ (0.020); $S_7/S_1 = 0.003$ (0.003).

For arbitrary values of the modulation depth $m$, the boundaries of the principal instability interval are shown by the first "tongue" of $T - m$ diagram in Figure 6.4.

In the presence of viscous friction the interval of instability of the state of rest in the equilibrium position shrinks. From Eq. (6.22) (with $\alpha = 0$) we find that over the threshold, when $m > 1/Q$, the following deviations $\Delta\omega$ of the frequency $\omega$ on both sides of its resonant value $\omega_{res}$ correspond to the boundaries of the interval:

$$\Delta\omega = \sqrt{(m\omega_0)^2 - 4\gamma^2} = \sqrt{m^2 - (1/Q)^2}\,\omega_0. \qquad (6.29)$$

From this equation we see that at the threshold (when $Q = 1/m$) the boundaries of the interval merge, and the interval of parametric instability disappears.

Outside the instability interval, the modulation of the moment of inertia causes only a few changes in the shape of those decaying natural oscillations that may have been excited.

## 6.2.2 Resonance of the Second Order

In contrast to the principal resonance, for which the energy supply due to the parameter modulation occurs even if we assume the torsional oscillations to be purely sinusoidal (see Eq. (6.8)), for the resonance of the second order a positive net energy delivery is possible only by virtue of the asymmetric distortions in the shape of the oscillations. These distortions are clearly seen in Figure 6.6. They provide the motion of the weights toward the axis of rotation ($v > 0$) to happen on average at a greater (in magnitude) angular velocity $\dot\varphi$ than the backward motion. The distortions can be described by the second harmonic component (frequency $2\omega$), whose contribution is proportional to the depth of modulation $m$. Hence the amount of energy delivered by modulation in conditions of the second parametric resonance is proportional not to $m$ (as at the principal resonance, see Eq. (6.12), p. 131), but only to $m^2$ (for small values of the modulation depth $m \ll 1$).

A system described by Mathieu's equation, Eq. (6.19), with a harmonic time dependence of the modulated parameter also has resonances of higher orders ($n >$

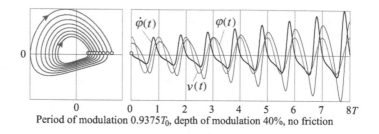

Period of modulation $0.9375T_0$, depth of modulation 40%, no friction

Figure 6.6: The phase trajectory of oscillations in conditions of the second parametric resonance (left), graphs of the angular velocity, angular deflection, and of the radial velocity of the weights (right).

1) near the values $\omega = 2\omega_0/n$. The width $\Delta\omega$ of these resonance bands (of the intervals of instability for the state $\varphi = 0$, $\dot\varphi = 0$) diminishes very quickly as the order $n$ of resonance is increased—as $m^n$. The index $s$ of the rate of the amplitude growth diminishes also as fast as does $\Delta\omega$ with the increase in $n$. Both of these properties make an experimental observation of parametric resonances of higher orders $n > 1$ at moderate values of $m$ very difficult.

In the case of sinusoidal motion of the weights, which is simulated in the relevant computer program, parametric resonance of order $n = 2$ (with $\omega \approx \omega_0$ or $T \approx T_0$, that is, when the period of modulation is approximately equal to the natural period) is extremely weak and narrow for small values of the depth of modulation. In order to find the boundaries of the second interval of parametric instability, we look for a solution of Eq. (6.19) that describes stationary oscillations near the value $\omega = \omega_0$. Considering terms up to the second order in the modulation depth $m$, we should include in this approximate solution the sinusoidal oscillations with the fundamental frequency[2] $\omega = \omega_0 + \varepsilon$ (the frequency of modulation) and the second harmonic with the frequency $2\omega$:

$$\varphi(t) = a_0 \cos\omega t + b_0 \sin\omega t + a_1 \cos 2\omega t + b_1 \sin 2\omega t. \tag{6.30}$$

An example of such stationary oscillation, which corresponds to the left boundary of the second instability interval in the absence of friction, is shown in Figure 6.7. We note the asymmetry of this oscillation: The angular excursion of the rotor on the right-hand side is smaller than on the left-hand side.

Obviously, for the same boundary there also exists one more asymmetric oscillation with greater excursion on the right-hand side. The phase orbits for these two similar oscillations are the mirror images of one another with respect to the ordinate axis.

---

[2]However, it may be convenient to consider the fundamental frequency of parametrically excited stationary oscillations to be always equal to one half of the frequency of modulation. Then the spectrum of oscillations in the case of resonance of an odd order includes only odd harmonics. The spectrum of stationary oscillations for resonance of an even order includes only even harmonics (the amplitude of the fundamental harmonic is zero).

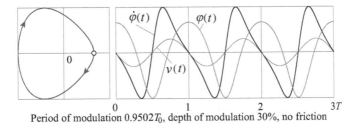

Period of modulation $0.9502T_0$, depth of modulation 30%, no friction

Figure 6.7: The phase trajectory of stationary oscillations occurring at the left boundary of the second instability interval (left), graphs of the angular velocity and angular displacement of the rotor, and of the radial velocity of the weights (right).

Substituting $\varphi(t)$ into Eq. (6.19), we transform there the products of sine and cosine functions into sums, keeping the terms with the frequencies $\omega$ and $2\omega$. Thus, for the coefficients $C_2$, $S_2$, and $C_4$, $S_4$ we obtain the following system of homogeneous equations:

$$\left(1 - \frac{\omega_0^2}{\omega^2}\right) C_2 + \frac{3}{4} m^2 \, C_2 + 2m \, S_4 - \frac{2\gamma}{\omega_0} \, S_2 = 0,$$

$$\left(1 - \frac{\omega_0^2}{\omega^2}\right) S_2 + \frac{1}{4} m^2 \, S_2 + 2m \, C_4 + \frac{2\gamma}{\omega_0} \, C_2 = 0, \qquad (6.31)$$

$$3 \, C_4 - 2m \, S_2 = 0, \qquad 3 \, S_4 + 2m \, C_2 = 0.$$

The last two equations of this system give us the expressions for the amplitudes $C_4$ and $S_4$ of the second harmonic in $\varphi(t)$ in terms of the depth of modulation $m$ and the amplitudes $C_2$ and $S_2$ of the principal harmonic:

$$C_4 = \frac{2}{3} m \, S_2, \qquad S_4 = -\frac{2}{3} m \, C_2. \qquad (6.32)$$

These relations mean essentially that the amplitude of the second harmonic in the stationary oscillations equals $\frac{2}{3} m$ times the amplitude of the principal harmonic. The ratio of the amplitudes of these harmonics is the same for both boundaries of the interval. However, for the left and right boundaries these harmonics add with different relative phases, creating a different shape of the resulting oscillations. Graphs of oscillations occurring at the right boundary of the second instability interval are shown in Figure 6.8.

We note that for the right boundary there also exists an oscillation with the opposite asymmetry of the angular velocity. The phase orbits for these two similar oscillations are the mirror images of one another with respect to the abscissa axis of the phase plane.

Substituting $C_4$ and $S_4$ from Eq. (6.32) into the first two equations of the system, Eq. (6.31), and taking into account that $\omega^2 = (\omega_0 + \varepsilon)^2 \approx \omega_0^2 + 2\omega_0\varepsilon$, we

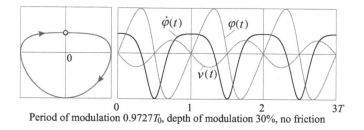

Period of modulation $0.9727T_0$, depth of modulation 30%, no friction

Figure 6.8: The phase trajectory of stationary oscillations occurring at the right boundary of the second instability interval (left), graphs of the angular velocity and angular displacement of the rotor, and of the radial velocity of the weights (right).

obtain the system of two homogeneous equations for $C_2$ and $S_2$:

$$\left(\frac{2\varepsilon}{\omega_0} - \frac{7}{12}m^2\right) C_2 - \frac{2\gamma}{\omega_0} S_2 = 0,$$

$$\frac{2\gamma}{\omega_0} C_2 + \left(\frac{2\varepsilon}{\omega_0} - \frac{13}{12}m^2\right) S_2 = 0. \tag{6.33}$$

A nontrivial solution to this system exists if its determinant equals zero. This condition determines the values of $\varepsilon = \omega - \omega_0$, which correspond to the boundaries $\omega_\pm$ of the second interval of instability:

$$\omega_\pm = \left(1 + \frac{5}{12}m^2 \pm \frac{1}{8}\sqrt{m^4 - (4/Q)^2}\right) \omega_0. \tag{6.34}$$

We note that even the lower boundary is displaced to a higher frequency from the value $\omega_0$. The boundaries of the interval merge at the threshold. From Eq. (6.34) we find the threshold conditions for the second parametric resonance:

$$m_{\min} = \frac{2}{\sqrt{Q}}, \qquad Q_{\min} = \frac{4}{m^2}, \qquad \omega_{\mathrm{res}} = \left(1 + \frac{5}{12}m^2\right)\omega_0. \tag{6.35}$$

Stationary oscillations occurring at the threshold of the second parametric resonance are illustrated by Figure 6.9.

In order to observe the mode of parametric regeneration (stationary oscillations at the threshold of the second parametric resonance) for a given modulation depth $m$ in the simulation experiment, we should choose the period of modulation and the quality factor according to Eq. (6.35), and properly set the initial conditions. For the threshold of the second resonance, Eqs. (6.33) give $S_2 = C_2$. Therefore,

$$\varphi(0) = C_2\left(1 + \frac{2}{3}m\right), \qquad \dot{\varphi}(0) = \omega_0 C_2\left(1 - \frac{4}{3}m\right). \tag{6.36}$$

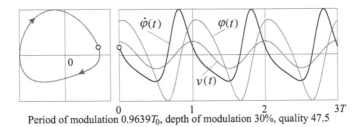

Period of modulation $0.9639T_0$, depth of modulation 30%, quality 47.5

Figure 6.9: The phase trajectory of stationary oscillations occurring at the threshold of the second instability interval (left), graphs of the angular velocity, angular displacement, and of the radial velocity of the weights (right).

To produce stationary oscillations, we can arbitrarily choose an initial angular displacement $\varphi(0)$, and enter an initial angular velocity $\dot\varphi(0) = \omega_0\varphi(0)(1 - 2m)$, as follows from Eq. (6.36). Or, equivalently, we can choose an arbitrary initial velocity $\dot\varphi(0)$, and enter an initial displacement $\varphi(0) = \dot\varphi(0)(1 + 2m)/\omega_0$.

In the absence of friction the width of the second interval of instability is proportional to the square of the depth of modulation: $\omega_+ - \omega_- = m^2\omega_0/4$. Equation (6.34) gives the following boundaries of the interval for zero friction:

$$\omega_+ = \left(1 + \frac{13}{24}m^2\right)\omega_0, \qquad \omega_- = \left(1 + \frac{7}{24}m^2\right)\omega_0. \qquad (6.37)$$

To find the frequencies corresponding to these boundaries with a greater precision, we should include more harmonics into the trial function $\varphi(t)$, Eq. (6.26). In the absence of friction it is more convenient to assume that the motion of the weights along the rod is described in Eq.(5.7) by $l(t) = l_0(1 + m\cos\omega t)$ instead of the sine function. In this case the stationary oscillations at the left boundary of the interval include only harmonics of the cosine type, and at the right boundary of the sine type.

The final (rather complicated) expressions for the periods of modulation and for the relative contributions of high harmonics at the boundaries of the instability interval show a very good agreement with the simulations. Below we cite the calculated values for the modulation depth $m = 0.3$ (30%). The corresponding experimental values are shown in the parenthesis:

Left (cosine-type) boundary: Period $T/T_0 = 0.9502$ (0.9502);

$C_4/C_2 = -0.203$ (−0.202); $C_6/C_2 = 0.038$ (0.039).

Right (sine-type) boundary: Period $T/T_0 = 0.9727$ (0.9727);

$S_4/S_2 = -0.207$ (−0.207); $S_6/S_2 = 0.039$ (0.039).

For arbitrary values of the modulation depth $m$ the calculated boundaries of this instability interval are shown by the second "tongue" of the $T - m$ diagram in Figure 6.4.

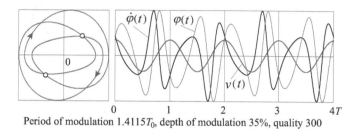

Period of modulation $1.4115T_0$, depth of modulation 35%, quality 300

Figure 6.10: The phase trajectory (left) and the time-dependent graphs (right) of stationary oscillations at the threshold of the third parametric resonance.

In order to observe stationary oscillations in the simulation experiment for the case when friction is zero, we should choose the period of modulation corresponding to one of these boundaries, and properly set the initial conditions. If $l(t) = l_0(1 + m \cos \omega t)$, for the left boundary we can choose arbitrarily an initial angular displacement $\varphi(0)$ and zero initial angular velocity. For the right boundary, vice versa, we arbitrarily choose an initial angular velocity $\dot{\varphi}(0)$ and zero initial displacement.

### 6.2.3 Resonances of the Third and Higher Orders

At parametric resonance of the third order, the flywheel makes three full oscillations during two cycles of modulation. Stationary oscillations at the threshold of parametric resonance of the third order are shown in Figure 6.10.

In order to find the boundaries of the third interval of parametric instability in the absence of friction, we assume that the weights move according to $l(t) = l_0(1 + m \cos \omega t)$, and use the trial function $\varphi(t)$ that includes the fundamental harmonic of the frequency $\frac{1}{2}\omega$ and several high odd-numbered harmonics of frequencies $\frac{3}{2}\omega, \frac{5}{2}\omega, \ldots$.

Stationary oscillations at the left boundary comprise only harmonics of cosine type, and at the right boundary only harmonics of sine type. After substituting the trial function into the differential equation

$$\frac{d}{dt}\left[(1 + m \cos \omega t)^2 \frac{d}{dt}\varphi\right] + \omega_0^2 \varphi = 0, \tag{6.38}$$

we equate to zero the coefficients of cosine (or sine) functions with frequencies $\frac{1}{2}\omega, \frac{3}{2}\omega, \frac{5}{2}\omega, \ldots$, and thus get a system of homogeneous equations for the coefficients $C_1, C_3, \ldots$ (or $S_1, S_3, \ldots$) of harmonic components in the trial function. The condition of existence of a nontrivial solution to this system yields an equation for the desired boundaries. This equation is the same as for the boundaries of the principal instability interval, but this time we look for its approximate solution in the vicinity of $\frac{3}{2}T_0$ (instead of $\frac{1}{2}T_0$). Third harmonic component (frequency $\frac{3}{2}\omega$) dominates the spectrum.

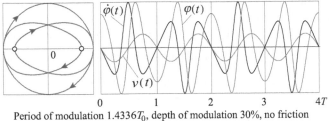

Period of modulation $1.4336T_0$, depth of modulation 30%, no friction

Figure 6.11: The phase trajectories and the time-dependent graphs of stationary oscillations at the left boundary of the third interval of parametric instability.

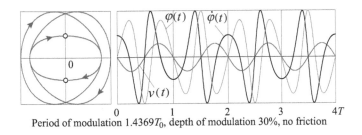

Period of modulation $1.4369T_0$, depth of modulation 30%, no friction

Figure 6.12: The phase trajectories and the time-dependent graphs of stationary oscillations at the right boundary of the third interval of parametric instability.

To increase precision, more harmonics should be included into the trial function $\varphi(t)$. We cite below the values of the period and of the relative contributions of different harmonics at stationary oscillations for the modulation depth $m = 0.3$, obtained by a calculation in which harmonics up to the thirteenth order were included. Such calculations can be fulfilled, say, with the use of *Mathematica* package by Wolfram Research Inc. The corresponding experimental values are shown in the parenthesis:

Left (cosine-type) boundary of the 3rd resonance: Period $T/T_0 = 1.4336 \,(1.4336)$;

$C_1/C_3 = 0.107 \,(0.110)$; $C_5/C_3 = -0.289 \,(-0.288)$; $C_7/C_3 = 0.065 \,(0.067)$.

Right (sine-type) boundary of the 3rd resonance: Period $T/T_0 = 1.4369 \,(1.4369)$;

$S_1/S_3 = 0.135 \,\,(0.136)$; $S_5/S_3 = -0.291 \,(-0.292)$; $S_7/S_3 = 0.066 \,(0.066)$.

Stationary oscillations at the left boundary of the third interval of parametric instability are shown in Figure 6.11, and at the right boundary of this interval in Figure 6.12.

Similar calculations allow us to find the periods of modulation at which resonances of higher orders occur. Corresponding ranges of parametric instability are

very narrow, that is, both their boundaries nearly coincide. We cite below the calculated values of the modulation periods and the spectral composition of stationary oscillations for the boundaries of the fourth and fifth resonances (at $m = 0.3$):

Left (cosine-type) boundary of fourth resonance: Period $T/T_0 = 1.9107$ (1.9107);

$C_2/C_4 = 0.219$ (0.220);  $C_6/C_4 = -0.377$ (−0.374);
$C_8/C_4 = 0.100$ (0.102);  $C_{10}/C_4 = -0.023$ (−0.021).

Right (sine-type) boundary of fourth resonance: Period $T/T_0 = 1.9112$ (1.9112);

$S_2/S_4 = 0.222$ (0.222);  $S_6/S_4 = -0.377$ (−0.377);
$S_8/S_4 = 0.100$ (0.100);  $S_{10}/S_4 = -0.023$ (−0.023).

Left (cosine-type) boundary of fifth resonance: Period $T/T_0 = 2.3872$ (2.3872);

$C_1/C_5 = 0.017$  (0.019);    $C_3/C_5 = 0.319$ (0.321);
$C_7/C_5 = -0.459$ (−0.466);  $C_9/C_5 = 0.124$ (0.146).

Right (sine-type) boundary of fifth resonance: Period $T/T_0 = 2.3873$ (2.3873);

$S_1/S_5 = 0.020$  (0.020);    $S_3/S_5 = 0.321$ (0.321);
$S_7/S_5 = -0.468$ (−0.468);  $S_9/S_5 = 0.142$ (0.144).

The spectral composition with the amplitudes of separate harmonics, phase
trajectories and time-dependent graphs of stationary oscillations at the boundaries
of the fourth and fifth intervals of parametric instability are shown in Figures 6.13
and 6.14, respectively.

Almost exact coincidence of both boundaries for the high-order intervals of
parametric instability means that at the period of modulation corresponding to
one of the intervals we can actually observe (in the absence of friction) not a
resonant growth but rather stationary (strongly non-harmonic) oscillations of a
constant (arbitrarily large) amplitude. From the graphs in Figures 6.13 and 6.14
we can conclude that at exact tuning to $n$-order resonance the oscillator completes
just the whole number $n$ of natural oscillations (of varying period and amplitude)
during exactly two cycles of modulation. The process is periodic at arbitrary initial
conditions, in contrast to the boundaries of low orders, for which special initial
conditions are required to provide periodic oscillations.

This behavior can be explained in terms of the familiar phenomenon of frequency modulation. For parametric resonances of high orders the weights move
along the rod of the exciter rather slowly: The period $T$ of their motion is large
compared to the period $T_0$ of natural torsional oscillations of the rotor ($T \gg T_0$).
A slow periodic variation of the moment of inertia means that the current natural
frequency of the oscillator is slowly modulated.

We see clearly in Figures 6.13 – 6.14 how oscillations slow down (the natural period increases) when the weights are moved towards the ends of the rotor,
and vice versa. Hence we can consider the motion of the rotor in conditions of a

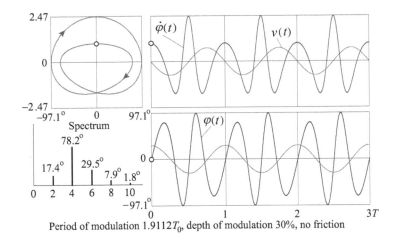

Period of modulation $1.9112 T_0$, depth of modulation 30%, no friction

Figure 6.13: The phase trajectory and time-dependent graph of the angular velocity (upper part), the spectrum and the graph of the angular displacement of the flywheel (lower part) for stationary oscillations at the right boundary of the fourth interval of parametric instability.

high-order parametric resonance as a frequency modulated oscillation,[3] in which the natural oscillation—the dominating harmonic component—plays the role of a carrier.

The spectral composition shown in Figures 6.13 and 6.14 gives convincing evidence in favor of this interpretation. The harmonic component with the frequency $n\omega/2 \approx \omega_0$ has the greatest amplitude (the carrier). The coefficients $C_{n-2}$ and $C_{n+2}$ of lateral spectral components with frequencies $(n\omega/2) \pm \omega$ have opposite signs and (for $n \gg 1$) are nearly equal in magnitude. This spectrum is characteristic of the frequency modulation.

## 6.3 Concluding Remarks

We have developed in this chapter a theoretical approach to the phenomenon of parametric resonance in a linear oscillator under a smooth modulation of the parameter, caused by periodic sinusoidal motion of masses along the rod of the flywheel. The analytical description is complemented by a computerized experimental investigation. A simple mathematical model of the physical system (based on a linear differential equation) is used. The model allows a complete quantitative description of the parametric excitation, which can be verified by the simulations.

---

[3]The amplitude of this oscillation is also slightly modulated with the same period as is the frequency. This means that the modulation is actually combined amplitude-frequency modulation, that is, not purely the frequency modulation.

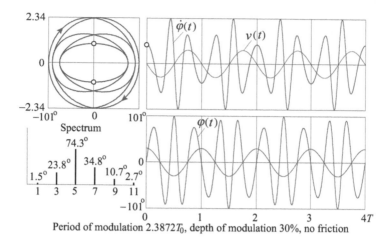

Figure 6.14: The phase trajectory and time-dependent graph of the angular velocity (upper part), the spectrum and the graph of the angular displacement of the flywheel (lower part) for stationary oscillations at the left boundary of the fifth interval of parametric instability.

Visualization of motion simultaneously with plotting the graphs of different variables and phase trajectories makes the simulation experiments very convincing and comprehensible. This investigation provides a good background for the study of more complicated nonlinear parametric systems like a pendulum whose length is periodically changed (a model of the playground swing), or a pendulum whose suspension point is driven periodically in the vertical direction. Such systems are investigated in Part II of the textbook.

## 6.4    Questions, Problems, Suggestions

### 6.4.1    Principal Parametric Resonance

#### 6.4.1.1*  Principal Resonance ($n = 1$) in the Absence of Friction.

(a) Let the period $T$ of the forced motion of the weights correspond to the principal resonance $n = 1$ ($T = T_0/2$). What expression describes the growth of the amplitude of parametrically excited oscillations when there is no friction? How does the rate of the growth in amplitude depend on the depth of modulation $m$ of the distance of the weights from the axis?

(b) What kind of initial conditions allows the growth of the amplitude from the very beginning of the modulation if the motion of the weights along the rod is given by Eq. (6.2)?

(c) For the initial conditions found in part (b) and in the absence of friction, after what number of periods $T$ of modulation does the amplitude increase by 5 times its initial value for $m = 10\%$?

(d) What is the difference in your answers if you set the initial deflection to be on the opposite side of the equilibrium position?

### 6.4.1.2* The Amplitude Growth under the Conditions of Principal Resonance without Friction.

(a) If the frequency of modulation is exactly tuned to the principal resonance ($n = 1$), what value of the modulation depth $m$ produces an increase in the amplitude of oscillation from 20 to 80 degrees during 12 mean natural periods $T_0$? Verify your answer experimentally.

(b) At what initial conditions does the amplitude of oscillation diminish during the early stages following the start of the simulation? Produce this case in a simulation experiment. Why does a decrease in the amplitude change after a while into a growth in the amplitude? Using the plots of the oscillations, explain the physical reason for the growth or decay in amplitude as they depend on the phase relationship between the initial natural oscillation and the forced variation in the moment of inertia. In what lapse of time does the initial decrease in amplitude turn into an unlimited increase?

### 6.4.1.3* The Threshold for Principal Resonance.

(a) Estimate the threshold (minimal) value of the modulation depth $m$ that produces a growth in the amplitude at parametric resonance ($T = T_0/2$) for a given value of the quality factor $Q$ (say, $Q = 15$). Verify your answer experimentally. Explain the features of the plots and of the phase trajectory in this case of parametric regeneration.

(b) Letting the frequency of modulation be exactly tuned to that of the principal resonance $n = 1$, estimate the threshold (minimal) value of the quality factor $Q$ that corresponds to stationary oscillations (i.e., to parametric regeneration) for some moderate value ($10 - 15$ percent) of the modulation depth $m$ of the distance of the weights from the axis.

(c) What initial conditions produce steady oscillations from the very beginning?

(d) Make your above-calculated estimation of the threshold $Q$-value more precise using a computer experiment. Why are there discrepancies between the theoretical predictions and your experimental results?

(e) Describe the character of the plots and of the phase trajectory under conditions of parametric regeneration and explain their features. Is the mode of stationary oscillations at the threshold stable with respect to small deviations in the properties of the system? Is it stable with respect to small deviations in the initial conditions?

(f) For the stationary oscillations at the threshold of principal parametric resonance, what is the ratio of the amplitude of the third harmonic to the amplitude of the fundamental harmonic? (The fundamental frequency in this case is one half of the frequency of modulation).

### 6.4.1.4* Amplitude Growth above the Threshold.

(a) For a given value of $m$ (say, for $m = 0.15$) calculate the minimal value of the quality factor $Q_{min}$ that corresponds to the threshold of parametric excitation. Enter the doubled value $Q = 2Q_{min}$. How many times is the initial amplitude of oscillation increased during the first 10 cycles of modulation when the modulation is exactly tuned to resonance? Does the answer depend on the initial conditions? Verify your answer by simulating the experiment on the computer.

(b) What is the amplitude of oscillations after the next 10 cycles of modulation? Why does the friction not restrict the growth of the amplitude of parametrically excited oscillations?

## 6.4.2    The Principal Interval of Parametric Resonance

### 6.4.2.1** The Principal Interval of Parametric Resonance in the Absence of Friction.

(a) Calculate the values of the modulation period $T$ that correspond to the boundaries of the instability interval at a given modulation depth $m$ and $\gamma = 0$ (for the approximation $m \ll 1$, say, for $m = 10\%$). How does the width of the interval depend on the depth of modulation?

(b) Calculate the boundaries of the principal interval of parametric instability up to terms of the second order in the depth of modulation $m$. Do the terms of the second order influence the width of the interval?

(c) What initial conditions should you enter for each boundary of the interval in order to observe steady oscillations of a constant amplitude from the very beginning of the simulation?

Input some initial deflection. What value of the initial angular velocity ought you to input at a given deflection in order to observe stationary oscillations?

(d) Verify experimentally your calculated approximate values of $T$ and of the initial angular velocity for either boundary of the instability interval, and try to find more precise values.

(e) Explain characteristic peculiarities of the graphs and the phase trajectories of stationary oscillations corresponding to each boundary of the instability interval.

(f) For these stationary oscillations, what is the ratio of the amplitude of the third harmonic to the amplitude of the fundamental harmonic?

### 6.4.2.2** Oscillations at a Boundary of the Instability Interval.

(a) Enter the value of the modulation period $T$ that corresponds to one of the boundaries of the instability interval containing the period of principal parametric resonance. Remember that at these boundaries stationary oscillations of constant amplitude are possible. If you then enter arbitrary initial conditions, the amplitude of oscillations at first either grows or decreases, and the phase trajectory is not a closed curve, as it should be for a periodic process. Explain why.

(b) Follow how the pattern of oscillations gradually approaches the shape that you should expect for the chosen boundary of the interval of instability. For a while the oscillations preserve this shape, but then the amplitude begins to grow or to decrease again, and the shape of oscillations changes again. Why?

### 6.4.2.3** The Threshold of Excitation within the Instability Interval.

(a) Choose a value $T$ of the period of modulation somewhere within the limits of the interval of instability, e.g., approximately halfway between the resonant value and one of the boundaries. Evaluate experimentally the growth of the amplitude without friction and calculate on this basis the threshold value $Q_{min}$ of the quality factor for parametric excitation occurring at the given values $T$ of the modulation period and $m$ of the modulation depth.

(b) Verify your result experimentally and obtain a more precise $Q$-value for the threshold. Compare the observed plots of these stationary oscillations with the plots of stationary (threshold) oscillations when the system is exactly tuned to resonance. What are the differences between the plots (and the phase trajectories) of stationary oscillations at the threshold within the interval of parametric excitation with friction and the plots (and the phase trajectories) of stationary oscillations at the boundaries of the instability interval without friction?

(c) How does friction change the width of the principal ($n = 1$) interval of instability? Estimate the values of the modulation period $T$ that correspond to the boundaries of the instability interval for some given values of $m$ and $Q$. Verify your results in a simulation experiment on the computer and try to determine the boundaries more exactly.

(d) For small values of the modulation depth $m \ll 1$, try to calculate the threshold value of the quality factor for some period of modulation $T$ lying within the principal interval of parametric instability up to terms of the second order in $m$. Compare your theoretical result with the threshold value that you have obtained experimentally in parts (a) and (b).

### 6.4.2.4 Oscillations outside the Interval of Parametric Resonance.

Enter a value $T$ of the modulation period lying somewhere outside the limits of the interval of instability for a given value of the modulation depth $m$. Convince yourself by simulating the experiment that for any initial conditions the oscillations eventually die out (even if the friction is very weak), and the oscillator comes to rest at the equilibrium position in spite of the forced periodic changes of the moment of inertia.

## 6.4.3 The Second Parametric Resonance

### 6.4.3.1** The Second Resonance ($n = 2$) in the Absence of Friction.

(a) For relatively small values of the modulation depth $m \ll 1$ and for zero friction, calculate the boundaries of the instability interval for resonance of the order $n = 2$ (when the period of modulation $T$ is close to the mean natural period $T_0$). How does the width of the interval depend on the depth of modulation $m$?

(b) For each boundary of the instability interval, what initial conditions produce stationary oscillations? Verify your calculated values experimentally.

(c) What is the ratio of the amplitude of the second harmonic to the amplitude of the fundamental harmonic in stationary oscillations at each of the boundaries of the instability interval? (We assume that in this case the fundamental frequency equals the frequency of modulation.)

### 6.4.3.2*** The Threshold for the Second Parametric Resonance

(a) Enter the period of modulation corresponding to parametric resonance of the second order. What initial conditions produce increasing oscillations from the beginning of the simulation?

Note the characteristic shape of the phase diagram at this resonance. Using the plots obtained in the simulation experiment, explain qualitatively the physical reasons for the observed growth of the amplitude. Compare the curves (and the phase diagram) for sinusoidal modulation with the corresponding curves for square-wave modulation (studied in the previous simulation program "Parametric Oscillations of Linear Torsion Pendulum").

(b) For a given modulation depth $m$, what is the threshold (minimal) value $Q_{min}$ of the quality factor for parametric resonance of the second order? What value of the period of modulation corresponds to the threshold?

(c) What initial conditions produce stationary oscillations for the threshold conditions from the beginning of the simulation?

(d) What is the ratio of the amplitude of the second harmonic to the amplitude of the fundamental harmonic in stationary oscillations at the threshold of this resonance?

# Part II

# Nonlinear Oscillations

# Chapter 7

# Free Oscillations of the Rigid Pendulum

**Annotation.** Various kinds of free (unforced) motion of the planar rigid pendulum in the gravitational field (including swinging with arbitrarily large amplitudes and complete revolutions) are investigated in this chapter both analytically and with the help of computerized simulations. The simulation experiments reveal many interesting peculiarities of this famous physical model and complement the analytical study of the subject in a manner that is mutually reinforcing. Chapter 7 also includes a set of theoretical and experimental problems to be solved by students, as well as various assignments that the instructor can offer students for possible individual work on their own.

## 7.1 Summary of the Theory

### 7.1.1 The Physical System

The most familiar example of a nonlinear mechanical oscillator is an ordinary pendulum in the gravitational field, that is, any rigid body that can swing and rotate about some fixed horizontal axis (a *physical pendulum*) or a massive small bob at the end of a rigid rod of negligible mass (a *simple pendulum*). We employ a rigid rod rather than a flexible string in order to examine complete revolutions of the pendulum as well as its swinging to and fro.

The simple pendulum is a famous physical model frequently encountered in textbooks and papers primarily due to its important role in the history of physics. This versatile model is useful and interesting not only in itself as the most familiar example of a nonlinear mechanical oscillator but more importantly because many problems in various branches of physics can be reduced to the differential equation describing the motion of a pendulum. The theory of solitons (solitary wave disturbances traveling in nonlinear media with dispersion), the problem of super-

Figure 7.1: Schematic diagram of the rigid pendulum on the computer screen.

radiation in quantum optics, and Josephson effects in weak superconductivity are the most important examples.

A very interesting treatment of the pendulum within a historical and cultural context can be found in the text of Gregory L. Baker and James A. Blackburn, "The Pendulum: a Case Study in Physics" [1]. This is a unique book in several ways. Firstly, it is a comprehensive quantitative study of one physical system, the pendulum, from the viewpoint of elementary and more advanced classical physics, modern chaotic dynamics, and quantum mechanics. In addition, pendulum analogs of superconducting devices are also discussed. Secondly, this book treats the physics of the pendulum in various aspects, showing, for example, that the pendulum has been intimately connected with studies of the earth's density, the earth's motion, and timekeeping.

A schematic diagram of the pendulum pictured on the screen by the simulation program is shown in Figure 7.1.

In the state of stable equilibrium the center of mass of the pendulum is located vertically below the axis of rotation. When the pendulum is deflected from this position through an angle $\varphi$, the restoring torque of the gravitational force is proportional to $\sin \varphi$. In the case of small angles $\varphi$ (i.e., for small oscillations of the pendulum) the values of the sine and of its argument nearly coincide ($\sin \varphi \approx \varphi$), and the pendulum behaves like a linear oscillator. In particular, in the absence of friction it executes *simple harmonic motion*. However, when the amplitude is large, the motion is oscillatory but no longer simple harmonic. In this case, a graph of the angular displacement versus time noticeably departs from a sine curve, and the period of oscillation noticeably depends on the amplitude.

If the angular velocity imparted to the pendulum at its initial excitation is great enough, the pendulum at first executes complete revolutions losing energy through friction, after which it oscillates to and fro.

In this chapter we describe a combined analytical and computerized approach to the eternal problem of the pendulum motion. Our study is based on the usage of the relevant simulation program of the software package *Physics of Oscillations*.

The simulations allow us to investigate interesting situations that are inaccessible in a real laboratory experiment. Special attention is devoted to cases in which the swing approaches 180°. Such oscillations are treated on the basis of a physically justified approach in which the cycle of oscillation is divided into several stages. The major part of the almost closed circular path of the pendulum is approximated by the limiting motion (see Section 7.1.6), while the motion in the vicinity of the inverted position is described on the basis of the linearized equation. The accepted approach provides additional insight into the dynamics of nonlinear physical systems. Full revolutions of the rigid planar pendulum are also investigated in detail.

## 7.1.2   The Differential Equation of Motion for a Pendulum

The equation of rotation of a solid about a fixed axis in the absence of friction in the case of a physical pendulum in a uniform gravitational field is:

$$J\ddot{\varphi} = -mga \sin \varphi. \tag{7.1}$$

Here $J$ is the moment of inertia of the pendulum relative the axis of rotation, $a$ is the distance between this axis and the center of mass, and $g$ is the acceleration of gravity. The left-hand side of Eq. (7.1) is the time rate of change of the angular momentum, and the right-hand side is the restoring torque of the force of gravity. This torque is the product of the force $mg$ (applied at the center of mass) and the lever arm $a \sin \varphi$ of this force. Dividing both sides of Eq. (7.1) by $J$ we have:

$$\ddot{\varphi} + \omega_0^2 \sin \varphi = 0, \tag{7.2}$$

where the notation $\omega_0^2 = mga/J$ is introduced.

For a simple pendulum $a = l$, $J = ml^2$, and so $\omega_0^2 = g/l$. For a physical pendulum, the expression for $\omega_0^2$ can be written in the same form as for a simple pendulum provided we define a quantity $l$ to be given by $l = J/(ma)$. It has the dimension of length, and is called the *reduced* or *effective* length of a physical pendulum. Since the differential equation of motion of a physical pendulum with an effective length $l$ is the same as that for a simple pendulum of the same length, the two systems are dynamically equivalent.

At small angles of deflection from stable equilibrium, we can replace $\sin \varphi$ with $\varphi$ in Eq. (7.2). Then Eq. (7.2) becomes the differential equation of motion of a linear oscillator (see Eq. (1.2) of Chapter 1). Therefore, the quantity $\omega_0$ in the differential equation of the pendulum, Eq. (7.2), has the physical sense of the angular frequency of infinitely small oscillations of the pendulum in the absence of friction.

In the presence of a torque due to viscous friction, we must add a term to the right-hand side of Eq. (7.2) that is proportional to the angular velocity $\dot{\varphi}$. Thus, with friction included, the differential equation of the pendulum assumes the form:

$$\ddot{\varphi} + 2\gamma\dot{\varphi} + \omega_0^2 \sin \varphi = 0. \tag{7.3}$$

We see that in our model a pendulum is characterized by two parameters: The *angular frequency* $\omega_0$ of small free oscillations, and the *damping constant*

$\gamma$, which has the dimensions of frequency (or of angular velocity). As in the case of a linear oscillator, it is convenient to use the dimensionless *quality factor* $Q = \omega_0/(2\gamma)$ (see Eq. (1.11) of Chapter 1) rather than the damping constant $\gamma$ to measure the effect of damping. At small free oscillations of the pendulum, the value $Q/\pi$ is the number of complete cycles during which the amplitude decreases by a factor of $e \approx 2.72$.

The principal difference between Eq. (7.3) for the pendulum and the corresponding differential equation of motion for a spring oscillator discussed in Chapter 1 is that Eq. (7.3) is a *nonlinear* differential equation. The difficulties in obtaining an analytical solution of Eq. (7.3) are caused by its nonlinearity. In the general case it is impossible to express the solution of Eq. (7.3) in elementary functions, although in the absence of friction the solution of Eq. (7.2) can be given in terms of special functions (elliptic integrals).

### 7.1.3    Dependence of the Period on the Amplitude

The nonlinear character of the pendulum is revealed primarily in dependence of the period of oscillations on the amplitude. To find an approximate formula for this dependence, we can expand $\sin\varphi$ in Eq. (7.3) into the power series. Keeping the first two terms, we obtain (for the conservative pendulum with $\gamma = 0$):

$$\ddot{\varphi} + \omega_0^2\varphi - \frac{1}{6}\omega_0^2\varphi^3 = 0. \tag{7.4}$$

An approximate solution to Eq. (7.4) can be searched as a superposition of the sinusoidal oscillation $\varphi(t) = \varphi_m\cos\omega t$ and its third harmonic $\epsilon\varphi_m\cos 3\omega t$. (We assume $t = 0$ to be the moment of maximal deflection). This solution is found in many textbooks (see, for example, [2]). The fractional contribution $\epsilon$ of the third harmonic equals $\varphi_m^2/192$, where $\varphi_m$ is the amplitude of the principal harmonic component whose frequency $\omega$ differs from the limiting frequency $\omega_0$ of small oscillations by a term proportional to the square of the amplitude:

$$\omega \approx \omega_0(1 - \varphi_m^2/16), \quad T \approx T_0(1 + \varphi_m^2/16). \tag{7.5}$$

The same approximate formula for the period can be obtained by expanding the exact solution expressed in terms of elliptic integrals (see, for example, [2], [3], or [4]) into a power series with respect to the amplitude $\varphi_m$.

Equation (7.5) shows that, say, for $\varphi_m = 30°$ (0.52 rad) the fractional increment of the period (compared to the period of infinitely small oscillations) equals 0.017 (1.7%). The fractional contribution of the third harmonic in this nonsinusoidal oscillation equals 0.14%, that is, its amplitude equals only 0.043°.

The relevant simulation program of the package PHYSICS OF OSCILLATIONS allows us to verify this approximate formula for the period. Table 7.1 gives the values of $T$ (for several values of the amplitude) calculated with the help of Eq. (7.5) and measured in the computational experiment. Comparing the values in the last two columns, we see that the approximate formula, Eq. (7.5), gives the value of the period for the amplitude of 45° with an error of only 0.13%. However,

Table 7.1. Dependence of period on the amplitude

| Amplitude $\varphi_{\mathrm{m}}$ | | $T/T_0$ (calculated) | $T/T_0$ (measured) |
|---|---|---|---|
| 30° | $(\pi/6)$ | 1.0171 | 1.0175 |
| 45° | $(\pi/4)$ | 1.0386 | 1.0400 |
| 60° | $(\pi/3)$ | 1.0685 | 1.0732 |
| 90° | $(\pi/2)$ | 1.1539 | 1.1803 |
| 120° | $(2\pi/3)$ | 1.2742 | 1.3730 |
| 135° | $(3\pi/4)$ | 1.3470 | 1.5279 |
| 150° | $(5\pi/6)$ | 1.4284 | 1.7622 |

for 90° the error is already 2.24%. The error does not exceed 1% for amplitudes up to 70°.

## 7.1.4   The Phase Portrait of the Pendulum

The evolution of the mechanical state of the pendulum during its entire motion can be graphically demonstrated very clearly by a *phase diagram*, i.e., a graph that plots the angular velocity $\dot{\varphi}$ (or the angular momentum $I\dot{\varphi}$) versus the angular displacement $\varphi$. If the motion of the physical system is *periodic*, the system returns to the same mechanical state after a full cycle, and the representative point, moving clockwise, generates a closed path in the phase plane. In general, the structure of a phase diagram tells us a great deal about the possible motions of a nonlinear physical system.

A general idea about the free motion of the pendulum resulting from various values of energy imparted to the pendulum is given by its *phase portrait*, i.e., the family of phase trajectories.

We can construct a phase portrait for a conservative system (e.g., for the pendulum), without explicitly solving the differential equation of motion of the system. The equations for phase trajectories follow directly from the law of the conservation of energy.

The potential energy $E_{\mathrm{pot}}(\varphi)$ of a pendulum in the gravitational field depends on the angle of deflection $\varphi$ measured from the equilibrium position:

$$E_{\mathrm{pot}}(\varphi) = mga(1 - \cos\varphi). \tag{7.6}$$

A graph of $E_{\mathrm{pot}}(\varphi)$ is shown in the upper panel of Figure 7.2. The potential energy of the pendulum has a minimal value of zero in the lower stable equilibrium position (at $\varphi = 0$), and a maximal value of $E_{\mathrm{max}} = 2mga$ in the inverted position (at $\varphi = \pm\pi$) of unstable equilibrium. This maximal value of the potential energy is assumed to be the unit of energy in Figure 7.2. The dashed line shows the parabolic potential well for a linear oscillator whose period is independent of the

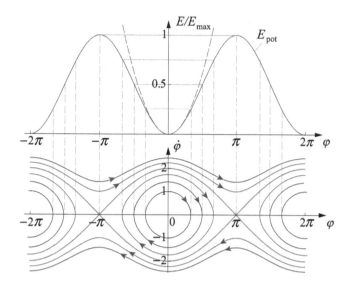

Figure 7.2: The potential well (graph of the potential energy $E_{pot}(\varphi)$) and the phase portrait of the conservative planar rigid pendulum. Closed phase trajectories that enclose the origin of the phase plane correspond to oscillations with different amplitudes. Trajectories passing over and below separatrix (see text for detail) correspond to counterclockwise and clockwise revolutions, respectively.

amplitude (and of the energy) and equals the period of infinitely small oscillations of the pendulum.

In the absence of friction, the total energy $E$ of the pendulum, i.e., the sum of its kinetic energy, $E_{kin}(\dot\varphi) = \frac{1}{2}J\dot\varphi^2$, and potential energy, $E_{pot}(\varphi)$, remains constant during the motion:

$$\frac{1}{2}J\dot\varphi^2 + mga(1 - \cos\varphi) = E. \tag{7.7}$$

This equation gives the relation between $\dot\varphi$ and $\varphi$, and therefore is the equation of the phase trajectory that corresponds to a definite value $E$ of total energy. It is convenient to express Eq. (7.7) in a slightly different form. Recalling that $mga/J = \omega_0^2$ and defining the quantity $E_0 = J\omega_0^2/2$ (the quantity $E_0$ has the physical sense of the kinetic energy of a body with the moment of inertia $J$, rotating with the angular velocity $\omega_0$), we rewrite Eq. (7.7):

$$\frac{\dot\varphi^2}{\omega_0^2} + 2(1 - \cos\varphi) = \frac{E}{E_0}. \tag{7.8}$$

If the total energy $E$ of the pendulum is less than the maximal possible value of its potential energy ($E < 2mga = E_{max} = 4E_0$), that is, if the total energy is less than the height of the potential barrier shown in Figure 7.2, the pendulum swings to and fro between the extreme deflections $\varphi_m$ and $-\varphi_m$. These angles correspond to the extreme points at which the potential energy $E_{pot}(\varphi)$ becomes equal to the total energy $E$ of the pendulum. If the amplitude is small ($\varphi_m \ll \pi/2$), the time dependence of oscillations is nearly sinusoidal, and the corresponding phase trajectory is nearly an ellipse. The elliptical shape of the curve follows from Eq. (7.8) if we substitute there the approximate expression $\cos\varphi \approx 1-\varphi^2/2$ valid for small angles $\varphi$:

$$\frac{\dot{\varphi}^2}{E\omega_0^2/E_0} + \frac{\varphi^2}{E/E_0} = 1. \tag{7.9}$$

This is the equation of an ellipse in the phase plane ($\varphi, \dot{\varphi}$). Its horizontal semiaxis equals the maximal deflection angle $\varphi_m = \sqrt{E/E_0}$. If the angular velocity $\dot{\varphi}$ on the ordinate axis is plotted in units of the angular frequency $\omega_0$ of small free oscillations, the ellipse (7.9) becomes a circle.

The shape of the closed phase trajectory gradually changes as the amplitude and the energy are increased. The width (along $\varphi$-axis) of the phase trajectory increases more rapidly than does its height as the total energy $E$ increases to $E_{max}$. The phase trajectory is stretched horizontally because for the same total energy the amplitude of oscillations in the potential well of the pendulum is greater than it is in the parabolic potential well of the linear oscillator. The greater the total energy $E$ (and thus the greater the amplitude $\varphi_m$), the greater the departure of the phase trajectory from an ellipse and the greater the departure of the motion from simple harmonic.

With the growth of the angular displacement, the restoring torque for the pendulum does not increase as rapidly as for the linear oscillator: The pendulum is a system with a "soft" restoring torque. The upper slopes of its potential well are not as steep as those of the parabola, and at large amplitudes the pendulum spends more time near the extreme points where its direction of motion is reversed.

The period, while independent of the amplitude for the linear oscillator, grows with the amplitude for the pendulum. The crests of the graph of $\varphi(t)$ are flattened, and those of the $\dot{\varphi}(t)$ graph are sharpened. These changes in the shape of time-dependent graphs of oscillations are clearly visible in Figure 7.3. We also note the increased period of these oscillations—the time marks on the abscissa axis of these graphs correspond to the period of infinitely small oscillations.

In the case of a linear oscillator whose potential well is parabolic, time dependencies of both potential and kinetic energies are sinusoidal, and their time average values are equal to one another.

As the extreme angular displacement approaches $180°$, the pendulum spends the greater part of its period near the inverted position, and so the potential energy of the pendulum is close to its maximal value $2mga$ most of the time. Only for the brief time during which the pendulum rotates rapidly through the bottom part of its circular path is the potential energy converted into kinetic energy. Crests of the graph of the potential energy $E_{pot}(t)$ become wider than the valleys between them (see the lower panel of Figure 7.3).

The opposite changes occur with the graph of the kinetic energy $E_{kin}(t)$. Although maximum values of both potential and kinetic energies are equal to the

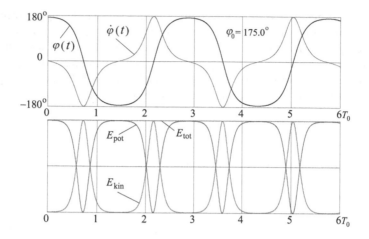

Figure 7.3: The graphs of large oscillations in the absence of friction (initial conditions: $\varphi(0) = \varphi_0 = 175°$, $\dot{\varphi}(0) = 0$, amplitude $\varphi_m = 175°$).

constant value of total energy $E_{tot}$, the average value of the potential energy during a complete cycle of this motion becomes considerably greater than that of the kinetic energy, in contrast to the case of small oscillations, for which the time average values of potential and kinetic energies are equal.

As the total energy imparted to the pendulum is increased so as to approach the value $2mga$ from below, the period of oscillations sharply increases and tends logarithmically to infinity. The shape of the curve of angular velocity versus time resembles a periodic succession of solitary impulses whose duration is close to the period $T_0$ of small oscillations. Time intervals between successive impulses are considerably greater than $T_0$. These intervals grow longer and longer as the total energy $E$ is changed so as to approach the maximal allowed potential energy $2mga$. Executing such swinging, the pendulum moves very rapidly through the bottom of its circular path and very slowly at the top, in the vicinity of the extreme points.

If $E > 2mga$, the kinetic energy and the angular velocity of the pendulum are non-zero even at $\varphi = \pm\pi$. In contrast to the case of swinging, now the angular velocity does not change its sign. The pendulum executes *rotation* in a full circle. This rotation is nonuniform. When the pendulum passes through the lowest point (through the position of stable equilibrium), its angular velocity is greatest, and when the pendulum passes through the highest point (through the position of unstable equilibrium), its angular velocity is smallest.

In the phase plane, rotation of the pendulum is represented by the paths that continue beyond the vertical lines $\varphi = \pm\pi$, repeating themselves every full cycle of revolution, as shown in Figure 7.2.

Upper paths lying above the $\varphi$-axis, where $\dot{\varphi}$ is positive and $\varphi$ grows in value, correspond to counterclockwise rotation, and paths below the axis, along which the representative point moves from the right to the left, correspond to clockwise rotation of the pendulum.

The phase trajectory that corresponds to the total energy $E = 2mga$ separates the central region of the phase plane which is occupied by the closed phase trajectories of oscillations from the outer region, occupied by the phase trajectories of rotations. This boundary is called the *separatrix*. The separatrix divides the phase plane of a conservative pendulum into regions that correspond to different types of motion (see Figure 7.2).

For a conservative system, the equation of a phase trajectory (e.g., Eq. (7.8) in the case of a pendulum) is always an even function of $\dot{\varphi}$, because the energy depends only on $\dot{\varphi}^2$. Consequently, the phase trajectory of a conservative system is symmetric about the horizontal $\varphi$-axis. This symmetry means that the motion of the system in the clockwise direction is mechanically the same as the motion in the counterclockwise direction. In other words, the motion of a conservative system is *reversible*: If we instantaneously change the sign of its velocity, the representative point jumps to the symmetric position of the same phase trajectory on the other side of the horizontal $\varphi$-axis. In the reverse motion the system passes through each spatial point $\varphi$ with the same speed as in the direct motion. Since changing the sign of the velocity ($\dot{\varphi} \rightarrow -\dot{\varphi}$) is the same as changing the sign of time ($t \rightarrow -t$), this property of a conservative system is also referred to as the *symmetry of time reversal*.

The additional symmetry of the phase trajectories of the conservative pendulum about the vertical $\dot{\varphi}$-axis (with respect to the change $\varphi \rightarrow -\varphi$) follows from the symmetry of its potential well: $E_{\text{pot}}(-\varphi) = E_{\text{pot}}(\varphi)$. (Unlike the symmetry about the $\varphi$-axis, this additional symmetry is not a property of all conservative systems.)

When we include friction in our model, motion of the pendulum becomes irreversible, and the above-discussed symmetry of its phase trajectories with respect to reflections in the coordinate axes of the phase plane vanishes. The influence of friction on the phase portrait we discuss below (see Section 7.4.1, p. 179).

The angles $\varphi$ and $\varphi \pm 2\pi, \varphi \pm 4\pi, \ldots$ denote the same position of the pendulum and thus are equivalent. Thus it is sufficient to consider only a part of the phase plane, e.g., the part enclosed between the vertical lines $\varphi = -\pi$ and $\varphi = \pi$ (see Figure 7.2). The cyclic motion of the pendulum in the phase plane is then restricted to the region lying between these vertical lines. We can identify these lines and assume that when the representative point leaves the region crossing the right boundary $\varphi = \pi$, it enters simultaneously from the opposite side at the left boundary $\varphi = -\pi$ (for a counterclockwise rotation of the pendulum).

We can imagine the two-dimensional phase space of a rigid pendulum not only as a part of the plane ($\varphi$, $\dot{\varphi}$) enclosed between the vertical lines $\varphi = +\pi$ and $\varphi = -\pi$, but also as a continuous surface. We may do so because opposing points on these vertical lines have the same value of $\dot{\varphi}$ and describe physically equivalent mechanical states. And so, taking into account the identity of the mechanical states of the pendulum at these points and the periodicity of the dependence of the restoring gravitational torque on $\varphi$, we can cut out this part of the phase plane and roll it into a cylinder so that the bounding lines $\varphi = +\pi$ and $\varphi = -\pi$ are joined. We can thus consider the surface of such a cylinder as the phase space of a rigid pendulum. A phase curve circling around the cylinder corresponds to a nonuniform rotational motion of the pendulum.

## 7.1.5   The Phase Portrait in the Simulation Program

Choosing the item of the computer program entitled "Phase Portrait and $T(E)$" (under the item "View" of the main menu), you can construct on the screen a whole family of phase trajectories of the pendulum. This phase portrait represents possible motions of the pendulum in the absence of friction. Different curves of this family correspond to various values of the total energy of the pendulum. In this experiment, simultaneously with drawing a phase trajectory, you can measure the corresponding period of oscillations or rotations. Doing this for various values of the energy imparted to the pendulum at the initial excitation, you can obtain the dependence $T(E)$ of the period on total energy $E$. Two modes of working with the computer are available in this section: demonstration and computation experiment.

To initiate the drawing of a phase curve, type in the desired value of the energy $E/E_{\max}$ into the text-box labeled "Energy value $E/E_{\max}$ for the next curve," and click on the **Go** command button. Each time you do so, the pendulum is set in motion from the equilibrium position by a momentary push. By clicking at the check-box "Demo mode," you can activate the demonstration mode, in which the energy imparted to the pendulum is increased by a fixed amount for each subsequent curve. You will obtain a phase portrait of the conservative pendulum formed by the family of phase curves for a set of values of the total energy differing by a fixed amount (similar to one shown in Figure 7.2).

During the construction of the phase curves, the motion of the representative point in the phase plane is compared with the graph of the potential energy versus the angle of deflection. This graph is displayed above the phase plane on the computer screen. At the extreme points of the phase diagram, where the curve crosses the $\varphi$-axis, vertical dotted lines are drawn that connect the corresponding points of the phase diagram and of the potential energy graph. (At these points the total energy equals the potential energy.)

Simultaneously with the phase trajectory, the simulation program draws graphs of the time-dependence of the kinetic energy and of the total energy (on the upper right-hand side of the screen). The graph of the total energy for a conservative system is clearly a horizontal straight line at a distance from the time-axis corresponding to the given value of the initial energy. For each value of the energy, the plotting of curves on the screen is terminated automatically when the pendulum completes one full cycle of its motion. Hence the lengths of these horizontal segments of the graphs of the total energy are equal to the period and give a very clear indication of the dependence of the period $T$ on the total energy $E$.

In the lower right-hand corner of the screen, a table is displayed that shows the values of period $T$ obtained in this computer simulation, for different values

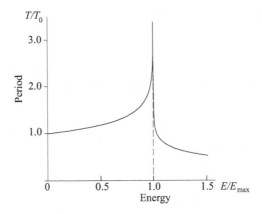

Figure 7.4: The period versus total energy.

of the total energy $E$. From this table you can form a quantitative judgment about the dependence of the period on the total energy $T(E)$.

It is also possible to draw a continuous curve through the ends of the horizontal segments of the graphs of the total energy that correspond to one cycle of oscillation or rotation. To do this, click on the relevant check-box "Plot $T(E)$" at the bottom of the panel. This curve gives a graph of dependence of the period $T$ of oscillations and revolutions on the total energy $E$ imparted to the pendulum. The axis of energy on this diagram is oriented vertically, the axis of period horizontally. The graph $T(E)$ with traditional orientation of axes—energy $E$ as abscissa, period $T$ as ordinate—is shown in Figure 7.4.

The initial almost linear growth of the period with $E$ corresponds to the approximate formula, Eq. (7.5). Indeed, Eq. (7.5) predicts a linear dependence of $T$ on $\varphi_m^2$, and for small amplitudes $\varphi_m$ the energy is proportional to the square of the amplitude.

When the energy approaches the value $E_m$, the period grows infinitely. We note the sharp unlimited growth of the period as the total energy $E$ of the pendulum approaches the maximal possible value $E_{max} = 2mga$ of the potential energy. Greater values of the energy correspond to the rotating pendulum. The period of rotation rapidly decreases with the energy.

## 7.1.6   The Limiting Motion along the Separatrix

The phase trajectory corresponding to a total energy $E$ that is equal to the maximal possible potential energy, namely $E_{pot}(\pi) = 2mga = E_{max}$, is of special interest. This boundary (the *separatrix*) divides the phase plane of a conservative pendulum into regions that correspond to different types of motion. The equation of the separatrix follows from Eq. (7.7) by setting $E = E_{max} = 2mga$, or from

Eq. (7.8) by setting $E = 4E_0 = 2J\omega_0^2$:

$$\dot{\varphi} = \pm 2\omega_0 \cos(\varphi/2). \tag{7.10}$$

The limiting motion of a conservative pendulum with total energy $E = 2mga$ is worthy of a more detailed investigation. In this case the representative point in the phase plane moves along the separatrix.

When the pendulum with the energy $E = 2mga$ approaches the inverted position at $\varphi = \pi$ or $\varphi = -\pi$, its velocity tends to zero, becoming zero at $\varphi = \pm\pi$. This state is represented in the phase plane by the saddle points $\varphi = \pi$, $\dot{\varphi} = 0$ and $\varphi = -\pi$, $\dot{\varphi} = 0$ where the upper and lower branches of the separatrix, Eq. (7.10), meet on the $\varphi$-axis. Both these points represent the same mechanical state of the system, that in which the pendulum is at rest in the unstable inverted position. The slightest initial displacement of the pendulum from this point to one side or the other results in its swinging with an amplitude that almost equals $\pi$, and the slightest initial push causes rotational motion (revolution) of the pendulum in a full circle. With such swinging, or with such rotation, the pendulum remains in the vicinity of the inverted position for an extended time.

For the case of motion along the separatrix, i.e., for the motion of the pendulum with total energy $E = 2mga = 4E_0$, there exists an analytical solution (in terms of elementary functions) for the angle of deflection $\varphi(t)$ and for the angular velocity $\dot{\varphi}(t)$. Integration of the differential equation Eq. (7.10) with respect to time (for the positive sign of the root) at the initial condition $\varphi(0) = 0$ yields:

$$-\omega_0 t = \ln \tan[(\pi - \varphi)/4], \tag{7.11}$$

and we obtain the following expression for $\varphi(t)$:

$$\varphi(t) = \pi - 4\arctan(e^{-\omega_0 t}). \tag{7.12}$$

This solution describes a counterclockwise motion beginning at $t = -\infty$ from $\varphi = -\pi$. At $t = 0$ the pendulum passes through the bottom of its circular path, and continues its motion until $t = +\infty$, asymptotically approaching $\varphi = +\pi$. A graph of $\varphi(t)$ for this motion is shown in Figure 7.5.

The second solution that corresponds to the clockwise motion of the pendulum (to the motion along the other branch of the separatrix in the phase plane) can be obtained from Eq. (7.12) by the transformation of time reversal, i.e., by the change $t \to -t$. Solutions with different initial conditions can be obtained from Eq. (7.12) simply by a shift of the time origin (by the substitution of $t - t_0$ for $t$).

Differentiating $\varphi(t)$ given by Eq. (7.12) with respect to time $t$, we find the following time dependence of the angular velocity $\dot{\varphi}(t)$ for the limiting motion of the pendulum:

$$\dot{\varphi}(t) = \frac{2\omega_0}{\cosh(\omega_0 t)} = \frac{4\omega_0}{e^{\omega_0 t} + e^{-\omega_0 t}}. \tag{7.13}$$

The graph of this function $\dot{\varphi}(t)$ has the form of an isolated impulse (see Figure 7.5). In Eq. (7.13) the origin $t = 0$ is chosen to be the instant at which the pendulum passes through the equilibrium position with the angular velocity $\dot{\varphi} = 2\omega_0$.

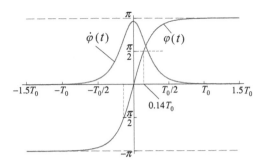

Figure 7.5: The graphs of $\varphi(t)$ and $\dot\varphi(t)$ for the limiting motion (total energy $E = E_{\max} = 2mga = 4E_0$).

This moment corresponds to the peak value of the impulse. The characteristic width of its profile, i.e., the duration of such a solitary impulse, is of the order of $1/\omega_0$. For time $t = \pm T_0/2$ on either side of the peak, Eq. (7.13) gives the angular velocity of only $\pm 0.17\,\omega_0$. Therefore the period $T_0 = 2\pi/\omega_0$ of small oscillations gives an estimate for the duration of the impulse on the velocity graph, that is, for the time needed for the pendulum to execute almost all of its circular path, from the vicinity of the inverted position through the lowest point to the other side of the inverted position.

Using the analytical expression for the time dependence of the angular deflection given by Eq. (7.12), we can calculate the time interval $\tau$ during which the pendulum moves from one horizontal position to the other, passing through the lower equilibrium position: $\tau = 0.28\,T_0$. During this time the kinetic energy of the pendulum is greater than its potential energy, so $\tau$ is the half-width of the solitary impulse of the kinetic energy for the motion under consideration.

The wings of the $\dot\varphi(t)$ profile for the limiting motion decrease exponentially as $t \to \pm\infty$. Actually, for large positive values of $t$, we may neglect the second term $\exp(-\omega_0 t)$ in the denominator of Eq. (7.13), and we find that:

$$\dot\varphi(t) \approx \pm 4\omega_0 e^{-\omega_0 t}. \qquad (7.14)$$

Thus, in the limiting motion of the representative point along the separatrix, when the total energy $E$ is exactly equal to the height $2mga$ of the potential barrier, the speed of the pendulum decreases steadily as it nears the inverted position of unstable equilibrium. The pendulum approaches the inverted position asymptotically, requiring an infinite time to reach it. The motion is not periodic.

The mathematical relationships associated with the limiting motion of a pendulum along the separatrix play an important role in the theory of solitons.

In the program that simulates the pendulum motion, there is a section "Spectrum of Oscillations." In this section a Fourier decomposition of periodic oscillations occurring in the absence of friction is performed. Oscillations with large amplitudes are of special interest because their spectrum is rich of harmonics. You

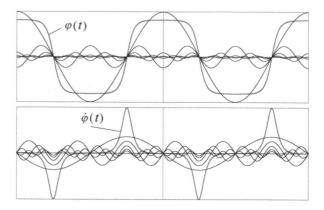

Figure 7.6: Graphs of $\varphi(t)$ (upper panel) and $\dot{\varphi}(t)$ (lower panel) for oscillations with the amplitude 179.885° in the absence of friction, and the graphs of their harmonics.

can obtain the graphs of the separate harmonic components as well as the graph of the partial sum of any number of harmonics. You can then compare these graphs with the graph of the actual motion, which is plotted by the program (in the same place on the screen) during the numerical simulation. Figure 7.6 shows the graph of $\varphi(t)$ and its harmonics (upper panel), and the graph of the angular velocity $\dot{\varphi}(t)$ with its harmonics (lower panel) for oscillations of the pendulum in the absence of friction with the amplitude 179.885°. The period $T$ of such oscillations equals more than five periods of oscillations with an infinitely small amplitude $T = 5.276\,T_0$).

## 7.2  Oscillations of the Pendulum with Extremely Large Amplitudes

The previous analysis of the limiting motion along the separatrix provides a means for understanding the character of oscillations with amplitudes approaching 180°, when total energy of the pendulum is slightly less than the height $2mga$ of the potential barrier, the potential energy of the inverted pendulum.

The old problem of large oscillations of a simple planar pendulum continues to attract attention of the academic community. Dozens of papers on the subject appeared during the last decade in journals—see, for example, [5]–[9] and references therein. In most of the papers various approximation schemes have been developed to express the large-angle pendulum period by simple formulae in terms of elementary functions. Each of the authors usually claims that the formula proposed by him is more simple and accurate when compared with other approximate formulae. A detailed comparison of several approximate expressions that have appeared in recent publications can be found in [10]. The common feature of all suggested approximation schemes can be reduced to a search for some empirical expression for the period $T(\varphi_m)$, which gives for large amplitudes $\varphi_m$ an acceptable numerical agreement with the values obtained from the exact formula given by the complete elliptic integral of the first kind, $K(q)$:

$$T(\varphi_m) = T_0 \frac{2}{\pi} K(\sin^2(\frac{\varphi_m}{2})), \qquad K(q) = \int_0^{\pi/2} \frac{dx}{\sqrt{1 - q\sin^2 x}}, \qquad (7.15)$$

where $T_0 = 2\pi/\omega_0 = 2\pi\sqrt{l/g}$ is the natural period and $\omega_0$ is the frequency of oscillations with infinitely small amplitude, $l$ is the effective length of the pendulum, and $g$ is the acceleration due to gravity.

The approximate expressions for the period that can be found in the literature (see [10] and references therein) give indefinitely increasing errors as the amplitude of the pendulum tends to $180°$. Moreover, all these exercises with various approximation schemes give little physical insight into the nonlinear dynamics of the pendulum behavior at large amplitudes.

In the present section we suggest a radically different approach to the problem of extremely large amplitudes. Our approach is based on physically clear presentation of large oscillations as consisting of several stages during which the motion can be described analytically with high precision in terms of elementary functions. The principal idea of our approach is very simple: The motion of the pendulum in the close vicinity of the inverted position can be described by a linear differential equation (if we choose as a variable the angle $\alpha = \pi - \varphi$, which the pendulum makes with the upper vertical line), while the remaining part of the pendulum's path (constituting nearly a full circle) is almost indistinguishable from the limiting motion (motion along the separatrix), for which a simple solution in elementary functions is available. Precision of the final (very simple) formula for the period, Eq. (7.22), p. 172, increases as the amplitude approaches $180°$.

## 7.2.1   Oscillations with Amplitudes Approaching $180°$

If the pendulum is released with zero initial velocity near the inverted position (say, at initial angle about $179°$), it slowly starts moving toward the down position with a small initial acceleration, because the torque of gravity, being proportional to the sine of deviation from the inverted position, is small. After the pendulum gains some speed, it rapidly makes almost a full circular path through the lower equilibrium position. When the pendulum occurs on the opposite side of the inverted position, its motion gradually slows down as it climbs up along the slope of the potential barrier to its summit. In the absence of friction the pendulum stops when its angular distance to the vertical becomes equal to the initial deviation. From this turning point all the motion repeats in the opposite direction, and after a period the pendulum occurs at the initial point with zero velocity.

To observe the simulation of oscillations discussed in this section, we should switch off the viscous friction (using the corresponding check-box on the "Parameters" panel), and choose appropriate initial conditions (initial angle about $179°$,

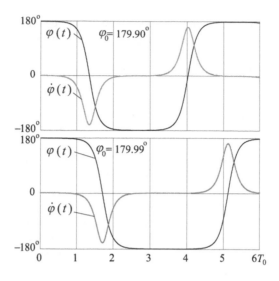

Figure 7.7: Graphs of $\varphi(t)$ and $\dot{\varphi}(t)$ for oscillations with the amplitude 179.90° and 179.99° in the absence of friction, obtained in the simulation experiment.

initial velocity zero). The program allows the user to plot the time dependencies of $\varphi(t)$ and $\dot{\varphi}(t)$, and to draw the phase trajectory simultaneously with the visualization of oscillations. Graphs of $\varphi(t)$ and $\dot{\varphi}(t)$ for oscillations with amplitudes 179.90° and 179.99° in the absence of friction are shown in Figure 7.7.

Comparing these graphs, we can see that for most of the angular excursion from $-\pi$ to $\pi$, these graphs for amplitudes 179.90° and 179.99° are nearly identical. We guess that for these stages of motion deflection angle $\varphi(t)$ and angular velocity $\dot{\varphi}(t)$ are characterized by almost the same time dependence as for the limiting motion along the separatrix, shown in Figure 7.5, p. 167. This dependence of $\varphi(t)$ on time is described (in elementary functions) by the simple expression (7.13), p. 166. Hence the duration of this stage of oscillation for all these cases of large amplitudes approaching 180° is about $T_0$ (the period of small oscillations) and can be calculated with high precision with the help of the same expression (7.13). The duration of the remaining stage, during which the pendulum lingers near the inverted position, depends critically on the amplitude $\varphi_{\mathrm{m}}$. This is clearly seen from comparison of the upper and lower panels of Figure 7.7. This duration increases indefinitely as $\varphi_{\mathrm{m}} \to 180°$. In order to calculate the duration of this stage for certain large amplitudes, we can make use of the linearized differential equation, applicable for small deviations from the inverted position. Next we will do this (see discussion on p. 171).

The closed phase trajectory of oscillatory motion with a large amplitude $\varphi_{\mathrm{m}}$ is shown in Figure 7.8. The largest part of the phase trajectory almost coincides with the separatrix. The representing point goes around the whole closed curve

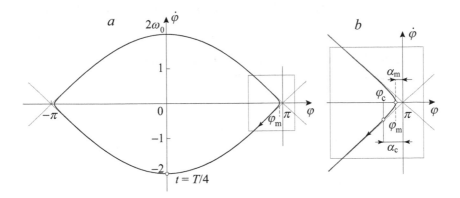

Figure 7.8: The phase trajectory of oscillatory motion with a large amplitude $\varphi_m$ (a) and its portion (increased) that corresponds to the motion of the pendulum in the vicinity of the inverted position (b).

during one period $T$ of oscillation. Next we consider one quarter of this curve, which starts in the phase plane at the initial point of maximal deflection $\varphi = \varphi_m$ and initial velocity $\dot\varphi(t) = 0$, and ends at the point $\varphi = 0$ (marked as $t = T/4$ in Figure 7.8a). To calculate this time $t = T/4$, we choose on this curve an arbitrary point $\varphi = \varphi_c$ located not far from the inverted position $\varphi = \pi$ (see Figure 7.8b), which divides the curve into two parts. The first part between $\varphi = \varphi_m$ and $\varphi = \varphi_c$ lies in the vicinity of the inverted position, so that duration $t_1$ of motion along this part can be calculated with the help of a linearized differential equation of motion (see below). The second part between $\varphi = \varphi_c$ and $\varphi = 0$ is almost indistinguishable from the separatrix, so that duration $t_2$ of motion along this part can be immediately expressed with the help of Eq. (7.12):

$$\omega_0 t_2 = -\ln\tan\frac{\pi - \varphi_c}{4} = -\ln\tan\frac{\alpha_c}{4} \approx \ln\frac{4}{\alpha_c}. \qquad (7.16)$$

Here we introduced the notation $\alpha_c = \pi - \varphi_c$ for the angle that the pendulum makes with the upward vertical line at $\varphi = \varphi_c$. When $\varphi_c$ is close to $\pi$, the angle $\alpha_c$ is small, so that in Eq. (7.16) we can assume $\tan(\alpha_c/4) \approx \alpha_c/4$. Therefore $\omega_0 t_2 \approx \ln(4/\alpha_c)$.

When considering the motion of the pendulum in the vicinity of the inverted position, we find it convenient to define the pendulum position (instead of the angle $\varphi$) by the angle $\alpha$ of deflection from the position of unstable equilibrium. This angle equals $\pi - \varphi$, so that $\varphi = \pi - \alpha$. Substituting angular acceleration $\ddot\varphi = -\ddot\alpha$ and $\sin\varphi = \sin\alpha$ in Eq. (7.2), we find the differential equation for the pendulum in terms of $\alpha$. Since near the inverted position $\alpha \ll 1$, we can replace in this equation $\sin\alpha$ by $\alpha$. Thus we get the following linear differential equation approximately valid for the pendulum's motion between $\varphi = \varphi_m$ and $\varphi = \varphi_c$:

$$\ddot{\alpha} - \omega_0^2 \alpha = 0. \tag{7.17}$$

The general solution to this linear equation can be represented as a superposition of two exponential functions of time $t$:

$$\alpha(t) = C_1 e^{\omega_0 t} + C_2 e^{-\omega_0 t}. \tag{7.18}$$

The initial conditions for the motion from $\varphi = \varphi_m$ to $\varphi = \varphi_c$ are $\alpha(0) = \alpha_m$ and $\dot{\alpha}(0) = 0$. Applying these conditions, we find the constants $C_1$ and $C_2$ in Eq. (7.18):

$$\alpha(t) = \frac{1}{2}\alpha_m (e^{\omega_0 t} + e^{-\omega_0 t}) = \alpha_m \cosh \omega_0 t. \tag{7.19}$$

To find the duration $t_1$ of motion from $\varphi = \varphi_m$ to $\varphi = \varphi_c$ (from $\alpha = \alpha_m$ to $\alpha = \alpha_c$), we substitute in Eq. (7.19) $\alpha(t_1) = \alpha_c$:

$$\alpha_c = \frac{1}{2}\alpha_m (e^{\omega_0 t_1} + e^{-\omega_0 t_1}) \approx \frac{1}{2}\alpha_m e^{\omega_0 t_1}. \tag{7.20}$$

We have omitted the second term in the right-hand side of Eq. (7.20). This is admissible if the arbitrary angle $\alpha_c$ (which divides the phase trajectory into two parts) is chosen to be large compared to $\alpha_m$. From Eq. (7.20) we get for $t_1$:

$$\omega_0 t_1 = \ln \frac{2\alpha_c}{\alpha_m}. \tag{7.21}$$

The desired period of oscillations $T$ is four times greater than the duration $t_1 + t_2$ of motion from $\varphi = \varphi_m$ to the lower equilibrium position $\varphi = 0$. Adding $t_1$ from Eq. (7.21) and $t_2$ from (7.16), we finally obtain the following expression for the period of oscillations with large amplitude $\varphi_m$ approaching $180°$:

$$T = 4(t_1 + t_2) = \frac{4}{\omega_0}\left(\ln \frac{2\alpha_c}{\alpha_m} + \ln \frac{4}{\alpha_c}\right) = \frac{2}{\pi}T_0 \ln \frac{8}{\alpha_m}. \tag{7.22}$$

(Here $\alpha_m = \pi - \varphi_m$.) We note that both $t_1$ and $t_2$ depend on the value $\alpha_c$ of the angle that we have chosen to divide the trajectory into one part that corresponds to the motion in the vicinity of the inverted position, and the other that almost merges with the separatrix. Nevertheless, this dependence on $\alpha_c$ disappears when we add $t_1$ and $t_2$: The final expression (7.22) for the period is independent of the arbitrarily chosen value of $\alpha_c$ (provided $\alpha_m \ll \alpha_c \ll 1$).

The approximation given by expression (7.22) is more accurate the closer the amplitude $\varphi_m$ to $180°$. Table 7.2 illustrates the precision of this simple expression for oscillations of extremely large amplitudes. The values of $T$ in the middle column are calculated on the basis of exact formula (7.15); the right column corresponds to the approximate expression (7.22).

We note that according to this table one cycle of the pendulum oscillation at large amplitudes covers several periods of small oscillations. As an assignment for students' activity, we suggest verifying the values cited in the table by direct measurements of the period in a simulation experiment using the relevant program.

Table 7.2. Period of large-amplitude oscillations

| Amplitude | | $T/T_0$ | $T/T_0$ |
|---|---|---|---|
| $\varphi_m$ | $(\alpha_m)$ | (exact value) | (approximate) |
| 175.000° | (5.000°) | 2.877664 | 2.876395 |
| 177.000° | (3.000°) | 3.202109 | 3.201597 |
| 179.000° | (1.000°) | 3.901065 | 3.900995 |
| 179.900° | (0.100°) | 5.366867 | 5.366866 |
| 179.990° | (0.010°) | 6.832737 | 6.832737 |
| 179.999° | (0.001°) | 8.298608 | 8.298608 |

## 7.2.2 Another Derivation of the Expression for the Period of Large Oscillations

In the above derivation of expression (7.22) we have arbitrarily chosen some small angle $\alpha_c$ for dividing the motion into stages described by different analytical time dependencies. Another way is to choose for this conventional boundary of the two stages, instead of the angular position $\alpha_c$, some arbitrary small angular velocity $\omega_c \ll \omega_0$, which the pendulum gains while moving from the turning point $\alpha_m$ at which its angular velocity is zero. To find the duration $t_3$ of this stage occurring in the vicinity of the inverted position, we can make use of the above-obtained solution (7.19) to the linearized Eq. (7.17), according to which

$$\dot{\alpha}(t) = \frac{1}{2}\alpha_m\omega_0(e^{\omega_0 t} - e^{-\omega_0 t}). \tag{7.23}$$

Substituting $\dot{\alpha}(t_3) = \omega_c$ in Eq. (7.23) and taking into account that $e^{-\omega_0 t_3} \ll e^{\omega_0 t_3}$, we find

$$\omega_0 t_3 = \ln \frac{2\omega_c}{\omega_0 \alpha_m}. \tag{7.24}$$

The further motion towards the equilibrium position is almost indistinguishable from the limiting motion. Hence the time dependence of the angular velocity $\dot{\alpha}(t) = -\dot{\varphi}(t)$ for this stage can be assumed to be the same as for the limiting motion; see Eq. (7.13). Therefore, for calculating the duration $t_4$ of this stage we can substitute $\dot{\varphi}(t_4) = \omega_c$ in (7.13) and take into account that $e^{-\omega_0 t_4} \ll e^{\omega_0 t_4}$. This yields

$$\omega_0 t_4 = \ln \frac{4\omega_0}{\omega_c}. \tag{7.25}$$

Adding $t_3$ from Eq. (7.24) and $t_4$ from (7.25), we finally obtain the same simple expression (7.22) for the period of oscillations with a very large amplitude $\varphi_m$:

$$T = 4(t_3 + t_4) = \frac{4}{\omega_0}\left(\ln \frac{2\omega_c}{\omega_0 \alpha_m} + \ln \frac{4\omega_0}{\omega_c}\right) = \frac{2}{\pi}T_0 \ln \frac{8}{\alpha_m}. \tag{7.26}$$

Again, the arbitrarily chosen angular velocity $\omega_c$ ($\omega_c \ll \omega_0$), which we have used to divide the motion on different stages, vanishes from the final expression (7.26).

## 7.3    Period of Revolutions and Large Oscillations

### 7.3.1    The Period of Fast Revolutions

If the total energy of the pendulum is considerably greater than the maximal value of its potential energy, we can assume all the energy of the pendulum to be the kinetic energy of its rotation. In other words, we can neglect the influence of the gravitational field on the rotation and consider this rotation to be uniform and occurring approximately with the angular velocity $\Omega$ received by the pendulum at the initial excitation. Thus for $E \approx J\Omega^2/2 \gg 2mga$ the asymptotic dependence of the period $T = 2\pi/\Omega$ on the initial angular velocity is the inverse proportion: $T \sim 1/\Omega$. To find the dependence $T(\Omega)$ more precisely, we need to take into account the variations in the angular velocity caused by gravitation. The angular velocity of the pendulum oscillates between the maximal value $\Omega$ in the lower position and the minimal value $\Omega_{\min}$ in the upper position. The latter can be found from the conservation of energy:

$$\Omega_{\min} = \sqrt{\Omega^2 - 4\omega_0^2} \approx \Omega \left(1 - 2\frac{\omega_0^2}{\Omega^2}\right). \tag{7.27}$$

For rapid rotation we can assume these oscillations of the angular velocity to be almost sinusoidal. Then the average angular velocity of rotation is approximately the half-sum of its maximal and minimal values. Thus,

$$\Omega_{\mathrm{av}} \approx \Omega \left(1 - \frac{\omega_0^2}{\Omega^2}\right), \quad T(\Omega) = \frac{2\pi}{\Omega_{\mathrm{av}}} \approx T_0 \frac{\omega_0}{\Omega} \left(1 + \frac{\omega_0^2}{\Omega^2}\right). \tag{7.28}$$

### 7.3.2    Relationship between the Periods of Revolutions and Large Oscillations

The most interesting peculiarities are revealed if we investigate the dependence of the period on energy in the vicinity of $E_{\mathrm{m}} = 2mga$. Measuring the period of oscillations for the amplitudes $179.900°$, $179.990°$, and $179.999°$, we see that duration of the impulses on the graph of the angular velocity very nearly remains the same, but the intervals between them become longer as the amplitude approaches $180°$: Experimental values of the period $T$ of such extraordinary oscillations are, respectively, $5.5\,T_0$, $6.8\,T_0$, and $8.3\,T_0$.

It is also interesting to compare the motions for two values of the total energy $E$, which differ slightly from $E_{\mathrm{max}}$ on either side by the same amount, e.g., for $E/E_{\mathrm{max}} = 0.9999$ and $E/E_{\mathrm{max}} = 1.0001$. In the phase plane, these motions occur very near to the separatrix, the first one inside (oscillations with the amplitude $178.9°$) and the second outside of the separatrix (very slow revolutions). Measuring the periods of these motions, we obtain the values $3.814\,T_0$ and $1.907\,T_0$, respectively: The period of oscillations is twice the period of rotation almost exactly. The graphs of $\varphi(t)$ and $\dot{\varphi}(t)$ for oscillations and revolutions of the pendulum whose energy equals $E_{\mathrm{max}} \mp \Delta E$ are shown respectively in the upper and lower parts of Figure 7.9.

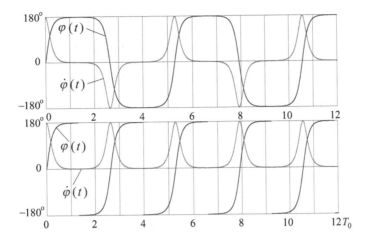

Figure 7.9: The graphs of $\varphi(t)$ and $\dot{\varphi}(t)$ for the pendulum excited at $\varphi = 0$ by imparting the initial angular velocity of $\dot{\varphi} = \omega_0(2 \mp 1 \cdot 10^{-6})$.

Next we suggest a theoretical approach, which can be used to calculate the period of oscillations and revolutions with $E \approx E_{\max}$. From the simulation experiments we can conclude that during the semicircular path, from the equilibrium position up to the extreme deflection or to the inverted position, both of the motions shown in Figure 7.9 almost coincide with the limiting motion (Figure 7.5, p. 167). These motions differ appreciably from the limiting motion only in the immediate vicinity of the extreme point or near the inverted position: In the first case ($E < E_{\max}$) the pendulum stops at this extreme point and then begins to move backwards, while in the limiting motion the pendulum continues moving for an unlimited time towards the inverted position; in the second case ($E > E_{\mathrm{m}}$) the pendulum reaches the inverted position during a finite time.

For the oscillatory motion under consideration, the representative point in the phase plane generates a closed path during one cycle, passing along both branches of the separatrix. The pendulum goes twice around almost the whole circle, covering it in both directions. On the other hand, executing rotation, the pendulum makes one circle during a cycle of revolutions, and the representative point passes along one branch of the separatrix (upper or lower, depending on the direction of rotation). To explain why the period of these oscillations is twice the period of corresponding revolutions, we must show that the motion of the pendulum with energy $E = E_{\max} - \Delta E$ from $\varphi = 0$ up to the extreme point requires the same time as the motion with the energy $E = E_{\max} + \Delta E$ from $\varphi = 0$ up to the inverted vertical position.

The major part of each of both motions under consideration occurs very nearly along the same path in the phase plane, namely, along the separatrix from the initial point $\varphi = 0$, $\dot{\varphi} \approx 2\omega_0$ up to some angle $\varphi_0$ whose value is close to $\pi$.

We choose this value $\varphi_0$ arbitrarily. To calculate the time interval required for this part of the motion, we can assume that the motion (in both cases) occurs exactly along the separatrix, and take advantage of the corresponding analytical solution, expressed by Eq. (7.12). Assuming $\varphi(t)$ in Eq. (7.12) to be equal to $\varphi_0$, we can find the time $t_0$ during which the pendulum moves from the equilibrium position $\varphi = 0$ up to the angle $\varphi_0$ (for both cases):

$$\omega_0 t_0 = -\ln\tan\frac{\pi - \varphi_0}{4} = -\ln\tan\frac{\alpha_0}{4}, \tag{7.29}$$

where we introduced the notation $\alpha_0 = \pi - \varphi_0$ for the angle that the pendulum at $\varphi = \varphi_0$ forms with the upward vertical line. When $\varphi_0$ is close to $\pi$, the angle $\alpha_0$ is small, so that in Eq. (7.29) we can assume $\tan(\alpha_0/4) \approx \alpha_0/4$. Therefore $\omega_0 t_0 \approx \ln(4/\alpha_0)$.

Next we shall consider in detail the subsequent part of the motion that occurs from this arbitrarily chosen angle $\varphi = \varphi_0$ towards the inverted position, and prove that the time $t_1$ required for the pendulum with the energy $E_{\max} + \Delta E$ (rotational motion) to reach the inverted position $\varphi = \pi$ equals the time $t_2$ during which the pendulum with the energy $E_{\max} - \Delta E$ (oscillatory motion) moves from $\varphi_0$ up to its extreme deflection $\varphi_m$, where the angular velocity becomes zero, and the pendulum begins to move backwards. We emphasize that these time intervals $t_1$ and $t_2$ are equal to one another only if $\Delta E$ is the same in both cases.

When considering the motion of the pendulum in the vicinity of the inverted position, we find it convenient to define its position (instead of the angle $\varphi$) by the angle $\alpha$ of deflection from this position of unstable equilibrium. This angle equals $\pi - \varphi$, and the angular velocity $\dot\alpha$ equals $-\dot\varphi$. The potential energy (measured relative to the lower equilibrium position) depends on $\alpha$ in the following way:

$$E_{\text{pot}}(\alpha) = mga(1 + \cos\alpha) \approx E_{\max}(1 - \alpha^2/4). \tag{7.30}$$

The latter expression is valid only for small values of $\alpha$, when the pendulum moves near the inverted position. Phase trajectories of motion with energies $E = E_{\max} \pm \Delta E$ near the saddle point (the origin in the new variables $\alpha$, $\dot\alpha$) can be found from the conservation of energy with the help of the approximate expression (7.30) for the potential energy:

$$\frac{1}{2}J\dot\alpha^2 + \frac{1}{4}E_{\max}\alpha^2 = \pm\Delta E, \text{ or } \frac{\dot\alpha^2}{\omega_0^2} - \alpha^2 = \pm4\varepsilon. \tag{7.31}$$

Here we use the notation $\varepsilon = \Delta E/E_{\max}$ for the small ($\varepsilon \ll 1$) dimensionless quantity characterizing the fractional deviation of energy $E$ from its value $E_{\max}$ for the separatrix. It follows from Eq. (7.31) that phase trajectories near the saddle point are hyperbolas whose asymptotas are the two branches of the separatrix that meet at the saddle point. Part of the phase portrait near the saddle point is shown in Figure 7.10. The curve *1* for the energy $E = E_{\max} + \Delta E$ corresponds to the rotation of the pendulum. It intersects the ordinate axis when the pendulum passes through the inverted position. The curve *2* for the energy $E = E_{\max} -$

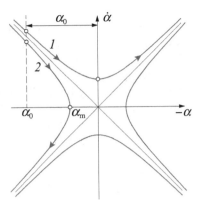

Figure 7.10: The phase curves near the saddle point.

$\Delta E$ describes the oscillatory motion. It intersects the abscissa axis at the distance $\alpha_m = \pi - \varphi_m$ from the origin (from the zero point of the axis). This point of intersection shows the extreme deflection in the oscillations.

The new variable $\alpha(t) = \pi - \varphi(t)$ near the point $\alpha = 0$ (for $\alpha \ll 1$) satisfies the linearized differential equation, Eq. (7.17), p. 172, which is valid for the pendulum motion in the vicinity of the inverted position. The general solution to this linear equation is given by a superposition of two exponential functions of time $t$, Eq. (7.18):

$$\alpha(t) = C_1 e^{\omega_0 t} + C_2 e^{-\omega_0 t}. \tag{7.32}$$

Next we separately consider the two cases of motion of the pendulum with the energies $E = E_{\text{max}} \pm \Delta E$.

1. Rotational motion ($E = E_{\text{max}} + \Delta E$) along the curve *1* from $\alpha_0$ up to the intersection with the ordinate axis. Let $t = 0$ be the moment of crossing the inverted vertical position: $\alpha(0) = 0$. Hence in Eq. (7.32), $C_2 = -C_1$. Then from Eq. (7.31) $\dot{\alpha}(0) = 2\sqrt{\varepsilon}\omega_0$, and $C_1 = \sqrt{\varepsilon}$. To determine duration $t_1$ of the motion, we assume in Eq. (7.32) $\alpha(t_1) = \alpha_0$:

$$\alpha_0 = \sqrt{\varepsilon}(e^{\omega_0 t_1} - e^{-\omega_0 t_1}) \approx \sqrt{\varepsilon}e^{\omega_0 t_1}. \tag{7.33}$$

(Here we can choose an arbitrary value of $\alpha_0$, although a small one, to be large compared to $\sqrt{\varepsilon}$, so that the condition $e^{-\omega_0 t_1} \ll e^{\omega_0 t_1}$ is fulfilled). Therefore $\omega_0 t_1 = \ln(\alpha_0/\sqrt{\varepsilon})$.

2. Oscillatory motion ($E = E_{\text{max}} - \Delta E$) along the curve *2* from $\alpha_0$ up to the extreme point $\alpha_m$. Let $t = 0$ be the moment of maximal deflection, when the phase curve intersects the abscissa axis: $\dot{\alpha}(0) = 0$. Hence in Eq. (7.32) $C_2 = C_1$. Then from Eq. (7.31) $\alpha(0) = \alpha_m = 2\sqrt{\varepsilon}$, and $C_1 = \sqrt{\varepsilon}$. To determine duration $t_2$ of this motion, we assume in Eq. (7.32) $\alpha(t_2) = \alpha_0$. Hence

$$\alpha_0 = \sqrt{\varepsilon}(e^{\omega_0 t_2} + e^{-\omega_0 t_2}) \approx \sqrt{\varepsilon}e^{\omega_0 t_2}, \tag{7.34}$$

and we find $\omega_0 t_2 = \ln(\alpha_0/\sqrt{\varepsilon})$.

We see that $t_2 = t_1$ if $\varepsilon = \Delta E/E_{\max}$ is the same in both cases. Therefore the period of oscillations is twice the period of rotation for the values of energy that differ from the critical value $E_{\max}$ on both sides by the same small amount $\Delta E$. Indeed, we can assume with great precision that the motion from $\varphi = 0$ up to $\varphi_0 = \pi - \alpha_0$ lasts the same time $t_0$ given by Eq. (7.29), since these parts of both phase trajectories very nearly coincide with the separatrix. In the case of rotation, the remaining motion from $\varphi_0$ up to the inverted position also lasts the same time as, in the case of oscillations, does the motion from $\varphi_0$ up to the utmost deflection $\varphi_m$, since $t_1 = t_2$.

The period of rotation $T_{\text{rot}}$ is twice the duration $t_0 + t_1$ of motion from the equilibrium position $\varphi = 0$ up to the $\varphi = \pi$. Using the above value for $t_1$ and Eq. (7.29) for $t_0$, we find:

$$T_{\text{rot}} = 2(t_0 + t_1) = \frac{2}{\omega_0} \ln \frac{4}{\sqrt{\varepsilon}} = \frac{1}{\pi} T_0 \ln \frac{4}{\sqrt{\varepsilon}}. \tag{7.35}$$

We note that an arbitrarily chosen angle $\alpha_0$ (however, $\sqrt{\varepsilon} \ll \alpha_0 \ll 1$), which delimits the two stages of motion (along the separatrix, and near the saddle point in the phase plane), falls out of the final formula for the period (when we add $t_0$ and $t_1$). For $\varepsilon = 0.0001$ (for $E = 1.0001 E_{\max}$) the above formula gives the value $T_{\text{rot}} = 1.907\, T_0$, which coincides with the cited above experimental result.

The period of oscillations $T$ is four times greater than the duration $t_0 + t_2$ of motion from $\varphi = 0$ up to the extreme point $\varphi_m$:

$$T = 4(t_0 + t_2) = \frac{4}{\omega_0} \ln \frac{4}{\sqrt{\varepsilon}} = \frac{2}{\pi} T_0 \ln \frac{8}{\alpha_m}. \tag{7.36}$$

For $\alpha_m \ll 1$ ($\varphi_m \approx \pi$) this formula agrees well with the experimental results: It yields $T = 5.37\, T_0$ for $\varphi_m = 179.900°$, $T = 6.83\, T_0$ for $\varphi_m = 179.990°$, and $T = 8.30\, T_0$ for $\varphi_m = 179.999°$. From the obtained expressions we see how both the period of oscillations $T$ and the period of rotation $T_{\text{rot}}$ tend to infinity as the total energy approaches $E_{\max} = 2mga$.

### 7.3.3   Mean Values of the Potential and Kinetic Energies

We can estimate the ratio of the values of the potential and kinetic energies, averaged over a period, if we take into account that most of the time the angular velocity of the pendulum is nearly zero, and for a brief time of motion the time dependence of $\varphi(t)$ is very nearly the same as it is for the limiting motion along the separatrix. Therefore we can assume that during an impulse the kinetic energy depends on time in the same way it does in the limiting motion. This assumption allows us to extend the limits of integration to $\pm\infty$. Since two sharp impulses of the angular velocity (and of the kinetic energy) occur during the period $T$ of oscillations, we can write:

$$\langle E_{\text{kin}} \rangle = \frac{J}{T} \int_{-\infty}^{\infty} \dot{\varphi}^2(t)\, dt = \frac{J}{T} \int_{-\pi}^{\pi} \dot{\varphi}(\varphi)\, d\varphi. \tag{7.37}$$

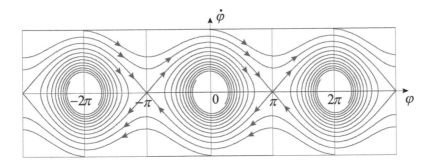

Figure 7.11: The phase portrait for revolutions and oscillations of the pendulum in the presence of friction ($Q = 18.12$).

The integration with respect to time is replaced here with an integration over the angle. The mean kinetic energy $\langle E_{kin} \rangle$ is proportional to the area $S$ of the phase plane bounded by the separatrix: $2\langle E_{kin} \rangle = JS/T$. We can substitute for $\dot{\varphi}(\varphi)$ its expression from the equation of the separatrix, Eq. (7.10):

$$\langle E_{kin} \rangle = \frac{J}{T} 2\omega_0 \int_{-\pi}^{\pi} \cos \frac{\varphi}{2} d\varphi = \frac{4}{\pi} J \omega_0^2 \frac{T_0}{T}. \qquad (7.38)$$

Taking into account that the total energy $E$ for this motion approximately equals $2mga = 2I\omega_0^2$, and $E_{pot} = E - E_{kin}$, we find:

$$\frac{\langle E_{pot} \rangle}{\langle E_{kin} \rangle} = \frac{2J\omega_0^2}{\langle E_{kin} \rangle} - 1 = \frac{\pi}{2} \frac{T}{T_0} - 1. \qquad (7.39)$$

For $\varphi_m = 179.99°$ the period $T$ equals $6.83\,T_0$, and so the ratio of mean values of potential and kinetic energies is 9.7 (compare with the case of small oscillations for which these mean values are equal).

## 7.4 The Influence of Friction

### 7.4.1 The Phase Portrait of the Pendulum in the Presence of Friction

When we take into account the small amount of friction inevitable in any real system, the phase portrait of the pendulum changes qualitatively (see Figure 7.11).

The closed phase trajectories corresponding to oscillations of a conservative system are transformed by friction into shrinking spirals that wind around a focus located at the origin of the phase plane. This focus represents a state of rest in the equilibrium position, and is an *attractor* of the phase trajectories: All phase trajectories of the damped pendulum spiral in toward the focus, forming an infinite

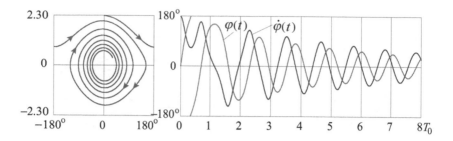

Figure 7.12: Phase diagram (left) and time dependent graphs of $\varphi(t)$ and $\dot{\varphi}(t)$ (right) for revolution and subsequent oscillation of the pendulum with friction ($Q = 18.1$) excited from the equilibrium position with the initial angular velocity $\Omega = 2.3\,\omega_0$.

number of loops. The phase curve that formerly passed along the upper branch of the separatrix does not now reach the saddle point $(\pi, \; 0)$. Instead it also begins to wind around the origin, gradually approaching it. Similarly, the lower branch crosses the abscissa axis $\dot{\varphi} = 0$ to the right of the saddle point $(-\pi, \; 0)$, and also spirals in towards the origin.

The inner part of the spiral, corresponding to small amplitudes, looks much like the phase trajectory of a linear oscillator experiencing viscous friction. The size of its gradually shrinking loops diminishes in a geometric progression.

## 7.4.2  Revolutions Followed by Oscillations

Typical graphs of the pendulum rotation followed by damped oscillations are shown in Figure 7.12 together with the corresponding phase diagram. The phase trajectory representing the counterclockwise rotation of the pendulum sinks lower and lower toward the separatrix with each revolution. When the pendulum has lost sufficient energy through friction so that it swings rather than rotates, its phase trajectory continues after crossing the separatrix as a gradually shrinking spiral, also winding around the origin.

When friction is weak, we can make some theoretical predictions for the motions whose phase trajectories pass close to the separatrix. For example, we can evaluate the minimal value of the initial velocity that the pendulum must be given in the lower (or some other) initial position in order to reach the inverted position, assuming that the motion occurs along the separatrix, and consequently that the dependence of the angular velocity on the angle of deflection is approximately given by the equation of the separatrix, Eq. (7.10).

The frictional torque is proportional to the angular velocity: $N_{\text{fr}} = -2\gamma J\dot{\varphi}$. Substituting the angular velocity from Eq. (7.10), we find

$$N_{\text{fr}} = \mp 4\gamma J\omega_0 \cos\frac{\varphi}{2} = \mp\frac{2mga}{Q}\cos\frac{\varphi}{2}. \qquad (7.40)$$

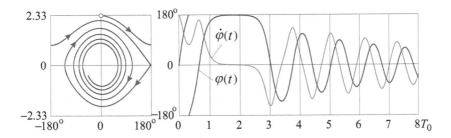

Figure 7.13: Revolution and subsequent oscillation of the pendulum with friction ($Q = 20$) excited from the equilibrium position with the initial angular velocity $\Omega = 2.3347\,\omega_0$.

Hence the work $W_{\text{fr}}$ of the frictional force during the motion from an initial point $\varphi_0$ to the final inverted position $\varphi = \pm\pi$ is:

$$W_{\text{fr}} = \int_{\varphi_0}^{\pm\pi} N_{\text{fr}}\,d\varphi = -4\frac{mga}{Q}\left(1 \mp \sin\frac{\varphi_0}{2}\right). \tag{7.41}$$

The necessary value of the initial angular velocity $\Omega$ can be found with the help of the conservation of energy, in which the work $W_{\text{fr}}$ of the frictional force is taken into account:

$$\Omega^2 = 2\omega_0^2\left[1 + \cos\varphi_0 + \frac{4}{Q}\left(1 \mp \sin\frac{\varphi_0}{2}\right)\right]. \tag{7.42}$$

For $\varphi_0 \neq 0$ the sign in Eq. (7.42) depends on the direction of the initial angular velocity. The exact value of $\Omega$ is slightly greater since the motion towards the inverted position occurs in the phase plane close to the separatrix but always outside it, with the angular velocity of a slightly greater magnitude. Consequently, the work of the frictional force during this motion is a little larger than the calculated value. For example, with $\varphi_0 = 0$ and the quality $Q = 20$, the above estimate yields $\Omega = \pm 2.098\,\omega_0$, but a more precise value of $\Omega$ determined experimentally by trial and error is $\pm 2.101\,\omega_0$.

Figure 7.13 shows the phase trajectory and the graphs of $\varphi(t)$ and $\dot{\varphi}(t)$ for a similar case in which the initial angular velocity is chosen exactly to let the pendulum reach the inverted position after a revolution.

## 7.5 Review of the Principal Formulas

The differential equation of motion for a planar rigid pendulum:

$$\ddot{\varphi} + 2\gamma\dot{\varphi} + \omega_0^2\sin\varphi = 0, \tag{7.43}$$

where $\omega_0$ is the frequency of small free oscillations:

$$\omega_0^2 = mga/J = g/l; \qquad l = J/ma. \tag{7.44}$$

Here $m$ is the mass of the pendulum, $a$ is the distance between the horizontal axis of rotation (the point of suspension) and the center of mass, $J$ is the moment of inertia about the same axis, $l$ is the reduced length of the physical pendulum, and $g$ is the acceleration of gravity.

The equation of a phase trajectory in the absence of friction:

$$\frac{\dot{\varphi}^2}{\omega_0^2} + 2(1 - \cos\varphi) = \frac{E}{E_0}, \tag{7.45}$$

where $E$ is the total energy, and

$$E_0 = \frac{1}{2}J\omega_0^2 = \frac{1}{2}mga = \frac{1}{4}(E_{\text{pot}})_{\text{max}}. \tag{7.46}$$

Here $(E_{\text{pot}})_{\text{max}} = 2mga$ is the maximal possible value of the potential energy of the pendulum, which is its potential energy when it is in the inverted vertical position.

The equation of the separatrix in the phase plane:

$$\dot{\varphi} = \pm 2\omega_0 \cos(\varphi/2). \tag{7.47}$$

The angular deflection and angular velocity for the motion of the pendulum that generates the separatrix in the phase plane are:

$$\varphi(t) = \pi - 4\arctan(e^{-\omega_0 t}), \quad \dot{\varphi}(t) = \pm\frac{2\omega_0}{\cosh(\omega_0 t)} = \pm\frac{4\omega_0}{(e^{\omega_0 t} + e^{-\omega_0 t})}. \tag{7.48}$$

## 7.6 Questions, Problems, Suggestions

### 7.6.1 Small Oscillations of the Pendulum

At small angles of deflection, when $\sin\varphi \approx \varphi$, the restoring torque of the force of gravity is approximately proportional to the angle of deflection from the position of stable equilibrium, and the pendulum behaves like a linear oscillator. In the absence of friction it executes simple harmonic motion. In the presence of weak friction, its motion can be considered as a nearly harmonic oscillation with a slowly decreasing amplitude.

**7.6.1.1 The Amplitude, Phase Trajectory, and Energy of Small Oscillations.** Select the case of no friction and use initial conditions which produce oscillations of small amplitude. For instance, let the initial deflection be 30° and the initial velocity be zero. In this case the amplitude will be 30°.

(a) What is the maximal value of the angular velocity in these oscillations? Verify your answer with a computer-simulated experiment.

(b) What initial angular velocity should you give the pendulum initially in the equilibrium position ($\varphi(0) = 0$) in order to excite oscillations of the same amplitude (of 30°)? Verify your answer with an experiment. Remember that the initial angular velocity you enter must be expressed in units of the frequency $\omega_0$ of small oscillations. What is the difference between these oscillations and oscillations excited by an initial deflection?

(c) Convince yourself that at small amplitudes the graphs of the angle of deflection versus time and of the angular velocity versus time have shapes which are close to that of a sine curve. Also convince yourself, that oscillations of the velocity lead the oscillations of the angular displacement in phase by a quarter period.

Compare the graphs of time dependence of the deflection angle and of the angular velocity with the motion of the representative point along the phase trajectory. What is the form of the phase trajectory for small oscillations? With what scale along the ordinate axis of the phase plane is the phase trajectory approximately a circle?

(d) What can you say about the time dependence of the kinetic and potential energies of the pendulum at small amplitudes? Prove that the time average values of kinetic and potential energy are approximately equal. If the amplitude equals 30 degrees, what is the ratio of total energy $E$ to the maximal possible value of potential energy $E/E_{\max}$?

### 7.6.1.2* Period of Small Oscillations.

For graphs of the time dependencies of the angle of deflection and of the angular velocity, the scale shown on the time axis is in the appropriate units for a given pendulum, namely in units of $T_0 = 2\pi/\omega_0 = 2\pi\sqrt{l/g}$), which is the period of small oscillations of the pendulum. That is, the duration between hatch marks on the time axis is $T_0$.

(a) Note that at small but finite amplitudes (say about 30°), the period of oscillations is a bit longer than $T_0$. You can make this observation either from the curves plotted on the screen or from readings of the timer. In the latter case, you can stop the timer by clicking the "Pause" button or by pressing the Spacebar at the moment when the pendulum completes a whole number of cycles. As a convenience in taking further readings, you may set the timer to zero during a pause in the simulation by clicking the "Reset Timer" button. Try to measure the period (in units of $T_0$) for several moderate values of the amplitude.

(b) In performing precise measurements of the period in the simulation experiments, which instants are better for starting and stopping the timer: When the pendulum passes through the equilibrium position or when it reaches the points of its greatest deflection? Give a convincing explanation of your answer.

(c) Compare the measurement of the period $T$ for a given amplitude $\varphi_0$ obtained from the simulation experiment with the value given by the theoretical approximation:

$$T = T_0(1 + \varphi_0^2/16), \tag{7.49}$$

in which the amplitude $\varphi_0$ is expressed in radians. Determine the maximal value of $\varphi_0$ for which Eq. (7.49) gives the value of $T$ to within one percent. Find the error the formula yields for an amplitude of 45 degrees.

### 7.6.1.3 Damping of Small Oscillations.

(a) Prove theoretically that weak viscous friction causes exponential damping of small free oscillations. At what value of the quality factor $Q$ does the amplitude halve during four complete cycles? Input the calculated value of $Q$ and verify it in a simulation experiment on the computer.

(b) Convince yourself that for large viscous friction for which the quality factor $Q$ is less than the critical value of 0.5, a disturbed pendulum returns to the equilibrium position without swinging. What is the principal qualitative difference of the phase trajectories for the cases of weak and strong damping?

## 7.6.2   Oscillations with Large Amplitudes

### 7.6.2.1 Comparison of the Pendulum with a Linear Oscillator. For large angular displacements from the equilibrium position, the nonlinearity of the dependence on the angle $\varphi$ of the restoring gravitational torque is more apparent. Because $\sin \varphi < \varphi$, the increase in the restoring torque with increasing angular deflection is not as large for a pendulum as it is for a linear oscillator. Therefore, a pendulum is referred to as a nonlinear oscillatory system with a "soft" restoring force.

(a) How do the differences between a pendulum and a linear oscillator reveal themselves in graphs of the time dependence of the angular deflection and the angular velocity? How do the differences reveal themselves in the phase trajectory? Give a qualitative physical explanation for the differences.

(b) What are the differences between the pendulum and a linear oscillator with respect to energy transformations? Compare the phase trajectory with the graph of potential energy versus deflection angle. The placing of the graphs on the computer screen (if you choose the item "View," "Energy Transformations" in the menu) is especially convenient for such comparison. Pay special attention to the position of the extreme points on the phase trajectory and on the potential well of the pendulum. For given initial conditions $\varphi(0) = \varphi_0$, $\dot{\varphi}(0) = \Omega$, what are the values of the potential energy and kinetic energy of the pendulum at the extreme points and at the equilibrium position?

### 7.6.2.2* Oscillations with Large Amplitudes.

(a) Study large oscillations of the pendulum experimentally in the absence of friction. Note the exact periodicity of these clearly non-sinusoidal oscillations of the dynamical variables in the conservative system.

When the amplitude exceeds $90°$, the graph of angular velocity versus time is nearly a saw-tooth with equilateral triangular teeth. Explain this shape.

The shape of a tooth in the corresponding graph of the angular deflection in this case is close to a parabola, in contrast to the sinusoidally shaped tooth associated with oscillations with small amplitudes. Explain this parabolic shape. Note the increase in the period with increasing amplitude. (Hatch marks on the time axis are separated by $T_0$, the period of small oscillations.)

(b) Note how the closed phase trajectories of the oscillating pendulum are stretched horizontally as the energy of the pendulum increases. Explain why these phase trajectories are different from the elliptical phase trajectories of a linear oscillator. To do so, use the shapes of the parabolic potential well of a linear oscillator and the sinusoidal potential well of the pendulum. Assume the curvature near the bottom to be the same for both potential wells: The period $T_0$ of small oscillations of the pendulum should be equal to the period of the linear oscillator. Remember that the latter period is independent of the energy.

Explain the increase of the period of the pendulum with increasing amplitude, comparing its potential well with that of a linear oscillator.

(c) At large amplitudes the pendulum passes rapidly through the vicinity of the equilibrium position (through the sinusoidal bottom of the potential well) and slowly climbs up the sinusoidal crest of the well, along its nearly horizontal upper slopes; then it slowly descends from them. So on the average the pendulum remains at large deflections longer than does a linear oscillator, whose parabolic potential well has steadily increasing slopes. Use the shapes of these potential wells to explain why, during a cycle, the time average values of the potential and kinetic energies of a pendulum are not equal to one another while those of the linear oscillator are.

(d)* Carefully study the interesting case of oscillations with an amplitude near $180°$. Set the initial deflection to be $179.999°$, and the initial velocity to be zero. After remaining near one side of the inverted position for a long time, the pendulum rapidly passes through the bottom of its path, and then remains for a long time again near the other side of the inverted position.

Compare the time during which the pendulum covers almost all its circular path (except a small vicinity of the extreme positions) with the period of small free oscillations of the pendulum. In other words, estimate the duration of a solitary impulse on the graph of angular velocity versus time. Or, equivalently, estimate the width of the nearly vertical portion of the nearly rectangular saw-tooth graph of the angular deflection versus time.

(e)* Try to discover what factor determines the width of this nearly rectangular tooth of the graph $\varphi(t)$, or, equivalently, what factor determines the time interval between successive impulses in the graph of angular velocity versus time. That is, try to discover the physical cause that determines the complete period of these extraordinary oscillations of the pendulum. (Hint: Set the initial deflection of the pendulum at the successive values $179.999°$, $179.990°$, and $179.900°$, each with an initial velocity of zero.

(f)** Try to evaluate theoretically the time interval needed for the pendulum to reach the extreme deflection of 179.99 degrees at excitation from rest in the lower stable equilibrium position. Use your results to estimate the period of oscillations with the amplitude 179.99 degrees. Compare your estimation of the period with the value of $T$ obtained in the simulation experiment.

(g) Note the character of energy transformations in the motion considered above. Total energy $E$ in this motion nearly equals the height $2mga$ of the potential barrier. It is the value of the potential energy of the pendulum in the inverted position ($\varphi = \pm\pi$). Since the pendulum spends most of its period near the inverted position (because the pendulum moves and accelerates very slowly while in the vicinity of the inverted position), the time averaged value of its potential energy, taken over a complete oscillation, is much greater than the mean value of its kinetic energy. In this case potential energy is converted into kinetic energy only for the short time during which the pendulum makes a rapid turn passing through the lower equilibrium position of minimal potential energy. Try to evaluate (to an order of magnitude) the ratio of the values of the potential energy to the kinetic energy, averaged over a period, during oscillations with an amplitude of $179.99°$.

### 7.6.2.3* Motion along the Separatrix.

(a) When you set the initial deflection to be almost 180 degrees and the initial velocity to be zero, the phase trajectory of the resulting motion nearly coincides with the separatrix $\dot\varphi = \pm 2\omega_0 \cos(\varphi/2)$. The point representing the mechanical state of the pendulum in the phase plane passes rapidly along the lower branch of the separatrix, remains for a long time at the left saddle point $(-\pi, \ 0)$, and then returns along the upper branch of the separatrix. What initial conditions should you choose in order to make the representative point move first along the upper branch of the separatrix and then along the lower one?

(b) What value of the initial angular velocity $\Omega$ (in units of $\omega_0$) must be initially given to the pendulum in its lower equilibrium position in order to make the representative point in the phase plane move along the separatrix? What value of the initial angular velocity should you input if the pendulum is to be initially deflected from the equilibrium position by an angle of $60°$? $90°$? $-90°$? $120°$? Verify your answers with simulation experiments.

(c) For the limiting motion along the separatrix, calculate the time interval $\tau$ during which kinetic energy of the pendulum is greater than its potential energy. In other words, for the pendulum making its circular path from one side of the inverted position to the other, find the lapse of time between the two instants at which the pendulum passes through the horizontal positions on either side of the lower equilibrium position. Express this time interval in units of the period $T_0$ of small oscillations. Verify your calculated value by the experiment on the computer.

### 7.6.2.4 Large Oscillations with Friction.

(a) Examine the influence of viscous friction on oscillations of large amplitude. Begin with rather weak friction ($Q \approx 20$). Note the gradual changes in the pattern of the graphs as friction slowly decreases the mechanical energy and the amplitude of the pendulum. In particular, note how the initial triangular saw-tooth curve of angular velocity, with its sharp nearly rectilinear teeth, as well as the initial curve of angular deflection with smooth parabolic crests, both evolve into the sinusoidal curves characteristic of the simple harmonic oscillator.

(b) Under the influence of viscous friction, the topologies of the phase trajectories of a pendulum change. Instead of closed curves corresponding to exactly periodic oscillations of a conservative pendulum, you see twisting spirals making an infinite number of gradually shrinking loops around the focus at the origin of the phase plane. Note how the form of the loops changes when they recede from the separatrix. Give a qualitative explanation for the observed changes. (For a linear oscillator experiencing viscous friction, the shrinking loops of the phase curve remain similar as the curve approaches the origin.)

(c) Using the program item "Energy Transformations," note how the rate at which energy is dissipated depends upon the position of the representative point in the potential well. At which part of a cycle does the rate of energy dissipation reach a maximum? Explain your answer.

(d)** Using the law of energy conservation, calculate the minimal value of the initial velocity that the pendulum must be given in the lower equilibrium position in order to reach the inverted position, for the case in which there is no friction and for the case in which the quality factor $Q = 20$. What must be the initial velocity in order to reach the inverted position if the pendulum is initially deflected by the angle 60 degrees? By 90 degrees?

## 7.6.3   The Rotating Pendulum

A pendulum makes a full revolution if its total energy exceeds the value $2mga$, the maximal value possible for its potential energy. The influence of the gravitational force makes this rotation in the vertical plane nonuniform: The angular velocity is a maximum (in the absence of friction) each time the pendulum passes through the lower, stable equilibrium position, and a minimum each time the pendulum passes through the upper, unstable equilibrium position.

### 7.6.3.1  The Angular Velocity at Revolutions.

(a) Select the case of the absence of friction. Calculate the minimal initial angular velocity needed to obtain a full revolution of the pendulum when it is initially at the lower equilibrium position. Note the character of the graph of angular velocity versus time: As the pendulum revolves, its angular velocity changes periodically (that is, the angular velocity oscillates in time), but the sign of the angular velocity does not change (that is, the curve does not intersect the time axis).

(b)* How does the period of these oscillations change if the initial angular velocity is increased? Calculate the minimal value of the oscillating angular velocity for a given value of the initial angular velocity. Find the asymptotic dependence of the period of rotation on the initial angular velocity $T(\Omega)$, valid for the values of total energy $E$, which are much greater than the potential energy of the inverted pendulum ($E \gg 2mga$).

(c) What initial conditions must be entered in order to obtain a phase trajectory located above the separatrix in the phase plane? ... located below the separatrix? ... coinciding with the upper or lower branch of the separatrix?

### 7.6.3.2*  The Period of Revolutions and Oscillations.

(a) It is especially interesting to compare the period of rotation with the period of oscillation of the conservative pendulum whose total energy $E$ is close to the maximal possible value of the potential energy $E_{\max} = 2mga$. In this case, the phase trajectories lie in the vicinity of the separatrix. Using the simulation experiment, perform the measurement of the period for two values of the total energy $E$, which are slightly different from $E_{max}$ by equal amounts on either side of $E_{max}$. For example, first let $E/E_{max} = 0.9999$ and then let $E/E_{max} = 1.0001$.

It is convenient to use the item "Phase Portrait of the Pendulum" of the computer program to carry out these experiments. For each value of the energy, the simulation of motion and the plotting of curves on the screen is terminated automatically when the pendulum completes one full cycle of its motion. So the final reading of the timer gives the value of the period (in units of the natural period $T_0$ of small oscillations) for the simulated motion.

(b) What is the ratio of the periods you have measured in these two cases? How can you explain this ratio?

(c) When the total energy $E$ of the pendulum is greater than the height $E_{max} = 2mga$ of the potential barrier, the period of rotation $T$ rapidly decreases as the energy is increased. The period tends to zero with the growth of the energy. What is the asymptotic behavior of $T(E)$ when $E$ tends to infinity?

### 7.6.3.3* Rotation of the Pendulum with Friction.

(a) Experimentally examine the rotation of the pendulum in the presence of weak viscous friction. Note the gradual approach of the phase trajectory to the separatrix. What is the value of the total energy of the pendulum at the moment when the phase trajectory crosses the separatrix? Note that before the crossing (while the pendulum is executing complete revolutions), the kinetic energy and the angular velocity of the pendulum are never zero.

(b)** Using the law of energy conservation, evaluate the minimal value of the initial velocity needed to obtain a complete revolution of the pendulum when it is initially in the lower position if the quality factor $Q = 15$. What value of the initial velocity is needed to obtain two revolutions of the pendulum? Verify your result in a simulation experiment. Try to improve the approximate theoretical value of the required initial velocity.

# Chapter 8

# Rigid Planar Pendulum under Sinusoidal Forcing

**Annotation.** In Chapter 8 several well-known and recently discovered counterintuitive regular and chaotic modes of the sinusoidally driven rigid planar pendulum are discussed and illustrated by computer simulations. The software supporting the investigation offers many interesting predefined examples that demonstrate various peculiarities of this famous physical model. Plausible physical explanations are suggested for some exotic and unexpected motions. The simulation program can also serve as an exploration-oriented tool for discovering new features of the driven pendulum and gives students an opportunity to perform mini-research projects on their own.

## 8.1 Regular Response of a Harmonically Driven Rigid Pendulum

### 8.1.1 Introduction

If we ask ourselves what is the most famous instrument in the history of physics, the first idea may be about the pendulum. We may expect that an ordinary pendulum subjected to periodic forcing will exhibit quite familiar behavior, which agrees well with our intuition. However, despite the apparent simplicity, this well-known nonlinear system can display a rich variety of rather complex, as-yet-unexplored modes of motion, which include various kinds of transient processes, single- and multiple-period stationary oscillations and complete revolutions, subharmonic and superharmonic resonance responses, bistability and multistability, intermittency, and transient and stationary chaos. Most of these modes delight the eye and certainly challenge our physical intuition. By slowly varying the control parameters of the system (the frequency and amplitude of the drive, and damping factor), we can observe various kinds of bifurcations manifesting transitions of the pendulum between dramatically different modes of behavior.

The seemingly simple situation of the forced pendulum is actually quite complex due to the subtle interplay between natural modes of the pendulum (these modes are described in detail in Chapter 7 and [11]) and the periodic driving force. The driven pendulum is interesting not only by virtue of its role in the history of physics, but, maybe more importantly, because it is isomorphic to many other physical systems, including rf-driven Josephson junctions and phase-locked voltage-controlled oscillators. The equation of motion of the dissipative, externally driven pendulum serves as a paradigmatic model of various low-dimensional nonlinear dynamical systems and plays an important role in explorations of bifurcational and chaotic phenomena. Mechanical analogues of such systems allow us to observe a direct visualization of their motion and thus can be very useful in gaining an intuitive understanding of complex phenomena.

Numerous nonlinear problems in relation to the forced pendulum are clearly presented in [12]. The nonlinear phenomena that can be predicted by analytical methods are described in [13]. Detailed reviews of the experimental and theoretical investigations of various regular and chaotic features of the system are available in the literature (see, for example, [14]–[15] and references therein).

An obvious way to understand the behavior of a nonlinear mechanical system is to observe a computer simulation of its motion. Sometimes the simulations can tell us much more than the equations can and thus contribute greatly in building our physical intuition. For this purpose we have developed an interactive simulation program, "Rigid pendulum driven by a sinusoidal torque," included in the software package "Nonlinear oscillations" that accompanies this textbook. The program illustrates the motion of the sinusoidally driven pendulum. It simulates all known modes of the pendulum behavior, and can also serve as a convenient tool for discovering new features of this seemingly inexhaustible system.

## 8.1.2   The Physical Model

In this chapter and in the relevant simulation program we consider an ordinary planar rigid pendulum, say, a weightless rigid rod with a massive bob (point mass) at one end (a simple or mathematical pendulum), or any other massive body (a physical pendulum) that can turn about a horizontal axis in a uniform gravitational field. Being excited, the pendulum can rotate in the vertical plane or swing about the stable equilibrium position in which its center of mass is below the axis. The period $T_0$ of infinitely small natural oscillations in the absence of friction is characteristic of the given pendulum and can serve as a convenient unit of time for the simulation. Natural oscillations gradually dampen due to friction whose braking torque is assumed in the model to be proportional to the angular velocity of the pendulum (viscous friction).

The momentary mechanical state of the pendulum is determined by its angular position $\varphi$, which is the angle of deflection from the vertical equilibrium position measured in radians (or degrees), and by the angular velocity $\dot{\varphi} = d\varphi/dt$ measured in the simulation program in units of the natural angular frequency $\omega_0$ of (undamped) infinitely small oscillations of the pendulum ($\omega_0 = 2\pi/T_0$). We assume that the pendulum is directly driven by an external sinusoidal torque with a definite frequency $\omega$ and some constant amplitude.

The differential equation of motion used here and in the computer program to simulate the damped driven pendulum is of the form

$$\ddot{\varphi} + 2\gamma\dot{\varphi} + \omega_0^2 \sin\varphi = \omega_0^2 \phi_0 \sin\omega t. \tag{8.1}$$

Here $\omega$ is the driving frequency, and $\gamma$ is the damping factor. To measure the viscous damping, we can use instead of $\gamma$ a more convenient dimensionless quantity $Q$—the quality factor that equals the ratio $\omega_0/2\gamma$.

The driving torque in the right-hand part of Eq. (8.1) is proportional to $\phi(t) = \phi_0 \sin\omega t$. This means that the dimensionless quantity $\phi(t)$ can be used as a convenient measure of the external torque. Its physical sense can be explained as follows. Imagine that some small constant (time-independent) external torque $\phi$ is exerted on the pendulum (instead of $\phi(t) = \phi_0 \sin\omega t$). This torque $\phi$ causes a static displacement $\varphi$ of the pendulum from the vertical. The sine of this angular displacement is proportional to the torque. Indeed, for the pendulum in equilibrium the time derivatives of $\varphi$ vanish ($\ddot{\varphi} = 0$ and $\dot{\varphi} = 0$), and we conclude from Eq. (8.1) that under a static torque $\phi$ the relation $\sin\varphi = \phi$ is valid. Hence the value $\phi = 1$ corresponds to the external torque, which is necessary to hold the pendulum stationary at horizontal position $\varphi = \pi/2$, the position of maximum restoring torque of gravity.

For a small enough value of a constant torque, the displacement is small ($\varphi \ll 1$), and we can assume $\sin\varphi \approx \varphi$. That is, $\varphi \approx \phi$. This means that the angular displacement of the pendulum under a small static torque just equals this torque measured in the assumed angular units. In the limit of a very low driving frequency (when $\omega \to 0$), the pendulum adiabatically follows the external torque, and the low frequency steady-state forced oscillation of the pendulum will occur just with the amplitude of the driving torque measured in these units (provided the amplitude is small enough so that the static displacement is proportional to the torque).

## 8.1.3 Behavior of the Pendulum under the Slow Varying Sinusoidal Torque Whose Amplitude is Close to 1

Graphs of potential energy $U(\varphi) \sim (1 - \cos\varphi - \phi\varphi)$ for the pendulum subjected to a static torque $\phi$ are shown in Figure 8.1 for several values of $\phi$. In the absence of external torque ($\phi = 0$) stable equilibrium positions—minima of $U(\varphi)$—are located at $\varphi = \pm 2\pi n$, $n = 0, 1, \ldots$. Natural oscillations of the hanging down pendulum can occur in any of the equivalent potential wells (say, about the midpoint $\varphi = 0$) with the frequency $\omega_0$.

The static torque $\phi$ causes a displacement of the equilibrium position to $\varphi = \arcsin\phi$. The pendulum can be in equilibrium under a static torque if $\phi < 1$ (curves $1$ and $2$ in Figure 8.1); at greater values of $\phi$ potential energy, $U(\varphi)$ has no

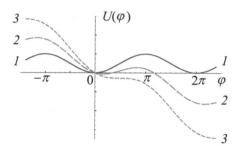

Figure 8.1: Potential energy $U(\varphi) \sim (1 - \cos\varphi - \phi\varphi)$ of the pendulum subjected to a static torque $\phi$. Curve $1 - \phi = 0$, curve $2 - \phi = 0.5$, curve $3 - \phi = 1$.

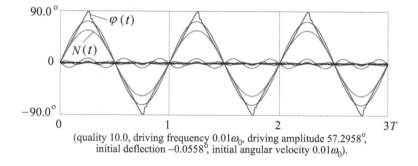

(quality 10.0, driving frequency $0.01\omega_0$, driving amplitude $57.2958°$, initial deflection $-0.0558°$, initial angular velocity $0.01\omega_0$).

Figure 8.2: Steady-state motion of the pendulum at $\phi_0 \approx 1$ under sinusoidal torque $N(t)$ of a low driving frequency ($\omega = 0.01\,\omega_0$). Harmonic components of $\varphi(t)$ are also shown.

minima (curve *3* in Figure 8.1), so that equilibrium (as well as an oscillatory motion) is impossible: The pendulum rotates. When $\phi \to 1$, the static displacement $\varphi \to \pi/2$ (tends to the horizontal position of the pendulum).

The case of a slow varying sinusoidal torque whose amplitude $\phi_0 \approx 1$ deserves special investigation. Figure 8.2 shows the time-dependent graph of the steady-state motion at $\omega = 0.01\,\omega_0$ under the driving torque whose amplitude $\phi_0$ slightly exceeds one radian. Period $T = 2\pi/\omega$ of the driving torque is chosen in this graph as an appropriate time unit. We note a linear dependence of $\varphi(t)$ on time $t$ when the external torque $N(t)$ increases with time sinusoidally from zero to its maximal value $\phi_0 = 1$.

Next we try to explain this counterintuitive behavior on the basis of the differential equation of the pendulum, Eq. (8.1). For a slow steady-state motion (at $\omega \ll \omega_0$), we can ignore the terms with the angular velocity and acceleration in the differential equation, Eq. (8.1), of the pendulum. In other words, the pendulum adiabatically follows the slow-varying external torque, remaining all the time in

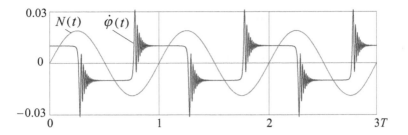

Figure 8.3: Angular velocity at the low-frequency ($\omega = 0.01\,\omega_0$) motion of the pendulum under the sinusoidal torque $N(t)$ whose amplitude $\phi_0 \approx 1$.

the equilibrium position (in the potential energy minimum), which is displaced from the vertical by the external torque.

The sine of this angular displacement $\varphi(t)$ is equal to the torque $\phi(t) = \phi_0 \sin\omega t$. This is evident from Eq. (8.1). Therefore for $\phi_0 = 1$ we get $\sin\varphi(t) = \sin\omega t$, and hence for the time interval $(0, T/4)$ the angle of deflection $\varphi(t) = \omega t$: When the external torque $\phi(t)$ increases sinusoidally, the equilibrium position $\varphi(t)$ is displaced linearly with time. The angular velocity of the pendulum in this slow uniform motion equals the driving frequency: $\dot{\varphi}(t) = \omega$.

This means that we can assume a linear function $\varphi(t) \approx \omega t$ for the zero-order solution to Eq. (8.1) in the time interval $(-T/4, T/4)$. Similarly, for the adjacent interval $(T/4, 3T/4)$ we can write $\varphi(t) \approx \pi/2 - \omega(t - T/4) = \pi - \omega t$. As a whole, this approximate steady-state periodic solution is characterized by a sawtooth pattern with equilateral triangle teeth.

The simulation shows that this rectilinear tooth shape is slightly distorted near each apex by rapid oscillations occurring after the external torque reaches a maximum and the direction of motion of the equilibrium position is reversed.

These rapid oscillations are especially pronounced in the angular velocity plot (Figure 8.3). The angular velocity $\dot{\varphi}$ is expressed here in units $\omega_0$ of the frequency of small undamped natural oscillations.

In order to investigate analytically the character of these oscillations, we assume that the angle of deflection for the time interval $(T/4, 3T/4)$ can be expressed as $\varphi(t) \approx \pi - \omega t + \delta(t)$, where the correction $\delta(t)$ to the zero-order function is small: $\delta(t) \ll 1$.

Differentiating Eq. (8.1) with respect to time, we obtain the following equation for the angular velocity $\dot{\varphi}(t) = \nu$:

$$\ddot{\nu} + 2\gamma\dot{\nu} + \omega_0^2 \cos\varphi(t)\,\nu = \omega\omega_0^2 \cos\omega t. \qquad (8.2)$$

In the left-hand part of this equation we can replace $\varphi(t)$ by its zero-order time-dependence $\varphi(t) = \pi - \omega t$, and substitute for $\cos\varphi(t)$ its approximate expression $\cos(\pi - \omega t) = -\cos\omega t$. Thus instead of (8.2) we get an approximate second-order

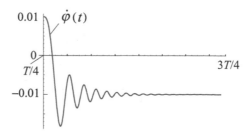

Figure 8.4: Approximate behavior of the angular velocity at the low-frequency steady-state motion of the pendulum.

linear homogeneous equation for the angular velocity $\dot{\varphi}(t) = \nu$:

$$\ddot{\nu} + 2\gamma\dot{\nu} - \omega_0^2 \cos\omega t\,(\nu + \omega) = 0. \tag{8.3}$$

We can conclude from Eq. (8.3) that, after the oscillations of the angular velocity $\nu$ damp out, and $\nu$ approaches a constant value, so that its time derivatives in Eq. (8.3) become negligible, this constant value equals $-\omega$: $\dot{\varphi} = \nu \to -\omega$.

During the preceding interval $(-T/4, T/4)$, oscillations of the angular velocity $\dot{\varphi}(t)$ have also damped out, and its constant value at the beginning of the interval $(T/4, 3T/4)$ approximately equals $\omega$.

In further calculations it is convenient to transfer the time origin to the initial moment of the interval $(T/4, 3T/4)$, that is, to replace $t \to (t + T/4)$, or $\omega t \to (\omega t + \pi/2)$ in Eq. (8.3):

$$\ddot{\nu} + 2\gamma\dot{\nu} - \omega_0^2 \sin\omega t\,(\nu + \omega) = 0. \tag{8.4}$$

Oscillations of the angular velocity at the beginning of the interval $(T/4, 3T/4)$ are approximately described by a solution to this homogeneous equation. We note that in Eq. (8.4) $\sin\omega t$ can be replaced by $\omega t$ for the time interval we are interested in. Even after this simplification Eq. (8.4) cannot be solved analytically. However, we can find numerically its particular solution for the given time interval with the help of any available mathematical package.

The initial conditions for the starting point of this interval follow from the known pattern of the velocity graph for the periodic steady-state motion shown in Figure 8.3: $\nu = \omega$, $\dot{\nu} = 0$.

The graph in Figure 8.4 shows the solution to Eq. (8.4) with these initial conditions obtained with the help of the "Mathematica" package ($\omega = 0.01\,\omega_0$, $Q = \omega_0/2\gamma = 0.1$).

Comparing this graph with those shown in Figure 8.3, we see that approximate expression, Eq. (8.4), indeed describes qualitatively the general character of oscillations of the angular velocity that take place after the motion of the equilibrium position is reversed.

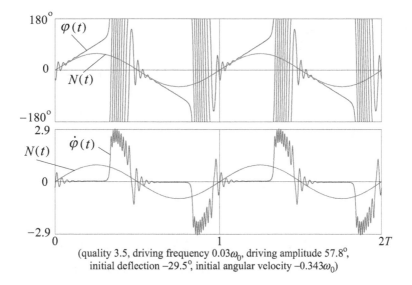

(quality 3.5, driving frequency $0.03\omega_0$, driving amplitude $57.8°$,
initial deflection $-29.5°$, initial angular velocity $-0.343\omega_0$)

Figure 8.5: Rotations and oscillations of the pendulum under a slow-varying sinusoidal external torque whose frequency equals $0.03\,\omega_0$ and whose amplitude
corresponds to $\phi_0$ slightly greater than 1.

If amplitude $\phi_0$ of the driving torque and its frequency $\omega$ are chosen to be
a little greater than in the case considered above, the pendulum at first follows
again the slowly varying torque, so that deflection angle $\varphi(t)$ also slowly increases
almost linearly with time up to the highest (nearly horizontal) position.

However, instead of reversing its slow motion alongside the equilibrium position when the latter starts to move back, the pendulum in this case escapes the
shallow potential well over its low right barrier and "slides down" along its bumpy
outer slope (see curves 2 and 3 in Figure 8.1). This means that the pendulum com-
mences a rapid, unidirectional, nonuniform rotation. Figure 8.5 shows the graphs
of angular position $\varphi(t)$ and angular velocity $\dot{\varphi}(t)$ time dependence for this extraordinary counterintuitive motion.

We can evaluate the average angular velocity $\langle\dot{\varphi}\rangle_{\mathrm{av}}$ of this rapid rotation by
equating the external torque $\phi_0$ at its maximum to the torque of viscous friction:
$\omega_0^2\phi_0 = 2\gamma\langle\dot{\varphi}\rangle_{\mathrm{av}}$, whence $\langle\dot{\varphi}\rangle_{\mathrm{av}} \approx Q\omega_0$. The average period $T_{\mathrm{rot}}$ of this rotation
can be estimated as $T_{\mathrm{rot}} = 2\pi/\langle\dot{\varphi}\rangle_{\mathrm{av}} \approx T_0/Q$, where $T_0 = 2\pi/\omega_0$ is the period
of the small natural oscillations.

When $t$ approaches $T/2$, the external torque $N(t)$ becomes smaller, the pendulum rotation gradually slows down, and finally (when the torque almost vanishes) the pendulum becomes trapped in the potential well, within which it executes damped natural oscillations near the equilibrium position, which moves
uniformly backward under the reversed external torque (see Figure 8.5). Then all
the above-described motion repeats in the opposite direction.

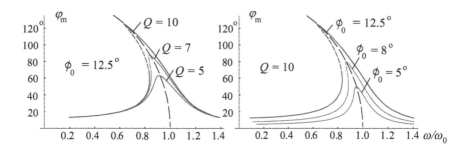

Figure 8.6: Approximate theoretical response-frequency curves of the pendulum: (a) constant driving amplitude $\phi_0 = 12.5°$; (b) constant quality factor $Q = 10$.

## 8.2 Steady-State Response-Frequency Curves of the Harmonically Driven Pendulum

### 8.2.1 Approximate Theoretical Resonance Curve

The steady-state response of a linear oscillator subjected to sinusoidal forcing is also a sinusoidal motion whose frequency equals the forcing frequency and whose amplitude depends on the frequency in a resonance manner. For the pendulum, as long as the driving amplitude is small and the damping is not too weak, the steady-state oscillation occurs with a small amplitude, so that the amplitude-frequency resonance curve is rather well approximated by the result for a harmonic oscillator.

For a stronger driving and/or weaker friction, the resonance curve of the pendulum bends toward lower frequencies and even folds, as shown in Figure 8.6. The nonlinear resonance curve can be approximated considerably well by the following heuristic approach [4]: We take the resonance curve of the harmonic oscillator and replace in it the natural frequency $\omega_0$ by $\omega_{\text{res}}(\varphi_{\text{m}})$. This approach assumes that the steady-state oscillation is still harmonic, i.e., sinusoidal, but its frequency depends on the amplitude $\varphi_{\text{m}}$.

For this dependence, $\omega_{\text{res}}(\varphi_{\text{m}})$, we can use the approximate frequency-amplitude relation of the pendulum, valid at moderate values of the amplitude $\varphi_{\text{m}}$ (see Chapter 7, Eq. (7.5), p. 158): $\omega_{\text{res}} \approx \omega_0(1 - \varphi_{\text{m}}^2/16)$. This approximate dependence of $\omega_{\text{res}}$ on the amplitude $\varphi_{\text{m}}$—the so-called *skeleton curve*—is shown by the thin dashed line in Figure 8.6. The resonance peak at its maximum is shifted to lower frequencies and acquires a shape typical for nonlinear systems with a "soft" restoring force. Over some critical value of the driving amplitude (for a given quality factor), the theoretical resonance curve becomes $S$–shaped with three solutions, only two of which are stable. This folding of the response-frequency curve (foldover effect) leads to bistability and hysteresis.

Within some interval of driving frequencies the pendulum oscillates either with a large amplitude or a small amplitude. In-between there is always an unsta-

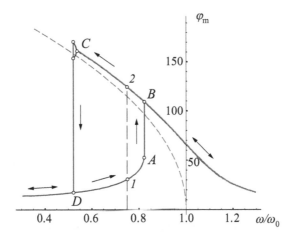

Figure 8.7: Hysteretic behavior of the amplitude-frequency dependence during the up and down sweeping of the drive frequency ($\phi_0 = 12.5°$, $Q = 10$).

ble solution (see the left overhanging slope of resonance peaks shown by dashed lines in Figure 8.6). Which of the two stable, periodic motions (limit cycles) eventually is established depends on the initial conditions.

A convenient traditional way to observe the nonlinear resonance response of the pendulum is to slowly vary ("sweep") the driving frequency from one side of the natural frequency through the resonance peak and to the other side in a process of continuous steady-state oscillations, while the amplitude of the driving torque is kept constant.

The pendulum responds differently depending on the direction of the frequency variation—there is an associated hysteresis characterized by abrupt jumps in the amplitude and phase of the steady-state response. When in the process of frequency sweeping an abrupt jump occurs from one slope of the folded resonance peak to the other, not only the amplitude of the steady-state oscillations changes considerably, but the whole mode undergoes a dramatic change.

Figure 8.7 shows the response-frequency curve (see [58]) obtained with the help of the simulation program. When we start the sweeping from low driving frequencies (and at the initial conditions of zero, with the pendulum resting in the equilibrium position), the observed steady-state response agrees perfectly well with the theoretical prediction: The forced oscillations occur almost in phase with the drive, and their amplitude grows gradually while the frequency is increased up to point $A$, which is characterized by a vertical tangent to the theoretical curve. Then an abrupt jump to point $B$ lying on the right slope of the resonance peak occurs. After this jump, the amplitude and phase again agree well with the theoretical prediction. In the process of further sweeping, the amplitude of the steady-state response gradually diminishes, and the pendulum oscillates in almost opposite phase with respect to the driving torque.

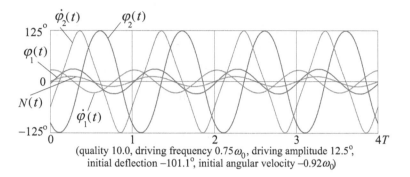

(quality 10.0, driving frequency $0.75\omega_0$, driving amplitude $12.5°$,
initial deflection $-101.1°$, initial angular velocity $-0.92\omega_0$)

Figure 8.8: Graphs of the angle and angular velocity for small-amplitude ($30°$) and large-amplitude ($125°$) oscillations at the same parameters of the pendulum and of the external torque ($\omega = 0.75\,\omega_0$, $\phi_0 = 12.5°$, and $Q = 10$). Initial conditions are indicated for the large-amplitude oscillations. These graphs correspond to points *1* and *2* respectively on the response-frequency diagram of Figure 8.7.

### 8.2.2  Autoresonance, Hysteresis, and Bistability

If we reverse the direction of the frequency sweep, the pendulum's response on the way back follows the same curve up to point $B$. However, the amplitude, instead of jumping down, continues to increase along the right slope of the theoretical resonance peak after we have passed through point $B$.

Figure 8.8 gives an example of this *bistability*, and of the hysteretic behavior of the pendulum at sweeping the driving frequency. Indeed, during the direct sweeping from left to right, the steady-state oscillations at point *1* are almost sinusoidal in shape (curves $\varphi_1(t)$ and $\dot{\varphi}_1(t)$ in Figure 8.8) and occur nearly in the same phase with the driving torque $N(t)$, while on the way back, at point *2* (at the same frequency and amplitude of the drive as at point *1*), the oscillations have a much greater amplitude, and these oscillations lag in phase behind the driving torque more than a quarter-period.

This phase relationship between the drive and the pendulum is characteristic of resonance, and is preserved during the slow sweeping of the drive frequency by virtue of the phase locking. Hence this phenomenon can be called *autoresonance*. On this branch of the frequency-response characteristic, the large-amplitude oscillations are no longer harmonic: The graph of $\dot{\varphi}_2(t)$ has a saw-toothed appearance, while the graph of $\varphi_2(t)$, though resembling a sinusoid, actually consists of nearly parabolic alternating segments.

Steady-state modes $\varphi_1(t)$ and $\varphi_2(t)$, coexisting at the same frequency and amplitude of the drive, differ in amplitude and in phase relationships with the driving torque—they correspond to different slopes of the resonance peak. The large amplitude mode $\varphi_2(t)$ has a much smaller basin of attraction than $\varphi_1(t)$.

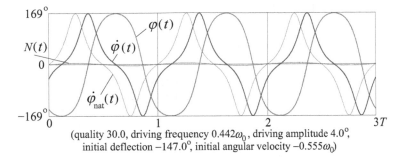

(quality 30.0, driving frequency $0.442\omega_0$, driving amplitude $4.0°$,
initial deflection $-147.0°$, initial angular velocity $-0.555\omega_0$)

Figure 8.9: Graphs of $\varphi(t)$ and $\dot{\varphi}(t)$ for the "bell-ringer" mode of forced oscillations. The graph of $\dot{\varphi}_{\mathrm{nat}}(t)$ for natural undamped oscillations ($\phi_0 = 0$) of the same amplitude (about 168°) is also shown for comparison.

### 8.2.3   Nonlinear Resonance and a "Bell-Ringer Mode"

As we continue to sweep the drive frequency $\omega$ from point $B$ towards lower values, we move to the left in Figure 8.7 almost along the skeleton curve, and can reach amplitudes that are considerably greater than the estimated theoretical maximum (see Figure 8.6).

During this sweeping, the phase relationships between the drive and the pendulum that are characteristic for resonance are automatically preserved (autoresonance) due to the phase locking. For weak damping, the steady-state amplitude at low drive frequencies can be very large (approaching 180°) at moderate and even quite small drive amplitudes.

Traces in Figure 8.9 give an example of such extraordinary motion of the pendulum, which Peters called a "bell-ringer mode" [18]. In this mode the pendulum passes rapidly through the lower equilibrium position, but "sticks" near the extreme points of oscillation, spending a very long time moving slowly in the vicinity of unstable equilibrium position (near the saddle point in the phase plane).

These large-amplitude oscillations give an example of nonlinear resonance: Their period (which equals the drive period) is very close to the period of natural oscillations of the same amplitude. For large amplitudes, this period can last several periods of small natural oscillations.

Actually, such forced oscillations in conditions of nonlinear resonance are very much like natural oscillations of the pendulum that occur at the corresponding large amplitude. To emphasize this similarity, we also show in Figure 8.9 by a thin line the graph of $\dot{\varphi}_{\mathrm{nat}}(t)$ for natural undamped oscillations ($\phi_0 = 0$) of the same amplitude (about 168°).

We can exploit the similarity between the bell-ringer mode and natural undamped oscillations for theoretical calculation of the amplitude at a given resonant drive frequency. It was shown in Chapter 7, Eq (7.22), p. 172 (and also in [11]) that for non-sinusoidal natural oscillations with an amplitude $\varphi_{\mathrm{m}}$ approaching 180° the period depends on the amplitude as follows:

$$T = \frac{2}{\pi} T_0 \ln \frac{8}{\pi - \varphi_m}, \quad \text{whence} \quad \varphi_m = \pi - 8 \exp\left(-\frac{\pi}{2}\frac{\omega_0}{\omega}\right). \quad (8.5)$$

In the simulation presented in Figure 8.9 the frequency of the drive $\omega = 0.442\,\omega_0$. Substituting this value in Eq. (8.5), we find for the corresponding amplitude of undamped natural oscillations: $\varphi_m = 167°$. This theoretical estimate is very close to the amplitude of the bell-ringer forced oscillations ($\varphi_m \approx 168°$) observed in the simulation experiment at the drive frequency $\omega = 0.442\,\omega_0$.

When the drive period is almost equal to the period of large natural oscillations, full synchronization between the motions (phase-locking) can occur. The pendulum lags in phase about a quarter period behind the periodic external torque $N(t)$. Due to the phase-locking, this small torque $N(t)$ at autoresonance is almost always directed in phase with the angular velocity $\dot{\varphi}(t)$ of the pendulum (see Figure 8.10) and therefore supplies the pendulum with energy needed to compensate for frictional losses and to maintain the constant amplitude of large non-sinusoidal nearly natural oscillations.

An alternative physical explanation of the "bell-ringer mode" is based on considering the motion of a particle in a time-dependent spatially periodic potential (see curve *1* in Figure 8.1) whose pattern is "rocking" slightly about the origin (point 0), so that the right barrier of the well lowers a bit and the left one rises when the external torque is directed to the right, and vice versa, after a half-period of the drive. Let us imagine that the particle in the well on its way from left to right slowly "climbs up" the slope of the right barrier and turns back approximately at the time when the potential pattern is horizontal (zero external torque), and passes back through the bottom of the well just after the moment at which the right barrier is at its maximal height. The duration of the particle motion back and forth in the non-parabolic well depends on the amplitude, and if this duration equals the period of "rocking" of the potential pattern, phase locking can occur and a steady-state process can eventually establish. The energy needed to overcome friction is supplied by the source that "rocks" the potential pattern (that is, by the periodic external torque).

This "bell-ringer" mode can certainly also be excited by carefully choosing proper starting conditions. However, to maintain the large-amplitude motion, the phase relation between the pendulum and the drive torque is critical. This means that for this mode the basin of attraction in the phase plane of initial conditions is rather small. Hence it is much easier to reach this mode experimentally by sweeping down the drive frequency as described above. After each step along this way, we must wait for transients to settle.

The spectrum of the bell-ringer large-amplitude oscillations, besides the fundamental harmonic whose period equals the driving period, also contains several harmonic components of higher orders. Their frequencies are odd integer multiplies of the fundamental frequency. Figure 8.10 shows the graphs of angular velocity $\dot{\varphi}(t)$ and its harmonics for the bell-ringer mode.

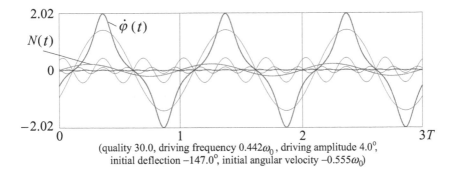

(quality 30.0, driving frequency $0.442\omega_0$, driving amplitude $4.0°$,
initial deflection $-147.0°$, initial angular velocity $-0.555\omega_0$)

Figure 8.10: Graphs of angular velocity $\dot{\varphi}(t)$ and its harmonics for the "bell-ringer" mode" of forced oscillations.

## 8.2.4   Symmetry-Breaking and Period-Doubling Bifurcations, Chaos and the Crisis

Further decreasing of the drive frequency brings the system to point $C$ of the amplitude-frequency characteristic (see Figure 8.7, p. 197) at which a symmetry-breaking bifurcation occurs: The pendulum's excursion to one side is greater than to the other side, for example, $173°$ versus $165°$. This spatial asymmetry of oscillations increases as we move further to lower frequencies. In the spectrum of such asymmetric steady-state oscillations, besides harmonics of odd orders, even-order harmonics of small amplitudes are present, including a zero-order component (constant mid-point displacement). Such asymmetric modes exist in pairs whose phase orbits (at the same drive frequency) are the mirror images of one another.

In this way, at a certain frequency, a period-doubling bifurcation occurs: In each subsequent cycle the maximal deflection to the same side slightly differs from the preceding one, but after two cycles the motion repeats exactly. This means that the period of this steady-state oscillation equals two drive periods. Period doubling breaks the original time-translational symmetry of the sinusoidally driven pendulum: Although the driving torque repeats exactly from cycle to cycle, the pendulum executes slightly different motions on alternate cycles. For this motion, the Poincaré section consists of two nearby points in the phase plane visited in alternation.

What happens after this period-doubling bifurcation in the simulation experiment depends strongly on details of the frequency diminution. If the sweeping occurs in very small steps, a whole cascade of close-set period-doubling bifurcations can be observed. Each bifurcation in this series doubles the period of motion and the number of fixed points in its Poincaré map.

This cascade of period-doubling bifurcations converges to a chaotic large-amplitude oscillation of the pendulum. During these chaotic oscillations of the bell-ringer type the maximal deflection is close to $180°$ and varies randomly from

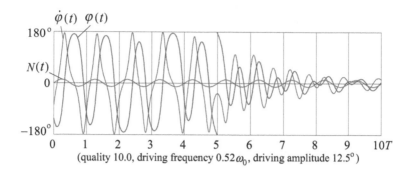

(quality 10.0, driving frequency $0.52\omega_0$, driving amplitude $12.5°$)

Figure 8.11: Transition from large-amplitude nonlinear oscillations (from the irregular "bell-ringer mode") to small-amplitude ordinary forced oscillations.

cycle to cycle whose duration equals approximately one driving period. Contrary to a complicated initial transient that leads eventually to a regular motion characterized by a fixed finite set of Poincaré sections, this chaotic regime persists indefinitely. The Poincaré map consists of two small nearby islands visited in alternation. Within each island the point bounces randomly from cycle to cycle. This chaotic state is stable in the sense that after a small perturbation the phase trajectory converges to the same region. Attracting regions in the phase space that correspond to chaotic regimes are called strange attractors because they are formed by fractals—geometric objects of non-integer dimensions. The fractal character of attractors is essential to the existence of persistent dynamical chaos.

The chaotic oscillatory regime following the bell-ringer mode exists in a very narrow interval of the driving frequencies, so that a slight perturbation can cause a crisis leading to an abrupt jump of the amplitude down to point $D$ (see Figure 8.7) located on the far left outskirt of the resonance peak. If the frequency sweeping is executed by steps that are not small enough, this jump can occur before the chaotic regime is established or even before the period-doubling bifurcation occurs.

Actually, this abrupt jump of the amplitude (the crisis) is presented by a long irregular transient (as in Figure 8.11), during which the motion of the pendulum undergoes a radical rearrangement.

Details of this transient are very sensitive to the character of perturbation (to the magnitude and timing of the frequency step). In particular, the initial stage of the transient may have the character of *intermittency*: During a long time the pendulum executes an asymmetric oscillation in which its excursion, say, to the left side is greater than to the right side. Then during several cycles the asymmetry changes to the opposite, that is, to prolonged oscillations with a greater maximal deflection to the right side. Such irregular interchanges of the two spatially asymmetric regimes are characterized by a time scale much longer than the cycle duration (the drive period), and can occur several times before the crisis.

The crisis leading to the jump down of the amplitude can be initiated, for example, if irregular amplitude variations lead the pendulum to cross the vertical (to make a full revolution), after which the pendulum gradually settles down to the low-frequency and low-amplitude regular (sinusoidal) steady-state oscillation for which the angular displacement $\varphi(t)$ almost exactly equals the driving torque: $\varphi(t) \approx \phi_0 \sin \omega t$. The simulation of motion in Figure 8.11 shows us what can happen during a transient that accompanies this amplitude jump.

If after this jump down of the amplitude at point $D$ (see Figure 8.7) we continue to sweep the frequency down, the amplitude and phase of steady-state oscillations again obey the theoretical response-frequency curve, just as they did while sweeping the frequency from left to right.

# 8.3 Subharmonic and Superharmonic Resonances

Steady-state forced oscillations of a large amplitude, resembling the bell-ringer mode, can also occur if the driving frequency is approximately three times greater than the natural frequency that corresponds to this large amplitude. An example of such oscillations is shown in Figure 8.12. Their graphs show clearly that the third harmonic of these steady-state oscillations has the frequency that equals the driving frequency, while the frequency of the fundamental harmonic equals one third of the driving frequency. In other words, one cycle of such non-harmonic oscillations of the pendulum covers three driving periods. Forced period-3 oscillations of large amplitude occurring under such conditions give an example of nonlinear third-order *subharmonic resonance*. Subharmonic resonances do not exist in linear systems.

Similarly to the bell-ringer mode (see Figure 8.10), the pendulum behaves here very much like during free (unforced) oscillations. The sinusoidal external torque, being synchronized with the third harmonic of these non-harmonic large-amplitude natural oscillations, compensates for frictional losses and maintains a constant angular excursion.

This synchronization (phase-locking) can occur only if at $t = 0$ (the time moment when the torque is switched on) large-amplitude natural oscillations already exist. This means that for given frequency of the external torque (lying within a definite interval) the third-order subharmonic resonance occurs only for initial conditions from a certain region (from the basin of attraction of this limit cycle). Different initial conditions cause the pendulum to eventually settle down into the low-amplitude period-1 antiphase oscillation that corresponds to the far-off high-frequency slope of the nonlinear resonance peak.

During slow reduction of the driving frequency under conditions of the third-order subharmonic resonance, a symmetry-breaking bifurcation takes place, after which the angular excursion to one side is greater than to the opposite side. Such spatially asymmetric modes exist in pairs whose phase orbits are the mirror images of one another. Further reduction of the driving frequency leads to a crisis: After a long transient the pendulum settles into the ordinary antiphase mode of forced oscillations that correspond to the right slope of the nonlinear resonance curve.

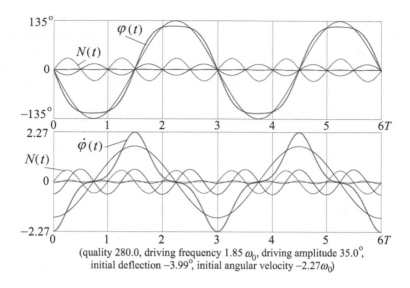

(quality 280.0, driving frequency 1.85 $\omega_0$, driving amplitude 35.0°, initial deflection −3.99°, initial angular velocity −2.27$\omega_0$)

Figure 8.12: Graphs of the angular deflection $\varphi(t)$, angular velocity $\dot\varphi(t)$ and their harmonics of oscillations at subharmonic resonance of the third order.

At subharmonic resonance of the fifth order (see the graph of $\dot\varphi(t)$ and its harmonics in Figure 8.13), one cycle of the pendulum's almost natural oscillation covers five driving periods: The external torque is synchronized with the fifth harmonic of a period-5 large-amplitude oscillation of the pendulum. On average, this phase-locking provides a surplus of energy transferred to the pendulum over the energy returned back to the source of the external torque, thus compensating for frictional losses.

Gradually reducing the driving frequency under conditions of the fifth-order subharmonic resonance, we can observe bifurcations of the symmetry-breaking and period-tripling, after which the period of steady-state forced oscillations equals 15 driving periods. The set of Poincaré sections consists of 15 fixed points in 5 groups visited by turn. Each group consists of 3 nearby points.

The subharmonic resonances discussed above occur at rather high drive frequencies, which are equal to an odd integer of the natural frequency. By contrast, superharmonic resonances can be excited at rather low drive frequencies: Synchronization of the drive with oscillations of the pendulum (phase locking) occurs if one period of the drive covers an odd integer number of natural periods.

The nature and origin of superharmonic resonances can be explained in the following way. Let us consider natural nonlinear oscillations of the pendulum in a potential well that slowly moves back and forth due to sinusoidally varying (with driving frequency $\omega$) external torque. Under certain conditions an integer number of natural cycles covers one period of the potential well motion. Figure 8.14 shows clearly that for the third-order superharmonic resonance just three natural cycles

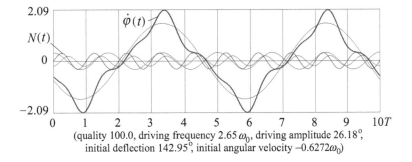

(quality 100.0, driving frequency 2.65 $\omega_0$, driving amplitude 26.18°,
initial deflection 142.95°, initial angular velocity −0.6272$\omega_0$)

Figure 8.13: Graphs of angular velocity $\dot{\varphi}(t)$ and its harmonics at subharmonic resonance of the fifth order.

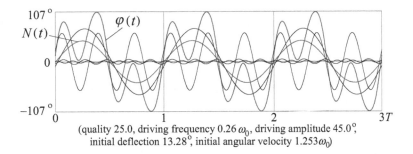

(quality 25.0, driving frequency 0.26 $\omega_0$, driving amplitude 45.0°,
initial deflection 13.28°, initial angular velocity 1.253$\omega_0$)

Figure 8.14: Plots of angular deflection $\varphi(t)$ and its harmonics at superharmonic resonance of the third order occurring under the sinusoidal external torque $N(t)$.

are executed during one period of the drive. In this case phase-locking of the potential well motion with natural oscillations can occur. By virtue of this synchronization, the external torque can continuously supply the pendulum with energy required to compensate for frictional losses and prevent damping of short-period natural oscillations of the pendulum in the moving potential well. As a result, a steady-state non-sinusoidal period-1 oscillation (its period equals that of the drive) is established, whose spectrum is distinguished by the considerable contribution of the third harmonic.

Depiction of such a motion on the screen with the help of the simulation program allows us to develop an intuitive feel for how nonlinear systems generate high harmonics of the sinusoidal input oscillation. The simulation tells us much more for understanding this phenomenon than the mathematical equations can do.

Superharmonic resonances are also accompanied by symmetry-breaking bifurcations and chaotic regimes. Examples of strange attractors that follow superharmonic resonances of the third and fifth order are shown in Figures 8.15$a$ and 8.15$b$, respectively.

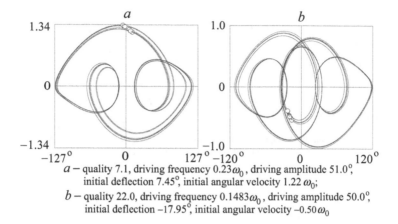

$a$ – quality 7.1, driving frequency $0.23\omega_0$, driving amplitude $51.0°$,
   initial deflection $7.45°$, initial angular velocity $1.22\,\omega_0$;

$b$ – quality 22.0, driving frequency $0.1483\omega_0$, driving amplitude $50.0°$,
   initial deflection $-17.95°$, initial angular velocity $-0.50\,\omega_0$

Figure 8.15: Phase diagrams (with Poincaré sections) of chaotic oscillations in the vicinity of superharmonic resonances of the third ($a$) and fifth ($b$) orders.

## 8.4   Other Extraordinary Regular Forced Oscillations

The original time-translational symmetry in the motion of the sinusoidally driven pendulum can be broken not only by the above-described period-doubling bifurcations: Under certain conditions regular oscillations of the pendulum have a period that covers some integer (other than two) number of the drive period.

Figure 8.16 shows the graphs of $\varphi(t)$ and $\dot\varphi(t)$ with their harmonics for a period-3 nonlinear oscillation in which the frequency of the third harmonic coincides with the driving frequency. The pendulum makes one oscillation during each drive period, but the swing differs from one cycle to the next as though the mid-point were moving with a period that is 3 times the drive period: Maximal angular excursion of the pendulum equals 171°, then 117°, and then 111°. After 3 cycles of the external torque all the motion repeats. The set of Poincaré sections consists of three fixed points visited in turn. If the initial conditions are chosen somewhere beyond the basin of attraction of this mode, the pendulum eventually settles into a coexisting simple mode—period-1 spatially symmetric oscillations with an amplitude of 142° that occurs in the opposite phase with respect to the driving torque.

When the driving frequency is slightly smaller than the natural frequency, rather counterintuitive steady-state modes can occur in which the motion of the pendulum resembles beats: The amplitude (and the frequency) of oscillation are not constant but instead vary slowly with a long period that equals an integer (odd and rather large) number of driving periods. An example of such steady-state *self-modulated oscillations* whose period equals 11 driving cycles is shown in Figure 8.17. We note the most surprising feature of this mode: The maximal

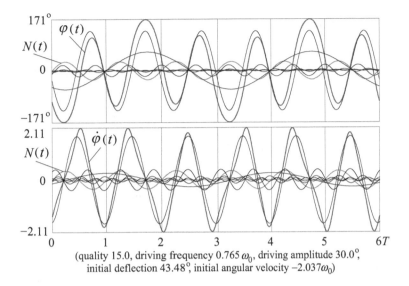

(quality 15.0, driving frequency $0.765\,\omega_0$, driving amplitude $30.0°$, initial deflection $43.48°$, initial angular velocity $-2.037\omega_0$)

Figure 8.16: Period-3 nonlinear oscillations. The frequency of the fundamental harmonic equals one third of the driving frequency.

deflections of the pendulum have no tendency to equalize in the course of time. Contrary to ordinary transient beats, in these oscillations the variations of amplitude and frequency do not fade: Once established, they continue forever.

We can suggest a simple physical explanation for this exotic mode of self-modulated oscillations. Let the phase of the sinusoidal driving torque be initially almost equal to the phase of (natural) oscillations of the angular velocity. That is, let us assume that the external torque varies with time in such a way that it is directed along the angular velocity during almost the entire period. In this case the energy is transferred to the pendulum, and the amplitude of oscillations gradually grows. But with the growing amplitude the natural period of the pendulum becomes longer. Therefore after a while the oscillations of the angular velocity accumulate some phase lag with respect to the driving torque. When this phase lag increases up to 180°, that is, the driving torque varies in the opposite phase with respect to the angular velocity, the energy flow is reversed. This causes the swing of oscillations to decrease.

Then after a while the phase relations again become favorable for supplying the energy to the pendulum, and the amplitude grows again. Thus the amplitude of oscillations is modulated with some (rather long) period.

A small amount of friction can stabilize the period of modulation. If this period equals an integer number of the driving periods, the phase-locking can occur. By virtue of this synchronization between the external drive and natural oscillations of the pendulum, the whole process of self-modulated oscillations becomes exactly periodic.

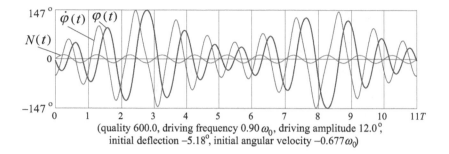

(quality 600.0, driving frequency $0.90\,\omega_0$, driving amplitude $12.0^\circ$,
initial deflection $-5.18^\circ$, initial angular velocity $-0.677\omega_0$)

Figure 8.17: Period-11 forced steady-state self-modulated oscillations.

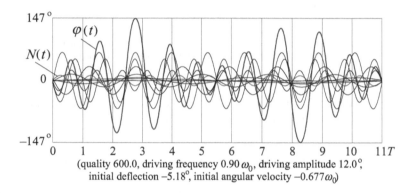

(quality 600.0, driving frequency $0.90\,\omega_0$, driving amplitude $12.0^\circ$,
initial deflection $-5.18^\circ$, initial angular velocity $-0.677\omega_0$)

Figure 8.18: Graph of period-11 steady-state forced self-modulated oscillations
and graphs of its harmonic components that constitute its spectrum.

The energy dissipation is compensated by a somewhat greater amount of energy being transferred to the pendulum on average (during a cycle of the modulation) compared to the backward transfer from the pendulum to the source of the
external drive.

Thin lines in Figure 8.18 show harmonic components (the spectrum) of these
period-11 self-modulated oscillations. The fundamental harmonic of this non-
sinusoidal oscillation has the frequency that equals (1/11)th of the driving frequency $\omega$. Its amplitude is an order of magnitude smaller than the amplitude of
the eleventh harmonic component whose frequency equals the driving frequency.
Besides this component, harmonics with frequencies 7/11, 9/11 and 13/11 of the
driving frequency contribute considerably to the resulting oscillation. The amplitudes of odd harmonics are listed in Table 8.1.

Another example of a periodic steady-state forced motion of a pendulum is
presented in Figure 8.19. During one period of the external torque the pendulum
makes 6 fast revolutions to one side, then its rotation slows down, and it makes 6

Table 8.1: Amplitudes of odd harmonics for period-11 oscillations

| No | Amplitude (radians) | Amplitude (degrees) | Velocity (units of $\omega_0$) |
|----|---------------------|---------------------|-------------------------------|
| 1  | 0.091               | 5.214               | 0.007                         |
| 3  | 0.137               | 7.850               | 0.039                         |
| 5  | 0.269               | 15.41               | 0.121                         |
| 7  | 0.585               | 33.52               | 0.374                         |
| 9  | 1.164               | 66.69               | 0.951                         |
| 11 | 0.883               | 50.59               | 0.885                         |
| 13 | 0.428               | 24.52               | 0.507                         |
| 15 | 0.013               | 0.745               | 0.018                         |

revolutions to the opposite side.

From the angular velocity graph of this extraordinary motion (see the lower panel of Figure 8.19) we suppose that the time dependence of $\dot{\varphi}(t)$ can be represented as a superposition of a slow periodic component (varying almost sinusoidally with the drive period $T$) and a small fast component distorting this slow variation. We suppose that the slow variation of $\dot{\varphi}(t)$ is caused by the slow varying external torque, while the additional fast oscillations of $\dot{\varphi}(t)$ appear by virtue of the gravitational force that influences the pendulum rotation. Hence to a first approximation this extraordinary behavior of the pendulum can be explained by neglecting the force of gravity. Omitting the last term (whose origin is related to gravity) in the left-hand part of Eq. (8.1), we get the following linear first-order equation for the angular velocity $\nu(t) = \dot{\varphi}(t)$:

$$\dot{\nu} + 2\gamma\nu = \omega_0^2 \phi_0 \sin \omega t. \tag{8.6}$$

The steady-state periodic solution to this equation can be represented as follows:

$$\nu(t) = -\nu_{\mathrm{m}} \cos(\omega t + \delta), \quad \nu_{\mathrm{m}} = \frac{\omega_0^2 \phi_0}{\sqrt{\omega^2 + 4\gamma^2}}, \quad \delta = \arctan \frac{2\gamma}{\omega}. \tag{8.7}$$

Hence the angular velocity $\nu(t)$ varies sinusoidally with drive frequency $\omega$ and amplitude $\nu_{\mathrm{m}}$ given by Eq. (8.7). Actually $\nu(t)$ corresponds to the slow component of $\dot{\varphi}(t)$ averaged over the period of fast rotation: $\nu(t) = \langle \dot{\varphi}(t) \rangle_{\mathrm{av}}$. According to Eq. (8.7), its amplitude $\nu_{\mathrm{m}} \approx \omega_0^2 \phi_0 / \omega$ equals $3.6\,\omega_0$ for the values $\phi_0 = 0.8$ and $\omega = 0.22\,\omega_0$ that were used in the simulation experiment shown in Figure 8.19, while the phase lag $\delta = \arctan(2\gamma/\omega) = \arctan(\omega_0/Q\omega) \approx 0.3$. These values agree rather well with the experiment. To evaluate the minimal period $\Delta t$ of fast rotation, we can divide the full angle $2\pi$ by the average angular velocity $\nu_{\mathrm{m}}$, whence $\Delta t/T = \omega/\nu_{\mathrm{m}} = 0.06$, which also agrees well with the experimental graph in Figure 8.19.

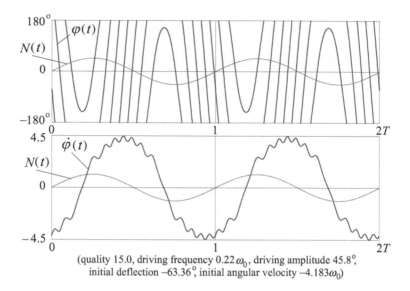

(quality 15.0, driving frequency $0.22\,\omega_0$, driving amplitude $45.8°$, initial deflection $-63.36°$, initial angular velocity $-4.183\omega_0$)

Figure 8.19: Stationary period-1 fast bidirectional revolutions of the pendulum.

We can evaluate the amplitude of fast oscillations of the angular velocity $\dot\varphi(t)$ on the basis of the energy conservation. Let $\dot\varphi_{\max}$ and $\dot\varphi_{\min}$ be the maximum and minimum values of $\dot\varphi(t)$ during the stage of fastest rotation. Kinetic energy of the rotating pendulum at the lowest point (which is proportional to $\dot\varphi^2_{\max}$) is greater than at the inverted position approximately by the difference in the potential energy at these points. From these considerations we find:

$$\dot\varphi_{\max,\,\min} = \nu_m\left(1 \pm \frac{\omega_0^2}{\nu_m^2}\right), \qquad \nu_m = \frac{\omega_0^2\phi_0}{\sqrt{\omega^2 + 4\gamma^2}} \approx \frac{\omega_0^2\phi_0}{\omega}. \qquad (8.8)$$

According to this estimate, the fractional difference $(\dot\varphi_{\max} - \dot\varphi_{\min})/\nu_m$ at $\phi_0 = 0.8$ and $\omega = 0.22\,\omega_0$ equals 0.15, again in a good agreement with the experiment.

An alternative physical explanation of this counterintuitive mode may be formulated by considering the motion of a particle in a time-dependent periodic potential (see curve *1* in Figure 8.1) whose lateral barriers are slowly rising and falling with time. Contrary to the similar approach in the explanation of the bell-ringer mode (see Section 8.2.3), now the potential pattern is "rocking" about the origin (point 0) with a large amplitude. After escaping the potential well by crossing its falling barrier, the particle starts to slide down along the bumpy slope crossing the barriers until the next barrier rises high enough to slow down the particle and to force its backward non-uniform motion.

To maintain this exotic steady-state periodic motion (to provide the phase locking), the phase relation between the pendulum and the periodic variation of the potential pattern (that is, the drive torque $N(t)$ time dependence) is critical.

This means that the mode can be excited only by choosing the initial conditions carefully. In other words, this limit cycle is characterized by a small basin of attraction.

## 8.5  Concluding Remarks

The dynamic behavior of the planar forced pendulum discussed in this chapter is richer in various modes than we might expect for such a simple physical system relying on our intuition. Its nonlinear large-amplitude motions can hardly be called simple. Variations of the parameters result in different regular and chaotic types of dynamical behavior. The simulation program "Rigid pendulum driven by a sinusoidal torque" offers many interesting predefined examples (besides those discussed above) that illustrate various peculiarities of this famous physical model in vivid computer simulations. Visualization of the motion simultaneously with plotting the graphs of different variables and phase trajectories makes the simulation experiments very convincing and comprehensible.

We have touched on only a small portion of the steady-state modes and regular motions of the sinusoidally driven rigid pendulum. The pendulum's dynamics exhibits a great variety of other counterintuitive rotational, oscillatory, and combined (both rotational and oscillatory) multiple-periodic stationary states (attractors), whose basins of attraction are sometimes characterized by a surprisingly complex (fractal) structure. Computer simulations also reveal intricate sequences of bifurcations, leading to numerous intriguing chaotic regimes. Most of these features remain beyond the scope of this chapter. With good reason we can say that this familiar and apparently simple physical system seems almost inexhaustible.

# Chapter 9

# Pendulum with a Square-Wave Modulated Length

**Annotation.** The phenomenon of parametric resonance in a rigid planar pendulum caused by a square-wave modulation of its length is investigated both analytically and with the help of a computer simulation. Characteristics of parametric resonance are found and discussed in detail. The role of nonlinear properties of the pendulum in restricting the resonant swinging is emphasized. The boundaries of parametric excitation as functions of the modulation depth and the quality factor are determined. Stationary oscillations at these boundaries and at the threshold conditions are investigated.

## 9.1   The Investigated Physical System

Periodic excitation of a physical system is called *parametric forcing* if it is realized by variation of some parameter that characterizes the system. In particular, a pendulum can be excited parametrically by a given vertical motion of its suspension point. This apparently simple physical system exhibits a surprisingly vast variety of possible regular and chaotic motions. Hundreds of texts and papers are devoted to investigation of the pendulum with vertically oscillating pivot: See, for example, [16], [17] and references therein. In the frame of reference associated with the pivot, such forcing of the pendulum is equivalent to periodic modulation of the gravitational field. A widely known curiosity in the behavior of an ordinary rigid planar pendulum whose pivot is forced to oscillate along the vertical line is the dynamic stabilization of its inverted position, occurring when the driving amplitude and frequency lie in certain intervals (see [17]–[21] and Chapter 10).

Another familiar method of parametric excitation that we explore in this chapter consists of a periodic variation of the length of the pendulum. In many text-books and papers (see, for example, [22]–[26]) such a system is considered as a simple model of a playground swing. Indeed, the swing can be treated as a phys-ical pendulum whose effective length changes periodically as the child squats at the extreme points, and straightens each time the swing passes through the equi-librium position. It is easy to illustrate this phenomenon of the swing pumping in a classroom by the following simple experiment. Let a thread with a bob hanging on its end pass through a little ring fixed in a support. You can pull by some small distance the other end of the thread that you are holding in your hand each time the swinging bob passes through the middle position, and release the thread to its previous length each time the bob reaches its extreme positions. These peri-odic variations of the pendulum's length with the frequency twice the frequency of natural oscillation cause the amplitude to increase progressively.

A remarkable description of an exotic example illustrating this mode of para-metric excitation can be found in [1], p. 27. In Spain, in the cathedral of a northern town Santiago de Compostela, there is a famous *O Botafumeiro*, a very large in-cense burner suspended by a long rope, which can swing through a huge arc. The censer is pumped by periodically shortening and lengthening the rope as it is wound up and then down around the rollers supported high above the floor of the nave. The pumping action is carried out by a squad of priests, called *tiraboleiros*, or ball swingers, each holding a rope that is a strand of the main rope that goes from the pendulum to the rollers and back down to near the floor. The *tiraboleiros* periodically pull on their respective ropes in response to orders from the chief verger of the cathedral. One of the more terrifying aspects of the pendulum's mo-tion is the fact that the amplitude of its swing is very large, and it passes through the bottom of its arc with a high velocity, spewing smoke and flames.

In this chapter we consider a pendulum with modulated length that can swing in the vertical plane in the uniform gravitational field (Figure 9.1). To allow ar-bitrarily large swinging and even full revolutions, we assume that the pendulum consists of a rigid massless rod (rather than a flexible string) with a massive small bob on its end. The effective length of the pendulum can be changed by shifting the bob up and down along this rod. Periodic modulation of the effective length by such mass redistribution can cause, under certain conditions, a growth of initially small natural oscillations. This phenomenon is called parametric resonance.

## 9.1.1   The Square-Wave Modulation of the Pendulum Length

In this chapter we are concerned with a periodic square-wave (piecewise constant) modulation of the pendulum length. The square-wave modulation provides an al-ternative (compared to the sinusoidal modulation) and may be a more straightfor-ward way to understand and describe quantitatively the phenomenon of parametric resonance. A relevant computer program [27] developed by the author simulates such a physical system and aids greatly in investigating the phenomenon.

In the case of the square-wave modulation, abrupt, almost instantaneous incre-ments and decrements in the length of the pendulum occur sequentially, separated

Figure 9.1: Modulation of the moment of inertia by periodic displacements of the massive bob up and down along the rod of the pendulum.

by equal time intervals. We denote these intervals by $T/2$, so that $T$ equals the period of the length variation (the period of modulation). It is easy to understand how the square-wave modulation can produce considerable oscillation of the pendulum if the period and phase of modulation are chosen properly.

For example, suppose that the bob is shifted up (toward the axis) at an instant at which the pendulum passes through the lower equilibrium position, when its angular velocity reaches a maximum value. While the weight is moved radially, the angular momentum of the pendulum with respect to the pivot remains constant. Thus the resulting reduction in the moment of inertia is accompanied by an increment in the angular velocity, and the pendulum gets additional energy. The greater the angular velocity, the greater the increment in energy. This additional energy is supplied by the source that moves the bob along the rod of the pendulum.

On the other hand, if the bob is instantly moved down along the rod of the swinging pendulum, the angular velocity and the energy of the pendulum diminish. The decrease in energy is transferred back to the source. In order that increments in energy occur regularly and exceed the amounts of energy returned, i.e., in order that, as a whole, the modulation of the length regularly feeds the pendulum with energy, the period and phase of modulation must satisfy certain conditions.

In particular, the greatest growth of the amplitude occurs if effective length of the pendulum is reduced each time the pendulum crosses the equilibrium position, and is increased back at greatest elongations, when the angular velocity is almost zero. Therefore this radial displacement of the bob into its former position causes nearly no decrement in the kinetic energy. The resonant growth of the amplitude occurs if two cycles of modulation are executed during one period of natural oscillations. This is the principal parametric resonance. The time history of such oscillations for the case of a very weak friction ($Q = 1500$) is shown in Figure 9.2 together with the square-wave variation of the pendulum length.

In a real system the growth of the amplitude at parametric resonance is restricted by nonlinear effects. In a nonlinear system like the pendulum, the natural

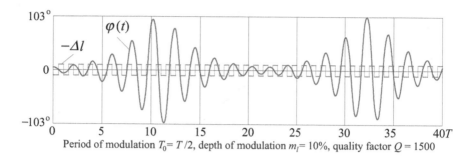

Period of modulation $T_0 = T/2$, depth of modulation $m_l = 10\%$, quality factor $Q = 1500$

Figure 9.2: Initial exponential growth of the amplitude of oscillations at parametric resonance of the first order ($n = 1$) under the square-wave modulation, followed by beats.

period depends on the amplitude of oscillations. As the amplitude grows, the natural period of the pendulum becomes longer. However, in the accepted model the drive period (period of modulation) remains constant. If conditions for parametric excitation are fulfilled at small oscillations and the amplitude is growing, the conditions of resonance become violated at large amplitudes—the drive slips out of resonance. The drive will then drift out of phase with the pendulum. The phase relationships between the modulation and oscillations of the pendulum change gradually to those favorable for the backward transfer of energy from the pendulum to the source of modulation. This causes gradual reduction of the amplitude. The natural period becomes shorter, and conditions for the growth of the amplitude restore. Oscillations of the pendulum acquire the character of beats, as shown in Figure 9.2. Due to friction these transient beats gradually fade, and the amplitude tends to a finite constant value.

Details of the process of resonant growth followed by a nonlinear restriction of the amplitude for a parametrically excited pendulum ($T = T_0/2$) with considerable values of the modulation depth and friction ($m_l = 15\%$, $Q = 5.0$) are shown in Figure 9.3. The vertical segments of the phase trajectory and of the $\dot{\varphi}(t)$ graph correspond to instantaneous increments and decrements of the angular velocity $\dot{\varphi}$ at the instants at which the bob is shifted up and down, respectively. The curved portions of the phase trajectory that spiral in toward the origin correspond to damped natural motions of the pendulum between the jumps of the bob. The initially fast growth of the amplitude (described by the expanding part of the phase trajectory) gradually slows down, because the natural period becomes longer. After reaching the maximum value of 78.3°, the amplitude alternatively decreases and increases within a small range, slowly approaching its final value of about 74°. The initially unwinding spiral of the phase trajectory simultaneously approaches the closed limit cycle, whose characteristic shape can be seen in the left panel of Figure 9.3.

It is evident that the energy of the pendulum is increased not only when two

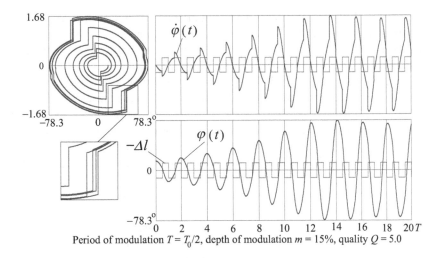

Period of modulation $T = T_0/2$, depth of modulation $m = 15\%$, quality $Q = 5.0$

Figure 9.3: The phase diagram ($\varphi - \dot{\varphi}$ plane) and time-dependent graphs of angular velocity $\dot{\varphi}(t)$ and angle $\varphi(t)$ for the process of resonant growth followed by nonlinear restriction of the amplitude.

full cycles of variation in the parameter occur during one natural period of oscillation, but also when two cycles occur during three, five, or any odd number of natural periods (resonances of odd orders). We shall see later that the delivery of energy, though less efficient, is also possible if two cycles of modulation occur during an even number of natural periods (resonances of even orders).

## 9.1.2 Conditions and Peculiarities of Parametric Resonance

There are several important differences that distinguish parametric resonance from the ordinary resonance caused by an external force exerted directly on the system. Variations of the length cannot take a resting pendulum out of equilibrium: In contrast to the direct forcing, parametric excitation can occur only if (even small) natural oscillations already exist. Parametric resonance is possible when one of the following conditions for the frequency $\omega$ (or for the period $T$) of a parameter modulation is fulfilled:

$$\omega = \omega_n = \frac{2\omega_0}{n}, \qquad T = T_n = \frac{nT_0}{2}, \qquad n = 1, 2, \ldots \qquad (9.1)$$

Parametric resonance is possible not only at the frequencies $\omega_n$ given in Eq. (9.1), but also in ranges of frequencies $\omega$ lying on either side of the values $\omega_n$ (in the ranges of instability). These intervals become wider as the depth of modulation is increased.

An important difference between parametric excitation and forced oscillations is related to the dependence of the growth of energy on the energy already stored in the system. While for a direct forced excitation the increment in energy during one period is proportional to the amplitude of oscillations, i.e., to the square root of the energy, at parametric resonance the increment in energy is proportional to the energy itself, stored in the system.

Energy losses caused by friction are also proportional to the energy already stored. In the case of direct forced excitation, energy losses restrict the growth of the amplitude because these losses grow with the energy faster than does the investment in energy arising from the work done by the external force. In the case of parametric resonance, both the investment in energy caused by the modulation of a parameter and the frictional losses are proportional to the energy stored, and so their ratio does not depend on the amplitude. Therefore, parametric resonance is possible only when a *threshold* is exceeded, that is, when the increment in energy during a period (caused by the parameter variation) is larger than the amount of energy dissipated during the same time. The critical (threshold) value of the modulation depth depends on friction. However, if the threshold is exceeded, the frictional losses of energy cannot restrict the growth of the amplitude. With friction, stationary oscillations of a finite amplitude eventually establish due to nonlinear properties of the pendulum.

## 9.2 The Threshold of Parametric Excitation

### 9.2.1 The Energy Supplied by the Square-Wave Modulation

We can use arguments employing the conservation laws to evaluate the modulation depth that corresponds to the threshold of the principal parametric resonance. Let the changes in the length $l$ of the pendulum occur between $l_1 = l_0(1 + m_l)$ and $l_2 = l_0(1 - m_l)$, where $m_l$ is the dimensionless *depth of modulation* (or modulation index). To calculate the change in total energy of the pendulum during a period, we should not worry about the potential energy. Indeed, after a period the pendulum occurs again in the vertical position with the bob at the same height, hence after a period its potential energy is the same. Thus we should calculate only the change in kinetic energy.

Next we calculate the fractional increment in energy $\Delta E/E$ during one cycle of modulation, namely, between two consecutive passages through the equilibrium position in opposite directions. At the first passage, the energy $E_1$ equals $v_1^2/2$ per unit mass of the bob, where $v_1$ is the bob's velocity. At this time moment the bob is shifted up, so the length of the pendulum changes from $l_0(1 + m_l)$ to $l_0(1 - m_l)$. During abrupt radial displacements of the bob along the pendulum rod, the angular momentum $L = J\dot{\varphi} = Ml^2\dot{\varphi}$ is conserved ($M$ is mass of the bob, $J = Ml^2$ is the moment of inertia about the pivot). Therefore the angular velocity changes at this moment from $\dot{\varphi}_1$ to $(1 + m_l)^2/(1 - m_l)^2\dot{\varphi}_1$. This means that the linear velocity $v$ of the bob changes from $v_1 = l_0(1 + m_l)\dot{\varphi}_1$ to $l_0(1 + m_l)/(1 - m_l)v_1$. Then the pendulum moves from the vertical $\varphi = 0$ up to the maximum deflection $\varphi_m$, whose value can be calculated using the energy conservation:

$$\frac{1}{2}v_1^2\left(\frac{1+m_l}{1-m_l}\right)^2 = g\,l_0(1-m_l)(1-\cos\varphi_{\mathrm{m}}). \tag{9.2}$$

When the frequency and phase of the modulation have those values that are favorable for the most effective delivery of energy to the pendulum, the abrupt backward displacement of the bob toward the end of the rod occurs at the instant when the pendulum attains its greatest deflection (more precisely, when the pendulum is very near it). At this instant the angular velocity of the pendulum is almost zero. Hence this action produces no change in the kinetic energy. At this time moment the bob is shifted down, and the length of the pendulum becomes $l_0(1+m_l)$. The pendulum starts its backward motion with zero velocity. Velocity $v_2$ in the equilibrium position, which is gained during this motion, again can be calculated, like in Eq. (9.2), on the basis of energy conservation:

$$\frac{1}{2}v_2^2 = g\,l_0(1+m_l)(1-\cos\varphi_{\mathrm{m}}). \tag{9.3}$$

From Eqs.(9.2)–(9.3) we find:

$$v_2^2 = v_1^2\left(\frac{1+m_l}{1-m_l}\right)^3, \qquad E_2 = E_1\left(\frac{1+m_l}{1-m_l}\right)^3, \tag{9.4}$$

where $E_2 = v_2^2/2$ is the kinetic energy (per unit mass) after a period $T$ of modulation. Hence

$$\frac{\Delta E}{E} = \frac{E_2}{E_1} - 1 = \left(\frac{1+m_l}{1-m_l}\right)^3 - 1 \approx 6\,m_l. \tag{9.5}$$

The last approximate expression in Eq. (9.5) is valid for small values of the modulation depth $m_l \ll 1$. That is, the fractional increment of total energy $\Delta E/E$ during one period $T$ of modulation approximately equals $6m_l$. The sequence of energy values $E_n$ at consecutive passages through the equilibrium position forms a geometric progression. A process in which the increment in energy $\Delta E$ during a period is proportional to the energy $E$ stored ($dE/dt \approx 6m_l E/T$) is characterized on average by the exponential growth of the energy with time:

$$E(t) = E_0 \exp(\frac{6m_l}{T}t) = E_0 \exp(\alpha t). \tag{9.6}$$

In this case of tuning to the principal resonance, the index of growth $\alpha$ is proportional to the depth of modulation $m_l$ of the pendulum length: $\alpha = 6m_l/T$.

## 9.2.2 Regime of Parametric Regeneration

When the modulation is exactly tuned to the principal resonance ($T = T_0/2$), the decrease of energy is caused almost only by friction. Dissipation of energy due to

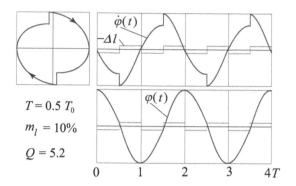

$$T = 0.5\, T_0$$

$$m_l = 10\%$$

$$Q = 5.2$$

Figure 9.4: The phase trajectory (left) and the time-dependent graphs of stationary oscillations (right) at the threshold condition $m_l \approx \pi/(6Q)$ for $T = T_0/2$. The square-wave modulation of the pendulum length is also shown.

viscous friction during an integer number of natural half-cycles (for $t = nT = nT_0/2$) is described by the following expression:

$$E(t) = E_0 \exp(-2\gamma t). \tag{9.7}$$

Comparing equations (9.6) and (9.7), we obtain the following estimate for the threshold (minimal) value $(m_l)_{\min}$ of the depth of modulation corresponding to the excitation of the principal parametric resonance:

$$(m_l)_{\min} = \frac{1}{3}\gamma T = \frac{1}{6}\gamma T_0 = \frac{\pi}{6Q}. \tag{9.8}$$

Here we introduced the dimensionless quality factor $Q = \omega_0/(2\gamma)$ to characterize viscous friction in the system.

The phase trajectory and the plots of time dependence of the angle and angular velocity of parametric oscillations of a small amplitude occurring at the threshold conditions, Eq. (9.8), are shown in Figure 9.4. We can see on the graphs and the phase trajectory only abrupt increments in the magnitude of the angular velocity occurring twice during the period of oscillation (when the bob is shifted upward). The downward shifts of the bob occur at instants when the angular velocity is almost zero. Therefore the corresponding decrements in velocity are too small to be visible on the graphs. This mode of steady oscillations (which have a constant amplitude in spite of the dissipation of energy) is called *parametric regeneration*. Computer simulations show that regime of parametric regeneration is stable with respect to small variations in initial conditions: At different initial conditions the phase trajectory and graphs acquire after a while the same characteristic shape. However, this regime is unstable with respect to variations of the pendulum parameters. If the friction is slightly greater or the depth of modulation slightly smaller than Eq. (9.8) requires, oscillations gradually damp in spite of the modulation. Otherwise, the amplitude grows.

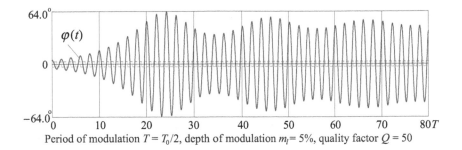

Period of modulation $T = T_0/2$, depth of modulation $m_l = 5\%$, quality factor $Q = 50$

Figure 9.5: Gradually fading beats of the amplitude of oscillations at parametric resonance of the first order ($n = 1$).

For the third resonance ($T = 3T_0/2$) the threshold value of the depth of modulation is three times greater than its value for the principal resonance: $(m_l)_{\min} = \pi/(2Q)$. In this instance two cycles of the parametric variation occur during three full periods of natural oscillations. Radial displacements of the pendulum bob again happen at the time moments most favorable for pumping the pendulum—up at the equilibrium position, and down at the extreme positions. The same investment in energy occurs during an interval that is three times longer than the interval for the principal resonance.

### 9.2.3 Transients over the Threshold

When the depth of modulation exceeds the threshold value, the energy of initially small oscillations during the first stage increases exponentially with time. For the principal parametric resonance this initial growth is shown in Figure 9.5. The growth of the energy again is described by Eq. (9.6). However, now the index of growth $\alpha$ is determined by the amount by which the energy delivered through parametric modulation exceeds the simultaneous losses of energy caused by friction: $\alpha = 6m_l/T - 2\gamma$. If the swing is small enough, the energy is proportional to the square of the amplitude. Hence the amplitude of parametrically excited oscillations initially also increases exponentially with time: $a(t) = a_0 \exp(\beta t)$. The index $\beta$ in the growth of amplitude is one half the index of the growth in energy. For the principal resonance, when the investment in energy occurs twice during one natural period of oscillation, we have $\beta = 3m_l/T - \gamma = 6m_l/T_0 - \gamma = 3m_l\omega_0/\pi - \gamma$.

If the threshold is exceeded, the amplitude grows, conditions of resonance are violated, and this causes a gradual reduction of the amplitude. Then the natural period becomes shorter, and oscillations of the pendulum acquire the character of beats. Due to friction, these transient beats gradually fade, and eventually steady-state oscillations of a finite amplitude establish (Figure 9.5). We note again that the growth of amplitude is restricted by nonlinear properties of the pendulum, namely, by the dependence of the natural period $T_0$ on the amplitude. For small

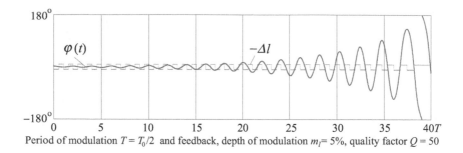

Period of modulation $T = T_0/2$ and feedback, depth of modulation $m_l = 5\%$, quality factor $Q = 50$

Figure 9.6: Parametric pumping of the pendulum with the usage of a feedback loop that provides the most effective delivery of energy to the pendulum.

and moderate values of the amplitude $\varphi_m$ this dependence is approximately given by $T_0(\varphi_m) \approx T_{\text{small}}(1 + \varphi_m^2/16)$, where $T_{\text{small}}$ is the period of infinitely small natural oscillations. In contrast with the ordinary resonance caused by direct periodic forcing in a linear isochronous system, friction alone cannot restrict the growth of the amplitude at parametric resonance. In an idealized linear system the amplitude of parametric oscillations over the threshold grows indefinitely (see [28]–[29]).

### 9.2.4   Parametric Swinging with the Feedback

In the above analysis we assumed that the period $T$ of modulation remains the same as the amplitude increases. At exact tuning to the principal resonance this period equals $T_0/2$, where $T_0$ is the period of small natural oscillations. When we apply the model of the pendulum with modulated length for explaining the pumping of a playground swing, we should take into account that the child on the swing may notice the lengthening of the natural period as the amplitude increases, and can react correspondingly, increasing the period of pumping to stay in phase with the swing.

This intuitive reaction may be considered as a kind of feedback loop: The child determines the time instants to squat and to stand depending on the actual position of the swing. We can include this feedback loop in our model by requiring that instantaneous upward shifts of the bob of the pendulum occur exactly at the time moments, at which the pendulum crosses the equilibrium position, and that backward shifts of the bob to the end of the rod occur exactly at extreme positions of the pendulum. Such manipulations provide the optimal control for the most effective and rapid pumping.

Figure 9.6 shows the graph of progressively growing oscillations occurring under this optimal control with a feedback. Initially the period of modulation $T$ satisfies conditions of the principal parametric resonance at small swing ($T = T_0/2$). We note how the period $T$ of the square-wave modulation increases with the amplitude due to the feedback. After the amplitude reaches 180°, the pendulum executes full revolutions.

Certainly, the priests that pump *O Botafumeiro* also use the feedback. They gradually increase the period of modulation as the amplitude grows, and then probably reduce the depth of modulation to the level sufficient to compensate for frictional losses and to maintain the desirable swing.

# 9.3 Parametric Autoresonance, Bifurcations, Multistability

## 9.3.1 Autoresonance

Is it possible to excite large oscillations of the pendulum at a small excess of the drive over the threshold without the feedback, that is, without appropriately adjusting the period and phase of modulation as the amplitude grows? It occurs that under certain conditions a spontaneous phase locking between the drive and the pendulum motion becomes possible: The pendulum can automatically adjust its amplitude to stay matched with the drive. By sweeping the drive period appropriately, we can control the amplitude of the pendulum. This phenomenon is called *autoresonance*. Autoresonance allows us to both excite and control a large resonant response in nonlinear systems by a small forcing.

We can start pumping the pendulum by modulating its length with period $T = T_0/2$, which corresponds to resonant condition at an infinitely small swing. Then, in the process of oscillations, we slowly increase the period of modulation. This can be done in small steps. After each increment of the period we wait for a while so that transients almost fade away. During this time the amplitude increases just to the amount that provides matching of an increased natural period of the pendulum with the new period of modulation. Thus in each step of this sweeping the pendulum remains locked in phase with the drive.

## 9.3.2 Bifurcations of Symmetry Braking and Period Doubling

To illustrate the phenomenon of parametric autoresonance in a computer simulation (Figure 9.7), we choose the following values for the pendulum parameters: Depth of modulation 5%, quality factor $Q = 20$. When the period of modulation is gradually increased from $T = 0.5T_0$ up to $T = 0.9T_0$, the pendulum swings with an amplitude of 153°. At $T = 0.90T_0$ a bifurcation of symmetry breaking occurs: The pendulum swings to one side through an angle of 161°, while its excursion to the other side is only 146°.

This asymmetry in the swing increases up to $T = 0.913T_0$, when a bifurcation of period doubling occurs: During two cycles of modulation the pendulum executes one asymmetric oscillation between the values 161.5° and 146.9°, while during the next two cycles the pendulum swings between 161.5° and 144.2°. We

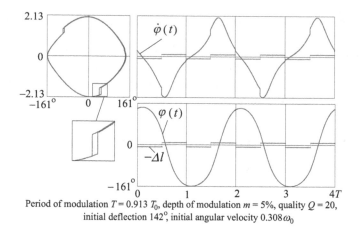

Period of modulation $T = 0.913\ T_0$, depth of modulation $m = 5\%$, quality $Q = 20$, initial deflection $142°$, initial angular velocity $0.308\,\omega_0$

Figure 9.7: Bifurcation of period doubling in parametric autoresonance. Phase diagram (left) and time-dependent graphs of the angular velocity $\dot\varphi(t)$ and angle of deflection $\varphi(t)$ from the equilibrium position (right).

note that the latter elongation is slightly smaller than in the preceding cycle. Then the process repeats.

Thus one period of the pendulum motion now covers four periods of excitation. These oscillations are illustrated in Figure 9.7. We note that the closed phase trajectory is formed by two nearby almost merging loops. Such asymmetric regimes exist (for the same values of $m_l$ and $Q$) in pairs, whose phase orbits are mirror images of one another.

Further increasing of the drive period by tiny steps causes a whole condensing cascade of nearby period doubling bifurcations, which ends at $T = 0.9148T_0$ by a crisis: Oscillations of the pendulum become unstable, finally it turns over the upper equilibrium, and then, after long irregular transient oscillations with gradually diminishing amplitude, the pendulum eventually comes to rest in the downward vertical position.

Stationary parametric oscillations of the pendulum with large amplitude that are locked in phase with the drive and occur at a rather small or moderate modulation (like those described above and shown in Figure 9.7), can be excited not only by slowly sweeping the drive period, but also by appropriate initial conditions. The system eventually comes to a certain periodic regime (limit cycle, or attractor), if initial conditions are chosen within the basin of attraction of this regime. In nonlinear systems different periodic regimes may coexist at the same values of parameters. This property is called multistability.

An example of multistability is shown in Figure 9.8. Curve *1* (upper side of Figure 9.8) describes stationary periodic oscillations of the pendulum with a finite amplitude corresponding to the principal parametric resonance. One period of these oscillations covers two cycles of excitation. Curves *2* and *3* (lower side

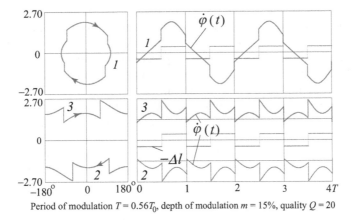

Period of modulation $T = 0.56T_0$, depth of modulation $m = 15\%$, quality $Q = 20$

Figure 9.8: Stationary periodic oscillations and rotations, occurring at the same values of the system parameters.

of Figure 9.8) correspond to period-1 unidirectional rotations of the pendulum in clockwise and counterclockwise directions, respectively. The pendulum makes one revolution during each period of modulation. One more attractor is represented by a single point at the origin of the phase plane, which describes the state of rest of the pendulum in the downward vertical position. Each of these different stationary modes, coexisting at the same values of all parameters of the pendulum and the drive, is characterized by a certain basin of attraction in the phase plane of initial states.

## 9.4 Quantitative Theory of Parametric Excitation

### 9.4.1 Differential Equation for Parametric Oscillations

Next we consider a more rigorous mathematical treatment of parametric resonance under square-wave modulation of the parameter. During the time intervals $(0,\ T/2)$ and $(T/2,\ T)$, the length of the pendulum is constant, and its motion can be considered as a free oscillation described by a corresponding differential equation. However, the coefficients in this equation are different for the adjacent time intervals $(0,\ T/2)$ and $(T/2,\ T)$:

$$\ddot{\varphi} + 2\gamma\dot{\varphi} + \omega_1^2 \sin\varphi = 0, \quad \omega_1 = \tfrac{\omega_0}{\sqrt{1+m}} \text{ for } 0 < t < T/2, \tag{9.9}$$

$$\ddot{\varphi} + 2\gamma\dot{\varphi} + \omega_2^2 \sin\varphi = 0, \quad \omega_2 = \tfrac{\omega_0}{\sqrt{1-m}} \text{ for } -T/2 < t < 0. \tag{9.10}$$

Here $\omega_0 = \sqrt{g/l_0}$ is the natural frequency of small oscillations for the pendulum with mean length $l_0$, and $\gamma$ is the damping constant characterizing the strength of viscous friction. For a slow pendulum traveling in air, the linear dependence of drag on velocity is a reasonable approximation. When damping is caused by the drag force exerted on the pendulum bob, and this force is proportional to the linear velocity of the bob, the frictional torque about the pivot is proportional to $l^2$. Since the moment of inertia is also proportional to $l^2$, the damping constant $\gamma$ in this model remains the same when the length of the pendulum changes, that is, its values in Eqs. (9.9) and (9.10) are equal.

At each instant $t_n = nT/2$ ($n = 1, 2, \ldots$) of an abrupt change in the length of the pendulum, we must make a transition from one of these equations (9.9)–(9.10) to the other. During each half-period $T/2$ the motion of the pendulum is a segment of some natural oscillation. An analytical investigation of parametric excitation can be carried out by fitting to one another known solutions to equations (9.9)–(9.10) for consecutive adjacent time intervals.

The initial conditions for each subsequent time interval are chosen according to the physical model in the following way. Each initial value of the angular displacement $\varphi$ equals the value $\varphi(t)$ reached by the oscillator at the end of the preceding time interval. The initial value of the angular velocity $\dot\varphi$ is related to the angular velocity at the end of the preceding time interval by the law of conservation of the angular momentum:

$$(1 + m_l)^2 \dot\varphi_1 = (1 - m_l)^2 \dot\varphi_2. \qquad (9.11)$$

In Eq. (9.11), $\dot\varphi_1$ is the angular velocity at the end of the preceding time interval, when the moment of inertia of the pendulum has the value $J_1 = J_0(1 + m_l)^2$, and $\dot\varphi_2$ is the initial value for the following time interval, during which the moment of inertia equals $J_2 = J_0(1 - m_l)^2$. The change in the angular velocity at an abrupt variation of the inertia moment from the value $J_2$ to $J_1$ can be found in the same way.

We may use here the conservation of angular momentum, as expressed in Eq. (9.11), because at sufficiently rapid displacement of the bob along the rod of the pendulum, the influence of the torque produced by the force of gravity is negligible. In other words, we can assume the pendulum to be freely rotating about its axis. This assumption is valid provided the duration of the displacement of the bob constitutes a small portion of the natural period.

Considering conditions for which equations (9.9)–(9.10) yield solutions with increasing amplitudes, we can determine the ranges of frequency $\omega$ near the values $\omega_n = 2\omega_0/n$, within which the state of rest is unstable for a given modulation depth $m_l$. In these ranges of parametric instability an arbitrarily small deflection from equilibrium is sufficient for the progressive growth of small initial oscillations.

## 9.4.2 The Mean Natural Period at Large Depth of Modulation

The threshold for the parametric excitation of the pendulum is determined above for the resonant situations in which two cycles of the parametric modulation occur during one natural period or during three natural periods of oscillation. The estimate obtained, Eq. (9.8), is valid for small values of the modulation depth $m_l$ of the pendulum length.

For large values of the modulation depth $m_l$, the notion of a natural period needs a more precise definition. Let $T_0 = 2\pi/\omega_0 = 2\pi\sqrt{l_0/g}$ be the period of oscillation of the pendulum when its massive bob is fixed in the middle position, for which the effective length equals $l_0$. The period is somewhat longer when the weight is moved further from the axis: $T_1 = T_0\sqrt{1 + m_l} \approx T_0(1 + m_l/2)$. The period is shorter when the weight is moved closer to the axis: $T_2 = T_0\sqrt{1 - m_l} \approx T_0(1 - m_l/2)$.

It is convenient to define the natural average period $T_{av}$ not as the arithmetic mean $\frac{1}{2}(T_1 + T_2)$, but rather as the period that corresponds to the arithmetic mean frequency $\omega_{av} = \frac{1}{2}(\omega_1 + \omega_2)$, where $\omega_1 = 2\pi/T_1$ and $\omega_2 = 2\pi/T_2$. So we define $T_{av}$ by the relation:

$$T_{av} = \frac{2\pi}{\omega_{av}} = \frac{2T_1T_2}{(T_1 + T_2)}. \tag{9.12}$$

Indeed, the period $T$ of the parametric modulation that is exactly tuned to any of the parametric resonances is determined not only by the order $n$ of the resonance, but also by the depth of modulation $m_l$. In order to satisfy the resonant conditions, the increment in the phase of natural oscillations during one cycle of modulation must be equal to $\pi$, $2\pi$, $3\pi$, ..., $n\pi$, .... During the first half-cycle the phase of oscillation increases by $\omega_1 T/2$, and during the second half-cycle by $\omega_2 T/2$. Consequently, instead of the approximate condition expressed by Eq. (9.1), we obtain:

$$\frac{\omega_1 + \omega_2}{2}T = n\pi, \quad \text{or} \quad T = T_n = n\frac{\pi}{\omega_{av}} = n\frac{T_{av}}{2}. \tag{9.13}$$

Thus, for a parametric resonance of some definite order $n$, the condition for exact tuning can be expressed in terms of the two natural periods, $T_1$ and $T_2$. This condition is $T = nT_{av}/2$, where $T_{av}$ is defined by Eq. (9.12). For small and moderate values of $m_l$ it is possible to use approximate expressions for the average natural frequency and period:

$$\omega_{av} = \frac{\omega_0}{2}\left(\frac{1}{\sqrt{1 + m_l}} + \frac{1}{\sqrt{1 - m_l}}\right) \approx \omega_0(1 + \frac{3}{8}m_l^2), \quad T_{av} \approx T_0(1 - \frac{3}{8}m_l^2). \tag{9.14}$$

The difference between $T_{av}$ and $T_0$ reveals itself in terms proportional to the square of the depth of modulation $m_l$.

# 9.5 Frequency Ranges for Parametric Resonance

## 9.5.1 Resonances of Odd Orders

To find the boundaries of the frequency ranges of parametric instability surrounding the resonant values $T = T_{av}/2$, $T = T_{av}$, $T = 3T_{av}/2$, ..., we can consider stationary oscillations of indefinitely small amplitude that occur when the period

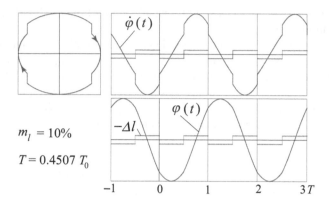

$m_l = 10\%$

$T = 0.4507\, T_0$

Figure 9.9: Phase trajectory and time-dependent graphs of stationary parametric oscillations at the lower boundary of the principal interval of instability (the period of modulation $T$ near $T_{\mathrm{av}}/2$).

of modulation $T$ corresponds to one of the boundaries. These periodic stationary oscillations can be represented as an alternation of natural oscillations with the periods $T_1$ and $T_2$.

## 9.5.2  Main Interval of Parametric Instability

We examine first the vicinity of the principal resonance occurring at $T = T_{\mathrm{av}}/2$. Suppose that the period $T$ of the parametric square-wave modulation is a little shorter than the resonant value $T = T_{\mathrm{av}}/2$, so that $T$ corresponds to the left boundary of the interval of instability. In this case a little less than a quarter of the mean natural period $T_{\mathrm{av}}$ elapses between consecutive abrupt increases and decreases of the pendulum length. Stationary regime with a constant swing in the absence of friction can be realized only if the abrupt increments and decrements of the angular velocity are equal in magnitude. The graphs of the angle $\varphi(t)$ and angular velocity $\dot\varphi(t)$ for this periodic stationary process have the characteristic symmetric patterns shown in Figure 9.9. The segments of the graphs of free oscillations (which occur within time intervals during which the length of the pendulum is constant) are alternating parts of sine or cosine curves with the periods $T_1$ and $T_2$. These segments are symmetrically truncated on both sides.

To find conditions at which such stationary oscillations take place, we can write the expressions for $\varphi(t)$ and $\dot\varphi(t)$ during the adjacent intervals in which the oscillator executes natural oscillations, and then fit these expressions to one another at the boundaries. Such fitting must provide a periodic stationary process.

We let the origin of time, $t = 0$, be the instant when the bob is shifted downward. The angular velocity is abruptly decreased in magnitude at this instant (see Figure 9.9). Then during the interval $(0,\ T/2)$ the graph describes a natural oscillation with the frequency $\omega_1 = \omega_0/\sqrt{1+m}$. Since the graph is symmetric with respect to time moment $T/4$, we can write the corresponding time dependencies of $\varphi(t)$ and $\dot{\varphi}(t)$ in the following form:

$$\varphi_1(t) = -A_1 \cos\omega_1(t-T/4), \quad \dot{\varphi}_1(t) = A_1\omega_1 \sin\omega_1(t-T/4), \quad 0 < t < T/2. \tag{9.15}$$

Similarly, during the interval $(-T/2,\ 0)$ the graph in Figure 9.9 is a segment of natural oscillation with the frequency $\omega_2 = \omega_0/\sqrt{1-m}$:

$$\varphi_2(t) = -A_2 \sin\omega_2(t+T/4), \quad \dot{\varphi}_2(t) = -A_2\omega_2 \cos\omega_2(t+T/4), \quad -T/2 < t < 0. \tag{9.16}$$

To determine the values of constants $A_1$ and $A_2$, we use the conditions that must be satisfied when the segments of the graph are joined together, and take into account the periodicity of the stationary process. At $t = 0$ the angle of deflection is the same for both $\varphi_1$ and $\varphi_2$, that is, $\varphi_1(0) = \varphi_2(0)$. The angular velocity at $t = 0$ undergoes a sudden change, which follows from the conservation of angular momentum: $(1 + m_l)^2\dot{\varphi}_1(0) = (1 - m_l)^2\dot{\varphi}_2(0)$, see Eq. (9.11). From these conditions of fitting the graphs we find the following equations for $A_1$ and $A_2$:

$$A_1 \cos(\omega_1 T/4) = A_2 \sin(\omega_2 T/4). \tag{9.17}$$

$$A_1(1 + m_l)^2\omega_1 \sin(\omega_1 T/4) = A_2(1 - m_l)^2\omega_2 \cos(\omega_2 T/4). \tag{9.18}$$

These homogeneous equations (9.17)–(9.18) for $A_1$ and $A_2$ are compatible only if the following condition is fulfilled:

$$(1 + m_l)^2\omega_1 \sin(\omega_1 T/4) \sin(\omega_2 T/4) = (1 - m_l)^2\omega_2 \cos(\omega_1 T/4) \cos(\omega_2 T/4). \tag{9.19}$$

This is the equation that determines period $T$ of modulation (for a given value $m_l$ of the depth of modulation) which corresponds to the left boundary of the interval of parametric instability. Next we rearrange Eq. (9.19) to the following form, which is more convenient for obtaining a numeric solution for the unknown variable $T$:

$$(q + 1) \cos(\omega_{\text{av}}T/2) = (q - 1) \cos(\Delta\omega T/4), \tag{9.20}$$

where $\omega_{\text{av}} = (\omega_1 + \omega_2)/2$, and $\Delta\omega = \omega_2 - \omega_1$. In Eq. (9.20) we have introduced a dimensionless quantity $q$ which depends on the depth of modulation $m_l$:

$$q = \left(\frac{1 + m_l}{1 - m_l}\right)^{3/2}. \tag{9.21}$$

To find the left boundary $T_-$ of the instability interval that contains the principal parametric resonance, we search for a solution $T$ to Eq. (9.20) in the vicinity of $T = T_0/2$. We replace $T$ in the argument of the cosine on the left-hand side of Eq. (9.20) by $T_{\text{av}}/2 + \Delta T$. Since $\omega_{\text{av}}T_{\text{av}} = 2\pi$, we can write the cosine as $-\sin(\omega_{\text{av}}\Delta T/2)$. Then Eq. (9.20) becomes:

$$\sin(\omega_{\text{av}}\Delta T/2) = -\frac{q - 1}{q + 1} \cos\frac{\Delta\omega(T_{\text{av}}/2 + \Delta T)}{4}. \tag{9.22}$$

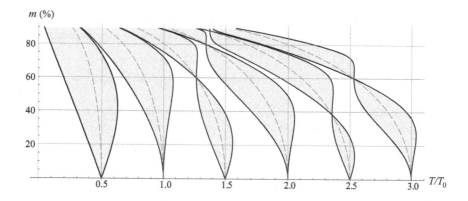

Figure 9.10: Intervals of parametric instability at square-wave modulation of the pendulum length in the absence of friction.

This equation for the unknown quantity $\Delta T$ can be solved numerically by iteration. We start with $\Delta T = 0$ as an approximation of the zeroth order, substituting it into the right-hand side of Eq. (9.22). Then the left-hand side of Eq. (5.22) gives us the value of $\Delta T$ to the first order. We substitute this first-order value into the right-hand side of Eq. (9.22), and on the left-hand side we obtain $\Delta T$ to the second order. This procedure is iterated until a self-consistent value of $\Delta T$ for the left boundary is obtained. Performing such calculations for various values of the modulation depth $m_l$, we obtain the whole left boundary $T_-(m_l)$ for the first interval of parametric instability. Below we explain how the right boundary of this interval can be calculated, as well as the boundaries of other intervals.

The intervals of instability in the plane $T - m_l$ for the first six parametric resonances, calculated numerically with the help of the above-described procedure, are shown in Figure 9.10. This is an analog of the Incze-Strutt diagram of parametric instability for a system that is described by the Mathieu equation, say, for a pendulum with vertical oscillations of the suspension point.

To observe stationary oscillations that correspond to the left boundary of the instability interval (see Figure 9.9) in the simulation, it is insufficient to choose for period $T$ of modulation a self-consistent solution to Eq. (9.22) for a given value of modulation depth $m_l$. After period $T$ is calculated, the initial conditions should also be chosen properly. This can be done on the basis of Eq. (9.15), according to which for an arbitrary initial displacement $\varphi(0)$ the initial angular velocity should have the value $\dot{\varphi}_1(0) = \omega_1 \tan(\omega_1 T/4)\varphi_1(0)$.

For the right boundary of the main interval of instability, the period $T$ of the parametric square-wave modulation is a little longer than the resonant value $T = T_{\mathrm{av}}/2$. In this case a little more than a quarter of the mean natural period $T_{\mathrm{av}}$ elapses between consecutive abrupt increases and decreases of the pendulum length. The graphs of the angle $\varphi(t)$ and angular velocity $\dot{\varphi}(t)$ for this periodic stationary process are shown in Figure 9.11.

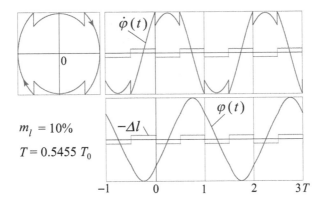

Figure 9.11: Stationary parametric oscillations at the upper boundary of the principal interval of instability (near $T = T_{av}/2$).

We can write the corresponding time dependencies of $\varphi(t)$ and $\dot{\varphi}(t)$ for the time interval $(0, \ T/2)$ in the following form:

$$\varphi_1(t) = B_1 \sin\omega_1(t - T/4), \quad \dot{\varphi}_1(t) = B_1\omega_1 \cos\omega_1(t - T/4), \quad 0 < t < T/2.$$
(9.23)

During the interval $(-T/2, \ 0)$ the graph in Figure 9.11 is a segment of natural oscillation with the frequency $\omega_2 = \omega_0/\sqrt{1 - m}$:

$$\varphi_2(t) = -B_2 \cos\omega_2(t + T/4), \quad \dot{\varphi}_2(t) = B_2\omega_2 \sin\omega_2(t + T/4), \quad -T/2 < t < 0.$$
(9.24)

Further calculations are similar to those for the left boundary already described after Eqs. (9.15)–(9.16). It occurs that $\Delta T$ for the right boundary is determined as a solution to the equation that differs from Eq. (9.22) by the opposite sign on its right-hand side. Solving it numerically by iterations for various values of $m_l$, we obtain the right boundary of the principal interval $(n = 1)$ of parametric instability, Figure 9.10.

To obtain approximate analytical solutions to Eq. (9.22) that are valid for small values of the modulation depth $m_l$, we can simplify the expression on its right-hand side by assuming that $q \approx 1 + 3m_l, q - 1 \approx 3m_l$. We may also assume the value of the cosine to be approximately 1. On the left-hand side of Eq. (9.22), the sine can be replaced by its small argument, in which $\omega_{av} = 2\pi/T_{av}$. This yields the following approximate expressions for both boundaries of the main interval that are valid up to terms to the second order in $m_l$:

$$T_{\mp} = \frac{1}{2}\left(1 \mp \frac{3m_l}{\pi}\right) T_{av} = \frac{1}{2}\left(1 \mp \frac{3m_l}{\pi} - \frac{3m_l^2}{8}\right) T_0.$$
(9.25)

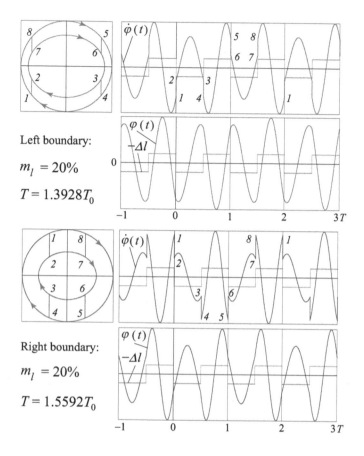

Figure 9.12: The phase trajectory and the graphs of the angular velocity and the deflection angle of stationary parametric oscillations at the left and right boundaries of the interval of instability near $T = 3T_{\mathrm{av}}/2$.

### 9.5.3  Third-Order Interval of Parametric Instability

The boundaries of the instability intervals that contain higher order parametric resonances can be determined in a similar way. At the third order resonance ($n = 3$) two cycles of variation of the pendulum length occur during approximately three natural periods of oscillation ($T \approx 3T_{\mathrm{av}}/2$). The phase trajectories and the time-dependent graphs of stationary oscillations at the left and right boundaries of the third interval are shown in Figure 9.12. The phase orbit of the periodic oscillation closes after two cycles of modulation. This orbit is formed by two concentric ellipses that correspond to small natural oscillations of the pendulum with frequencies $\omega_1$ and $\omega_2$. The representative point moves clockwise along this orbit, jumping from one ellipse to the other each time the bob is shifted along the pendulum rod. The numbers in Figure 9.12 make it easier to follow how the representative point describes this orbit: Equivalent points of the phase orbit and the graph of angular velocity are marked by equal numbers.

Considering conditions at which the graphs of natural oscillations with frequencies $\omega_1$ and $\omega_2$ on the left boundary fit one another for adjacent time intervals and produce the periodic process shown in Figure 9.12, we get the same equations (9.17)–(9.18) for $A_1$ and $A_2$, as well as Eq. (9.22) for the period of modulation. Actually, this is true for all intervals of parametric instability of odd orders. Similarly, for the right boundary we get the same equations for $B_1$ and $B_2$ as in case $n = 1$, and also Eq. (9.22) with the opposite sign for determination of the corresponding period of modulation $T$. However, if we are interested in the third interval, we should search for a solution to these equations in the vicinity of $T = 3T_{\mathrm{av}}/2$, as well as for any other interval of odd order $n$ in the vicinity of $T = nT_{\mathrm{av}}/2$. The boundaries of intervals of the third and fifth orders, obtained by a numerical solution, are also shown in Figure 9.10.

For small values of the depth of modulation $m_l$, we can find approximate analytical expressions for the lower and the upper boundaries of the third interval that are valid up to quadratic terms in $m_l$:

$$T_{\mp} = \frac{3}{2}\left(1 \mp \frac{m_l}{\pi}\right)T_{\mathrm{av}} = \frac{3}{2}\left(1 \mp \frac{m_l}{\pi} - \frac{3m_l^2}{8}\right)T_0, \quad m_l \ll 1. \qquad (9.26)$$

In this approximation, the third interval has the same width $(3m_l/\pi)T_0$ as does the interval of instability in the vicinity of the principal resonance. However, this interval is distinguished by greater asymmetry: Its central point is displaced to the left of the value $T = \frac{3}{2}T_0$ by $\frac{9}{16}m_l^2 T_0$.

### 9.5.4 Parametric Resonances of Even Orders

For small and moderate square-wave modulation of the pendulum length, parametric resonance of the order $n = 2$ (one cycle of the modulation during one natural period of oscillation) is relatively weak compared to the above-considered resonances $n = 1$ and $n = 3$. In the case in which $n = 2$ the abrupt shifts of the bob induce both an increase and a decrease of the energy only once during each natural period. The growth of oscillations occurs only if the increase in energy at the instant when the bob is shifted up is greater than the decrease in energy when the bob is shifted down. This is possible only if the bob is shifted up when the angular velocity of the pendulum is greater in magnitude than it is when the bob is shifted down. For $T \approx T_{\mathrm{av}}$, these conditions can be fulfilled only because there is a (small) difference between the natural periods $T_1$ and $T_2$ of the pendulum, where $T_1 = T_0\sqrt{1 + m_l}$ is the period with the bob shifted down and $T_2 = T_0\sqrt{1 - m_l}$ is the period with the bob shifted up. This difference in the natural periods is proportional to $m_l$.

The growth of oscillations at parametric resonance of the second order is shown in Figure 9.13. We note the asymmetric character of oscillations at $n = 2$ resonance: The angular excursion of the pendulum to one side is greater than to

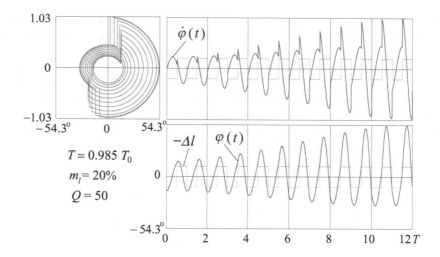

Figure 9.13: The phase trajectory and the graphs of angular velocity $\dot{\varphi}(t)$ and angle $\varphi(t)$ of oscillations corresponding to parametric resonance of the second order $n = 2$ ($T = T_{\mathrm{av}}$).

the other. In this case, the investment in energy during a period is proportional to the *square* of the depth of modulation $m_l$, while in the cases of resonances with $n = 1$ and $n = 3$ the investment in energy is proportional to the first power of $m_l$. Therefore, for the same value of the damping constant $\gamma$ (the same quality factor $Q$), a considerably greater depth of modulation is required here to exceed the threshold of parametric excitation. The growth of the amplitude again is restricted by the nonlinear properties of the pendulum.

The interval of instability in the vicinity of $n = 2$ resonance (for small values of $m_l$) is considerably narrower compared to the corresponding intervals of $n = 1$ and $n = 3$ resonances. Its width is also proportional only to the square of $m_l$.

To determine the boundaries of this interval of instability, we can consider, as is done above for other resonances, stationary oscillations for $T \approx T_0$ formed by alternating segments of free oscillations with the periods $T_1$ and $T_2$. The phase trajectory and the graphs of the angular velocity $\dot{\varphi}(t)$ and the angle $\varphi(t)$ of such stationary periodic oscillations for one of the boundaries are shown in Figure 9.14. During oscillations occurring at the boundary of the instability interval, the abrupt increment and decrement in the angular velocity exactly compensate each other.

To describe these stationary oscillations with small amplitude, we can use the following expressions for $\varphi(t)$ and $\dot{\varphi}(t)$ in the interval $(0, -T/2)$ (see Figure 9.14):

$$\varphi_1(t) = -A_1 \cos \omega_1(t-T/4), \quad \dot{\varphi}_1(t) = A_1 \omega_1 \sin \omega_1(t-T/4), \quad 0 < t < T/2, \tag{9.27}$$

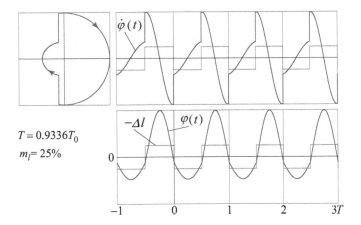

$T = 0.9336T_0$

$m_l = 25\%$

Figure 9.14: Stationary parametric oscillations at the left boundary of the interval of instability of the second order $n = 2$ (near $T = T_{av} \approx T_0$).

and during the interval $(-T/2, 0)$:

$$\varphi_2(t) = A_2 \cos \omega_2(t + T/4), \quad \dot{\varphi}_2(t) = -A_2\omega_2 \sin \omega_2(t + T/4), \quad -T/2 < t < 0.$$
(9.28)

The conditions for joining the graphs at $t = 0$ are the same as for other resonances, namely, at $t = 0$ we require $\varphi_1(0) = \varphi_2(0)$, and the angular velocity undergoes a sudden change, which follows from the conservation of angular momentum (see Eq. (9.11)). These conditions yield the following equations for $A_1$ and $A_2$:

$$A_1 \cos(\omega_1 T/4) = -A_2 \cos(\omega_2 T/4),$$
(9.29)

$$A_1(1 + m_l)^2 \omega_1 \sin(\omega_1 T/4) = A_2(1 - m_l)^2 \omega_2 \cos(\omega_2 T/4).$$
(9.30)

These homogeneous equations (9.29)–(9.30) for $A_1$ and $A_2$ are compatible only if the following condition is fulfilled:

$$(1 + m_l)^2 \omega_1 \sin(\omega_1 T/4) \cos(\omega_2 T/4) = -(1 - m_l)^2 \omega_2 \sin(\omega_2 T/4) \cos(\omega_1 T/4).$$
(9.31)

This is the equation that determines period $T$ of modulation (for a given value $m_l$ of the depth of modulation) which corresponds to the left boundary of the second interval of parametric instability. We transform Eq. (9.31) to the following form, which is convenient for a numeric solution by iteration:

$$(q + 1) \sin(\omega_{av}T/2) = (q - 1) \sin(\Delta\omega T/4),$$
(9.32)

where $q$ depends on the depth of modulation $m_l$ according to Eq. (9.21). Next we replace $T$ in the argument of the sine on the left-hand side of Eq. (9.32) by $T_{av} + \Delta T$. Since $\omega_{av}T_{av} = 2\pi$, we can write this sine as $-\sin(\omega_{av}\Delta T/2)$. Then

Eq. (9.32) becomes:

$$\sin(\omega_{\text{av}} \Delta T/2) = -\frac{q-1}{q+1} \sin \frac{\Delta\omega(T_{\text{av}} + \Delta T)}{4}. \tag{9.33}$$

This equation for $\Delta T$ can be solved numerically by iteration with the help of the above-described procedure. Its self-consistent solutions for various values of the modulation depth $m_l$ give the left boundary of the $n = 2$ instability interval. After period $T$ for this boundary is calculated, the initial conditions that provide stationary oscillations can be chosen on the basis of Eq. (9.27), according to which, for an arbitrary initial displacement $\varphi(0)$, the initial angular velocity should have the value $\dot{\varphi}_1(0) = \omega_1 \tan(\omega_1 T/4)\varphi_1(0)$.

The right boundary of the second interval is given by an equation, which differs from Eq. (9.33) by the opposite sign on its right-hand side. Both boundaries are shown on the diagram in Figure 9.10 together with intervals of higher even orders, which are obtained with the help of similar numeric calculations.

We note how the intervals of even resonances ($n$ =2, 4, 6) are narrow at small values of the modulation depth $m_l$ in contrast to the intervals of odd orders. With the growth of $m_l$ the even intervals expand and become comparable with the intervals of odd orders.

For small and moderate values of the depth of modulation $m_l \ll 1$, an approximate analytical expression for both boundaries of the second interval of instability can be found as a solution to Eq. (9.33) (and to an equation with the opposite sign for the other boundary):

$$T_{\mp} = \left(1 \mp \frac{3}{4}m_l^2\right) T_{\text{av}} = T_0 + \left(\mp\frac{3}{4} - \frac{3}{8}\right) m_l^2 T_0, \tag{9.34}$$

i.e., $T_- = T_0(1 - \frac{9}{8}m^2)$, $T_+ = T_0(1 + \frac{3}{8}m_l^2)$. As mentioned above, the width of this interval of instability $T_+ - T_- = \frac{3}{2}m_l^2 T_0$ is proportional to the square of the modulation depth.

### 9.5.5   Intersections of the Boundaries at Large Modulation

Figure 9.10 shows that at certain values of $m_l$ both boundaries of intervals with $n > 2$ coincide (we may consider that they intersect). This means that at these values of $m_l$ the corresponding intervals of parametric instability disappear. Such values of $m_l$ correspond to the natural periods of oscillation $T_1$ and $T_2$, whose ratio is 2 : 1, 3 : 1, and 3 : 2.

For the first intersection (ratio 2 : 1) exactly one half of the natural oscillation with period $T_1$ is completed during the first half of the modulation cycle (see Figure 9.15). On the phase diagram, the representing point traces a half of the smaller ellipse (1 — 2), and then abruptly jumps down to the larger ellipse (2 — 3). During the second half of the modulation cycle the oscillator executes exactly a whole natural oscillation with period $T_2 = T_1/2$, so that the representing point passes in the phase plane along the whole larger ellipse (3 — 4), and then jumps up to the smaller ellipse along the same vertical segment (4 — 5).

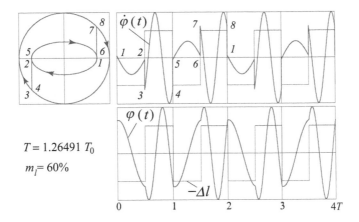

$T = 1.26491\, T_0$

$m_l = 60\%$

Figure 9.15: The phase trajectory and time-dependent graphs of angular velocity $\dot{\varphi}(t)$ and angle $\varphi(t)$ for stationary oscillations at the intersection of both boundaries of the third interval.

During the next modulation cycle the representing point first generates the other half of the smaller ellipse ($5 — 6$), and then again the whole larger ellipse ($7 — 8$). Therefore during any two adjacent cycles of modulation the representing point passes once along the closed smaller ellipse and twice along the larger one, returning finally to the initial point of the phase plane. We see that such an oscillation is periodic for arbitrary initial conditions. This means that for the corresponding values of the modulation depth $m_l$ and the period of modulation $T$ the growth of amplitude is impossible even in the absence of friction (the instability interval vanishes).

Similar explanations can be suggested for other cases in Figure 9.10 in which the boundaries of the instability intervals intersect.

## 9.6 Intervals of Parametric Excitation in the Presence of Friction

When there is friction in the system, the intervals of the period of modulation that correspond to parametric instability become narrower, and for strong enough friction (below the threshold) the intervals disappear. Above the threshold, approximate values for the boundaries of the first interval are given by Eq. (9.25) provided we substitute for $m_l$ the expression $\sqrt{m_l^2 - (m_l)_{\min}^2}$ with the threshold value $(m_l)_{\min} = \pi/(6Q)$ defined by Eq. (9.8). The proof can be found in the following Section 9.6.1. For the third interval, we can use Eq. (9.26), substituting $\sqrt{m_l^2 - (m_l)_{\min}^2}$ for $m_l$, with $(m_l)_{\min} = \pi/(2Q)$. When $m_l$ is equal to the corresponding threshold value $(m_l)_{\min}$, the interval of parametric resonance

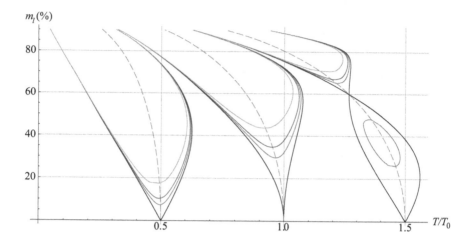

Figure 9.16: Intervals of parametric excitation at square-wave modulation of the pendulum length without friction, for $Q = 7$, $Q = 5$, and for $Q = 3$.

disappears.

The boundaries of the second interval of parametric resonance in the presence of friction are approximately given by Eq. (9.34) provided we substitute for $m_l^2$ the expression $\sqrt{m_l^4 - (m_l)_{\min}^4}$ with the threshold value $(m_l)_{\min} = \sqrt{2/(3Q)}$, which corresponds to the second parametric resonance (see Section 9.6.1).

The diagram in Figure 9.16 shows the boundaries of the first three intervals of parametric resonance for $Q = 3$, $Q = 5$, and $Q = 7$ (and also in the absence of friction). We note the "island" of parametric resonance of the third order ($n = 3$) at $Q = 7$. This resonance disappears when the depth of modulation exceeds 48% and reappears when $m_l$ exceeds approximately 66%.

In the presence of friction, for any given value $m_l$ of the depth of modulation, only several first intervals of parametric resonance (where $m_l$ exceeds the threshold) can exist. We note that in case the equilibrium of the system is unstable due to modulation of the parameter, parametric resonance can occur only if at least small oscillations are already excited. Indeed, when the initial values of $\varphi$ and $\dot{\varphi}$ are exactly zero, they remain zero over the course of time. This behavior is in contrast to that of resonance arising from direct forcing, when the amplitude increases with time even if initially the system is at rest in the equilibrium position (if the initial conditions are zero).

## 9.6.1   Boundaries of Instability for Resonances of Odd Orders

Stationary oscillations occurring at the left boundary of the instability interval in the vicinity of the principal parametric resonance in the presence of friction are shown in Figure 9.17 (compare with Figure 9.9). Twice during the full cycle of

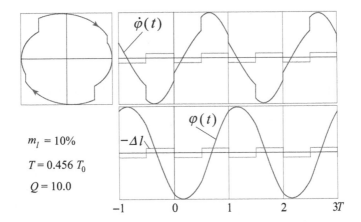

$m_l = 10\%$

$T = 0.456\,T_0$

$Q = 10.0$

Figure 9.17: Stationary oscillations in the presence of friction at the left boundary of the principal instability interval.

modulation the angular velocity abruptly increases, and twice it decreases. The increments are greater than the decrements, so that as a whole the energy received by the pendulum exceeds the energy given away. This surplus compensates for the dissipation of the energy that occurs at natural oscillation during the intervals between the abrupt displacements of the bob along the rod of the pendulum.

To find conditions at which such stationary oscillations take place, we can write the expressions for $\varphi(t)$ and $\dot\varphi(t)$ during the adjacent intervals when the pendulum executes damped natural oscillations, and then fit these expressions to one another at the boundaries. Contrary to the frictionless pendulum (see Figure 9.9, p. 228), now the phase trajectory is not symmetric with respect to the ordinate axis (Figure 9.17). We choose as the time origin $t = 0$ the instant when the bob is shifted down, and the angular velocity decreases in magnitude. Then during the interval $(0,\ T/2)$ the graph describes a damped natural oscillation with the frequency $\omega_1 = \omega_0/\sqrt{1 + m_l}$. We can represent this motion as a superposition of damped natural oscillations of sine and cosine type with some constants $A_1$ and $B_1$:

$$\begin{aligned}\varphi_1(t) &= (A_1 \sin\omega_1 t + B_1 \cos\omega_1 t)\, e^{-\gamma t},\\ \dot\varphi_1(t) &\approx (A_1\omega_1 \cos\omega_1 t - B_1\omega_1 \sin\omega_1 t)\, e^{-\gamma t}.\end{aligned} \tag{9.35}$$

The latter expression for $\dot\varphi(t)$ is valid for relatively weak friction ($\gamma \ll \omega_0$). To obtain it, we differentiate $\varphi(t)$ with respect to the time, considering the exponential factor $e^{-\gamma t}$ to be approximately constant. Indeed, at weak damping the main contribution to the time derivative originates from the oscillating factors $\sin\omega_1 t$ and $\cos\omega_1 t$ in the expression for $\varphi(t)$.

Similarly, during the interval $(-T/2,\ 0)$ the graph in Figure 9.17 is a segment of damped natural oscillation with the frequency $\omega_2$:

$$\varphi_2(t) = (A_2 \sin \omega_2 t + B_2 \cos \omega_2 t)\, e^{-\gamma t},$$
$$\dot{\varphi}_2(t) \approx (A_2\omega_2 \cos \omega_2 t - B_2\omega_2 \sin \omega_2 t)\, e^{-\gamma t}. \tag{9.36}$$

To determine the values of constants $A_1$, $A_2$, and $B_1$, $B_2$, we use the conditions that must be satisfied when the segments of the graph are joined together, and take into account the periodicity of the stationary process. At $t = 0$ the angle of deflection is the same for both $\varphi_1$ and $\varphi_2$, that is, $\varphi_1(0) = \varphi_2(0)$. From this condition we get $B_2 = B_1$. We later denote these equal constants by $B$. The angular velocity at $t = 0$ undergoes a sudden change, which follows from the conservation of angular momentum: $(1+m_l)^2 \dot{\varphi}_1 = (1-m_l)^2 \dot{\varphi}_2$, see Eq. (9.11). This condition gives us the following relation between $A_1$ and $A_2$: $A_2 = qA_1 = qA$ (further on we denote $A_1$ as $A$), where the factor $q$ depends on modulation depth $m_l$ according to Eq. (9.21).

For stationary periodic oscillations, corresponding to the principal resonance, as well as to all resonances of odd orders $n = 1,\ 3,\ \ldots$ in Eq. (9.13), the conditions of periodicity are:

$$\varphi_1(T/2) = -\varphi_2(-T/2), \quad (1+m)^2 \dot{\varphi}_1(T/2) = -(1-m)^2 \dot{\varphi}_2(-T/2). \tag{9.37}$$

Substituting $\varphi$ and $\dot{\varphi}$ in Eq. (9.37), we obtain the system of homogeneous equations for the unknown quantities $A$ and $B$:

$$(pS_1 - qS_2)A + (p\,C_1 + C_2)B = \quad 0,$$
$$q(p\,C_1 + C_2)A - (p\,qS_1 - S_2)B = \quad 0, \tag{9.38}$$

where $p = \exp(-\gamma T)$. In Eq. (9.38) the following notations are used:

$$C_1 = \cos(\omega_1 T/2), \quad C_2 = \cos(\omega_2 T/2),$$
$$S_1 = \sin(\omega_1 T/2), \quad S_2 = \sin(\omega_2 T/2).$$

The homogeneous system of Eqs. (9.38) for $A$ and $B$ has a non-trivial (non-zero) solution only if its determinant is zero:

$$2qC_1C_2 - (1+q^2)S_1S_2 + q(p+1/p) = 0. \tag{9.39}$$

This condition for the existence of a non-zero solution to Eqs. (9.38) gives us an equation for the unknown variable $T$, which enters Eq. (9.39) as the arguments of sine and cosine functions in $S_1$, $S_2$ and $C_1$, $C_2$, and also as the argument of the exponent in $p = e^{-\gamma T}$. The desired boundaries of the interval of instability $T_-$ and $T_+$ are given by the roots of Eq. (9.39). To find approximate solutions $T$ to this transcendental equation, we transform it into a more convenient form. We first represent in Eq. (9.39) the products $C_1C_2$ and $S_1S_2$ as follows:

$$C_1C_2 = \frac{1}{2}\left(\cos \frac{\Delta\omega T}{2} + \cos \omega_{\mathrm{av}} T\right), \qquad S_1S_2 = \frac{1}{2}\left(\cos \frac{\Delta\omega T}{2} - \cos \omega_{\mathrm{av}} T\right). \tag{9.40}$$

Then, using the identity $\cos\alpha = 2\cos^2(\alpha/2) - 1$, we reduce Eq. (9.39) to the following form:

$$(q+1)\cos(\omega_{\text{av}}T/2) = \pm\sqrt{(q-1)^2\cos^2(\Delta\omega T/4) - q(p+1/p - 2)}. \quad (9.41)$$

To find the boundaries of the interval that contains the principal resonance, we should search for a solution $T$ of Eq. (9.41) in the vicinity of $T = T_0/2 \approx T_{\text{av}}/2$. If for a given value of the quality factor $Q$ ($Q$ enters $p = e^{-\gamma T}$) the depth of modulation $m_l$ exceeds the threshold value, Eq. (9.41) has two solutions that correspond to the desirable boundaries $T_-$ and $T_+$ of the instability interval. These solutions exist if the expression under the radical sign in Eq. (9.41) is positive. Its zero value corresponds to the threshold conditions:

$$\frac{(q-1)^2}{q}\cos^2(\Delta\omega T/4) = p + \frac{1}{p} - 2. \quad (9.42)$$

To evaluate the threshold value of $Q$ for small values of the modulation depth $m_l \ll 1$, we may assume here $q \approx 1 + 3m_l$ (see Eq. (9.21)), and $\cos(\Delta\omega T/4) \approx 1$. On the right-hand side of Eq. (9.42), in $p = e^{-\gamma T}$, we can consider $\gamma T \approx \gamma T_0/2 = \pi/(2Q) \ll 1$, so that $p + 1/p - 2 \approx (\gamma T)^2 = (\pi/2Q)^2$. Thus, for the threshold of the principal parametric resonance we obtain

$$Q_{\min} \approx \frac{\pi}{6m_l} \qquad (m_l)_{\min} \approx \frac{\pi}{6Q}. \quad (9.43)$$

At the threshold the expression under the radical sign in Eq. (9.41) is zero. Both its roots (the boundaries of the instability interval) merge. This occurs when the cosine on the left-hand side of Eq. (9.41) is zero, that is, when its argument equals $\pi/2$:

$$\omega_{\text{av}}\frac{T}{2} = \frac{\pi}{2}, \qquad \text{or} \qquad T = \frac{\pi}{\omega_{\text{av}}} = \frac{1}{2}T_{\text{av}},$$

so that the threshold conditions (9.43) correspond to exact tuning to resonance, when $T = T_{\text{av}}/2$.

To find the boundaries $T_-$ and $T_+$ of the instability interval, we represent $T$ in the argument of the cosine function on the left-hand side of Eq. (9.41) as $T_{\text{av}}/2 + \Delta T$. Since $\omega_{\text{av}}T_{\text{av}} = 2\pi$, we can write this cosine as $-\sin(\omega_{\text{av}}\Delta T/2)$. Then Eq. (9.41) becomes:

$$\sin(\omega_{\text{av}}\Delta T/2) = \mp\frac{1}{q+1}\sqrt{(q-1)^2\cos^2\frac{\Delta\omega(\frac{1}{2}T_{\text{av}} + \Delta T)}{4} - q\frac{(p-1)^2}{p}}. \quad (9.44)$$

For zero friction $p = 1$, and Eq. (9.44) coincides with Eq. (9.21). The diagram in Figure 9.16 is obtained by numerically solving this equation for $\Delta T$ by iteration. Boundaries of the instability for intervals of higher odd orders $n = 3, 5, \ldots$ are calculated similarly by representing $T$ in Eq.(9.41) as $nT_{\text{av}}/2 + \Delta T$. They are also shown in Figure 9.16 for several values of the quality factor $Q$. For large values of the modulation depth $m_l$ these boundaries almost merge with the corresponding boundaries in the absence of friction.

To find an approximate solution of Eq. (9.44) that is valid for small values of the modulation depth $m_l \ll 1$ up to terms to the second order in $m_l$, we can simplify the expression under the radical sign on the right-hand side of Eq. (9.38), assuming $q \approx 1 + 3m_l$, $(q - 1)^2 \approx 9m_l^2$, and the value of the cosine function to be 1. The last term of the radicand can be represented as $(\pi/6Q)^2 \approx (m_l)_{min}^2$. On the left-hand side the sine can be replaced with its small argument, where $\omega_{av} = 2\pi/T_{av}$. Thus we obtain:

$$\frac{\Delta T}{T_{av}} \approx \mp \frac{1}{2\pi}\sqrt{m_l^2 - (m_l)_{min}^2}, \text{ or } T_{\mp} = \frac{T_{av}}{2}\left(1 \mp \frac{1}{\pi}\sqrt{m_l^2 - (m_l)_{min}^2}\right).$$
(9.45)

For the case of zero friction $(m_l)_{min} = 0$, and these approximate expressions for the boundaries of the instability interval reduce to Eq. (9.25). For the threshold conditions $m_l = (m_l)_{min}$, and both boundaries of the interval merge, that is, the interval disappears.

After the substitution of one of the roots $T_-$ or $T_+$ of Eq. (9.25) into (9.38), both equations for $A$ and $B$ become equivalent and allow us to find only the ratio $A/B$. Nevertheless, these oscillations have a definite shape, which is determined by the ratio of the amplitudes $A$ and $B$ of the sine and cosine functions whose segments form the typical pattern of the stationary parametric oscillation (see Figures 9.9 and 9.11).

### 9.6.2  Resonances of Even Orders

To describe stationary oscillations occurring on the boundaries of instability intervals of even orders, we can use the same expressions for $\varphi(t)$ and $\dot\varphi(t)$, Eqs. (9.35) and (9.36). The conditions of joining the graphs at $t = 0$ are also the same. However, the conditions of periodicity at the instants $-T/2$ and $T/2$ for resonances of even orders differ from Eqs. (9.37) by the opposite sign. This yields, instead of Eq. (9.41), the following equation for the boundaries of instability intervals:

$$(q + 1)\sin(\omega_{av}T/2) = \pm\sqrt{(q - 1)^2 \sin^2(\Delta\omega T/4) - q(p + 1/p - 2)}. \quad (9.46)$$

For the interval of the second order, we should search for its solution $T$ in the vicinity of $T_0 \approx T_{av}$. If for a given value of the quality factor $Q$ ($Q$ enters $p = e^{-\gamma T}$) the depth of modulation $m_l$ exceeds the threshold value, Eq. (9.46) has two solutions that correspond to the boundaries $T_-$ and $T_+$ of the instability interval. These solutions exist if the expression under the radical sign in Eq. (9.46) is positive. Its zero value corresponds to the threshold conditions, that is, to $(m_l)_{min}$ for a given $Q$ or $Q_{min}$ for a given $m_l$:

$$\frac{(q - 1)^2}{q}\sin^2(\Delta\omega T_{av}/4) = \frac{(p - 1)^2}{p}. \quad (9.47)$$

The threshold conditions fulfil at exact tuning to second resonance, when $T = T_{av}$. To estimate the threshold value of $Q$ for small values of the modulation depth $m_l$, we may assume here $q \approx 1 + 3m_l$, $\sin(\Delta\omega T/4) \approx \Delta\omega T_0/4$, and $\Delta\omega \approx m_l\omega_0$. On the right-hand side of Eq. (9.47), in $p = e^{-\gamma T}$, we can consider $\gamma T \approx \gamma T_0 = \pi/Q \ll 1$, so that $p + 1/p - 2 = (p - 1)^2/p \approx (\gamma T)^2 = (\pi/Q)^2$. Thus, for the threshold of the second parametric resonance we obtain:

$$Q_{\min} \approx \frac{2}{3m_l^2}, \qquad (m_l)_{\min} \approx \sqrt{\frac{2}{3Q}}. \tag{9.48}$$

To find the boundaries $T_-$ and $T_+$ of the second instability interval, we represent $T$ in the argument of the sine function on the left-hand side of Eq. (9.46) as $T_{av} + \Delta T$. Since $\omega_{av}T_{av} = 2\pi$, we can write this sine as $- \sin(\omega_{av}\Delta T/2)$. Then Eq. (9.46) becomes:

$$\sin \frac{\omega_{av}\Delta T}{2} = \mp \frac{1}{q+1} \sqrt{(q-1)^2 \sin^2 \frac{\Delta\omega(T_{av} + \Delta T)}{4} - q\frac{(p-1)^2}{p}}. \tag{9.49}$$

This form of the equation is convenient for numerical solution by iteration. For the zero friction $p = 1$, and Eq. (9.49) coincides with Eq. (9.33). To obtain an approximate solution to Eq. (9.49), valid for small values of the modulation depth $m_l$ up to the terms of the second order of $m_l$, we can simplify the expression under the radical sign on the right-hand side of Eq. (9.49), assuming $q \approx 1 + 3m_l$, $(q - 1)^2 \approx (3m_l)^2$, and $\sin[\Delta\omega(T_{av} + \Delta T)/4] \approx \Delta\omega T_{av}/4 = \pi m_l/2$. The last term of the radicand can be represented as $(2/3Q)^2 \approx (m_l)^4_{\min}$. On the left-hand side the sine can be replaced by its small argument, where $\omega_{av} = 2\pi/T_{av}$. Thus for the boundaries of the second instability interval we get:

$$\frac{\Delta T}{T_{av}} \approx \mp\frac{3}{4}\sqrt{m_l^4 - (m_l)^4_{\min}}, \quad \text{or} \quad T_{\mp} = \left(1 \mp \frac{3}{4}\sqrt{m_l^4 - (m_l)^4_{\min}}\right) T_{av}. \tag{9.50}$$

## 9.7 Concluding Remarks

We have shown in this chapter that a pendulum whose length is subject to square-wave modulation by mass reconfiguration gives a very convenient example in which the phenomenon of parametric resonance in a nonlinear system can be clearly explained physically with all its peculiarities. The threshold of parametric excitation is easily determined on the basis of energy considerations.

In a linear system, if the threshold of parametric excitation is exceeded, the amplitude of oscillations increases exponentially with time. In contrast to forced oscillations, linear viscous friction is unable to restrict the growth of the amplitude at parametric resonance. In real systems like the pendulum, the growth of the amplitude is restricted by nonlinear effects that cause the natural period to depend on the amplitude. During parametric excitation the growth of the amplitude causes an increment in the natural period of the pendulum. The system slips out of resonance, the swing becomes smaller, and conditions of resonance restore. These transient beats fade out due to friction, and oscillations of finite amplitude eventually establish.

Computer simulations aid substantially in understanding the restriction of the amplitude growth over the threshold caused by nonlinear properties of the pendulum. The simulations illustrate the phenomenon of parametric autoresonance, stationary periodic oscillatory and rotational regimes that are possible due to the phase locking between the drive and the pendulum. The simulations also reveal bifurcations of symmetry breaking and intriguing sequences of period doubling. The boundaries of parametric instability for a pendulum with the square-wave modulated length are investigated quantitatively by rather modest mathematical means.

# Chapter 10

# Rigid Pendulum with Oscillating Pivot

## 10.1 Introductory Notes

Familiar and apparently simple dynamical systems for which our intuition may seem to be well developed can behave in very complicated and even irregular ways, in spite of the quite simple and exact nature of governing physical laws and the deterministic character of relevant differential equations.

Various kinds of motion of the pendulum whose axis is driven periodically in the vertical direction are of special interest. Depending on the frequency and amplitude of this constrained oscillation of the suspension point, this seemingly simple system exhibits a rich variety of nonlinear phenomena characterized by amazingly different types of motion. Some modes of such a parametrically forced pendulum are quite simple indeed and agree well with our intuition, while others are very complicated and counterintuitive.

When the external frequency is approximately twice the natural frequency of the pendulum, the lower state of equilibrium becomes unstable, and the system leaves it executing oscillations whose amplitude increases progressively, provided the driving amplitude exceeds some threshold value. This well-known phenomenon is called *parametric resonance*. In contrast to the case of ordinary resonance caused by a direct influence of some periodic external force, over the threshold friction is unable to restrict the growth of parametrically excited oscillations. The growth of the amplitude is restricted because the period of natural oscillations increases with the amplitude due to nonlinear properties of the pendulum. Parametric resonance is possible when two driving cycles occur during approximately one, two, three and any other integer number of natural periods. With increasing friction, parametric resonances of higher orders become weaker and disappear.

Besides the principal parametric resonance, excited when two driving cycles occur during approximately one natural oscillation, parametric resonances of higher orders are possible when approximately two driving cycles occur during two, three and any other integer number of natural periods. At small (and moderate) driving amplitudes, parametrically excited oscillations in all these cases are very much like the natural ones — their frequency is close to the natural frequency of the pendulum. The forced oscillation of the pivot at resonant conditions supplies the pendulum with energy needed to compensate for frictional losses, thus preventing these almost natural oscillations from damping. With increasing friction, parametric resonances of higher orders become weaker and disappear.

Another possible kind of regular motion is a synchronized non-uniform unidirectional rotation in a full circle with a period that equals the period of the constrained motion of the axis or an integer multiple of this period. More complicated regular modes of the parametrically forced pendulum are formed by combined rotational and oscillating motions synchronized with the pivot. Different competing modes can coexist at the same values of the driving amplitude and frequency. Which mode is activated depends on the starting conditions.

Behavior of the pendulum whose axis is forced to oscillate with a frequency from a certain interval (and with large enough amplitude) can be irregular, chaotic. The pendulum makes several revolutions in one direction, then swings for a while with permanently changing amplitude, then rotates again in the former or in the opposite direction, and so forth. At first sight such essentially unpredictable, random behavior contradicts the well-known uniqueness of solution to a differential equation of motion with given initial conditions. Within the scope of classical mechanics, which naturally includes the concept of mechanical determinism, chaotic behavior of simple dynamical systems considered admissible only as a result of external random perturbations of the system, i.e., as something introduced from the outside, from the environment. Discovery of random behavior and intrinsic irregular, chaotic oscillations in deterministic dynamical systems of different nature (physical, chemical, biological) is one of the most prominent recent scientific sensations. It is remarkable that such a simple mechanical system as a pendulum whose pivot is forced to oscillate regularly can exhibit at some conditions a chaotic behavior, illustrated by a strange attractor in the phase plane. Chaotic modes of the parametrically driven pendulum have been intensively investigated over past decades (see, for example, [30]–[35] and references therein).

## 10.2   Kapitza's Pendulum — Dynamic Stabilization

Another well-known interesting feature in the behavior of a rigid pendulum whose suspension point is constrained to vibrate with a high frequency along the vertical line is the dynamic stabilization of its inverted position. When the frequency and/or the amplitude of these vibrations are large enough (the necessary conditions are determined below), the inverted pendulum shows no tendency to turn down. Moreover, at small and moderate deviations from the vertical inverted position the pendulum tends to return to it. Being deviated, the pendulum executes relatively slow oscillations about the vertical line on the background of rapid os-

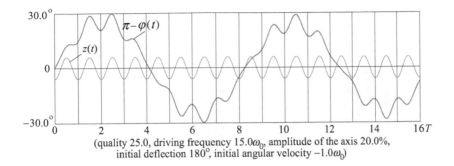

(quality 25.0, driving frequency $15.0\omega_0$, amplitude of the axis 20.0%, initial deflection 180°, initial angular velocity $-1.0\omega_0$)

Figure 10.1: Graphs of the angular deflection from the inverted (upside-down) position for the pendulum whose pivot oscillates at a high frequency along the vertical line.

cillations of the suspension point. An example of the graphs of such oscillations of the inverted pendulum obtained in the computer simulation of the motion is shown in Figure 10.1. We note how the rapid vibrations with the frequency of the pivot superimpose on the slow oscillation of the pendulum and distort its smooth shape. In the presence of friction these slow oscillations gradually damp away, and the pendulum eventually comes to the vertical inverted position.

This type of dynamic stability was first pointed out by Stephenson almost a century ago [36]. In 1951 such extraordinary behavior of the pendulum was explained and investigated experimentally in detail by Pjotr Kapitza [37], and the corresponding physical device is now widely known as "Kapitza's pendulum." Simple hand-made devices are often used in lectures to show this fascinating phenomenon of classical mechanics. An old electric shaver's mechanism (or a jig saw) can serve perfectly well to force the pivot of a light rigid pendulum vibrating with a high enough frequency and sufficient amplitude to make the inverted position stable (Figure 10.2). The hand holds the shaver in the position that provides the vertical direction of the pivot oscillations. If the rod is turned into the inverted vertical position, it remains there until the axis is vibrating. When the rod is slightly deflected to one side and released, it oscillates slowly about the inverted position. Such a demonstration inevitably evokes vivid response, astonishing those students who see it for the first time.

Below is a citation from the paper of Kapitza [38] published in the Russian journal *Uspekhi*:

> Demonstration of oscillations of the inverted pendulum is very impressive. Our eyes cannot follow the fast small movements caused by vibrations of the pivot, so that behavior of the pendulum in the inverted position seems perplexing and even astonishing ... When we carefully touch the rod of the pendulum trying to deviate it from the vertical, the finger feels the resistance produced by the vibrational

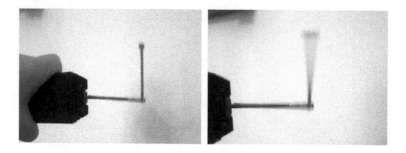

Figure 10.2: Demonstration of the inverted pendulum's dynamic stabilization by vertical vibration of the pivot (left), and small slow oscillations about the inverted position (right).

torque. After acquaintance with the experiment on dynamic stabilization of the inverted pendulum we reasonably conclude that this phenomenon is as much instructive as the dynamic stabilization of a gyroscope, and should be necessarily included in lecture demonstrations on classical mechanics.

After Kapitza, this simple but very curious and intriguing system attracted the attention of many researchers, and the theory of the phenomenon may seem to be well elaborated (see, for example, [39]). Nevertheless, more and more new features in behavior of this inexhaustible system are reported regularly. Many related papers were published in the *American Journal of Physics* ([40]–[49]).

However, in the abundant literature on the subject (a vast bibliography can be found in [50]) it is almost impossible to find a simple and clear interpretation of this interesting phenomenon. Understanding the dynamic stabilization of an inverted pendulum is certainly a challenge to our intuition. The principal aim of this chapter is to present a quite simple qualitative physical explanation for the phenomenon. We also focus on an approximate quantitative theory (leading to the well-known concept of the effective potential for the slow motion of the pendulum) that can be developed on the basis of the suggested approach to the problem. Finally, we show that the loss of dynamic stability at large amplitudes of the pivot is closely related to the commonly known conditions of parametric instability of the non-inverted pendulum.

## 10.3 The Physical Model of the Investigated System

For simplicity we consider a light rigid rod of length $l$ with a heavy small bob of mass $m$ on its end and assume that the rod has zero mass. Let the axis of the pendulum be forced to execute a given harmonic oscillation along the vertical line with a frequency $\omega$ and an amplitude $a$, i.e., let the motion of the axis be described by the following equation:

$$z(t) = a \sin \omega t \quad \text{or} \quad z(t) = a \cos \omega t. \tag{10.1}$$

Depending on the problem under consideration, either sine or cosine time dependence may be more convenient for calculations.

The force of inertia $F_{in}(t)$ exerted on the bob in the non-inertial frame of reference also has a sinusoidal dependence on time:

$$F_{in}(t) = -m\ddot{z}(t) = ma\omega^2 \sin \omega t \quad \text{or} \quad F_{in}(t) = ma\omega^2 \cos \omega t. \tag{10.2}$$

This force of inertia is directed downward during the time intervals for which $z(t) < 0$, i.e., when the axis is below the middle point of its oscillations. We see this directly from the equation for $F_{in}(t)$, Eq. (10.2), whose right-hand side depends on time exactly as the $z$-coordinate of the axis. Therefore during the corresponding half-period of the oscillation of the pivot this additional force is equivalent to some strengthening of the force of gravity. During the other half-period the axis is above its middle position ($z(t) > 0$), and the action of this additional force is equivalent to some weakening of the gravitational force. When the frequency and/or amplitude of the pivot are large enough (when $a\omega^2 > g$), for some part of the period the apparent gravity is even directed upward.

On the basis of this approach, taking into account the periodic variations of the apparent gravity, we can easily explain, say, the physical reason for the ordinary parametric swinging of the pendulum, when its pivot is driven vertically with a frequency approximately twice the frequency of natural oscillations.

## 10.4 Parametric Resonance

Taking into account the modulation of the apparent gravitational force in the non-inertial reference frame associated with the pivot, we can easily understand the reason for the growth of oscillations in conditions of parametric resonance. The principal parametric resonance takes place if approximately two cycles of modulation occur during one period of natural oscillations. Indeed, let us consider the time interval during which the pendulum moves from the utmost deflection toward the lower equilibrium position (a quarter of the natural period). During this time interval let the pivot in its constrained oscillation be below the midpoint, $z(t) < 0$ (a half-period of the pivot oscillation). Due to the additional apparent gravity during this interval the pendulum gains a greater speed than it would have gained in the absence of the pivot's motion.

During the further motion of the pendulum away from the equilibrium position, the pivot is above its midpoint ($z(t) > 0$, Figure 10.3), so that the force of inertia reduces the apparent gravity. Thus the pendulum reaches a greater angular displacement than it would have reached otherwise. During the second half-period of the pendulum's motion the swing increases again, and so on, until the stationary motion is established due to violation of the resonance conditions at large swing.

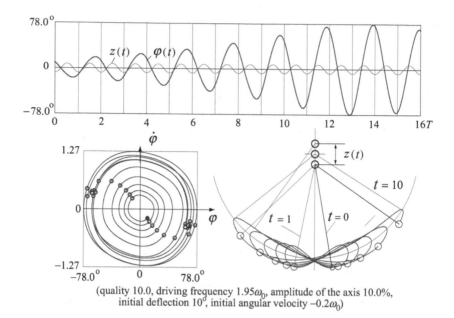

(quality 10.0, driving frequency $1.95\omega_0$, amplitude of the axis 10.0%,
initial deflection $10^\circ$, initial angular velocity $-0.2\omega_0$)

Figure 10.3: The time history $\varphi(t)$, phase trajectory (with Poincarè sections), and the spatial trajectory of the pendulum bob in conditions of parametric resonance. The period $T$ of the pivot oscillations is used as a unit of time.

The most effective growth of the amplitude occurs when the frequency of the pivot oscillation is twice the natural frequency of the pendulum, and if the pivot, moving upward, crosses its middle position at the moment at which the pendulum crosses its equilibrium position.

The growth of the energy is provided by the work done by the source that makes the pivot oscillate. The swing of the pendulum increases if the gain of energy during the period exceeds frictional losses, that is, when the amplitude of the pivot is greater than some threshold value.

In contrast to the ordinary resonance (which is excited by an external force whose frequency equals the natural frequency of the pendulum), friction cannot restrict the growth of amplitude at parametric resonance if the threshold is exceeded.

The growth of the pendulum swing in conditions of parametric resonance is restricted due to nonlinear properties of this system. Indeed, the natural period of the pendulum increases when the angular excursion becomes greater. This growth of the natural period at the fixed period of the pivot oscillations leads to violation of resonance conditions as the swing grows. The amplitude starts to diminish, and the resonance conditions restore, so that the amplitude grows again, and so on. Due to friction these transient beats gradually fade away, and periodic oscillations of a finite swing finally establish.

The lower left panel of Figure 10.3 shows the phase trajectory of the pendulum at resonance conditions. The Poincarè sections correspond to the time instants at which the oscillating pivot reaches its lowest position. During the transient process the expanding phase trajectory approaches the closed curve, that is, approaches the limit cycle that corresponds to steady-state periodic oscillations. During this transient the Poincarè sections condense to the two fixed points in the phase plane. The lower right panel of Figure 10.3 shows the corresponding trajectory of the pendulum's bob in space. The simulation is based on a numerical integration of the exact differential equation, Eq. (10.11), p. 257, for the momentary angular deflection $\varphi(t)$. This equation includes the torque of the force of gravity and the instantaneous value of the torque exerted on the pendulum by the force of inertia $F_{in}(t)$ that depends explicitly on time $t$.

Excitation of the principal parametric resonance at finite amplitudes of the pivot oscillation is possible not only when two driving cycles occur during exactly one natural oscillation of the pendulum (or, generally, during an integer number $n$ of natural periods), but rather in intervals of the pivot frequencies $\omega$ in the vicinity of $\omega = 2\omega_0$ and, generally, near $\omega = 2\omega_0/n$. The intervals of parametric instability are characterized by "tongues" in the parameter's plane. These "tongues" are discussed in more detail in Section 10.10.2, p. 281.

# 10.5 Physical Reasons for Stability of the Inverted Pendulum

In the case of oscillations of the axis with high enough frequency, the mean value of the force of inertia, averaged over the short period of these oscillations, is zero, but the value of its *torque* about the axis, averaged over the period, is not zero. Next we show why. This non-zero mean torque of the force of inertia explains the pendulum stabilization in the inverted position.

Let us begin with the case in which the rod of the pendulum is oriented horizontally, i.e., at the right angle $\psi = \pi/2$ to the direction of oscillations of the axis (Figure 10.4a).

To better understand the influence of the force of inertia upon the system, we first forget for a while about the force of gravity. If the bob has zero initial velocity, in the inertial reference frame in the absence of gravity, it stays practically at the same level while the axis $A$ oscillates between the extreme points *1* and *2* and the rod turns down and up through a (small) angle, as shown in the upper panel of Figure 10.4a. (For the sake of clarity the amplitude of the pivot and hence this angle are exaggerated in this figure.)

In the non-inertial frame of reference associated with the oscillating axis, the same motion of the rod is shown in the lower panel of (Figure 10.4a): The bob of the pendulum moves up and down along an arc of a circle and occurs in positions *1* and *2* (lower panel of Figure 10.4a) at the instants at which the oscillating axis reaches its extreme positions *1* and *2* respectively (upper panel of Figure 10.4a). In position *1* the force of inertia $F_1$ exerted on the bob is directed downward, and

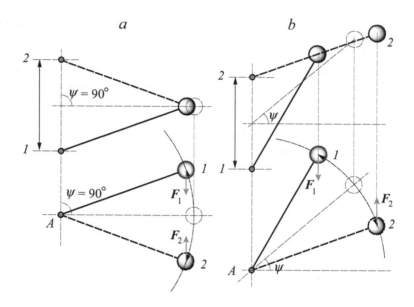

Figure 10.4: The forces of inertia exerted on the pendulum in the non-inertial reference frame at the extreme positions *1* and *2* of the oscillating axis *A*.

in the position *2* the force $F_2$ of the same magnitude is directed upward. The arm of the force in the positions *1* and *2* is the same. Therefore the torques of the force of inertia in positions *1* and *2* are equal and opposite. The same is true for all pairs of symmetric intermediate positions of the pivot.

Hence the torque of this force, averaged over the period of oscillations, is zero. In the absence of gravity this orientation of the pendulum (perpendicularly to the direction of oscillations) corresponds to a dynamic equilibrium position (an unstable one, as we shall see later).

Now let us consider the case in which on average the rod is deflected through an arbitrary angle $\psi$ from the direction of oscillations, and the axis oscillates between extreme points *1* and *2*, as shown in the upper panel of Figure 10.4*b*. In the non-inertial frame of reference associated with the oscillating axis, the bob moves at these oscillations between the points *1* and *2* in the lower panel of Figure 10.4*b* along an arc of a circle whose center coincides with the axis *A* of the pendulum. We note that the rod has the same simultaneous orientations in both reference frames at the instant *1* as well as at the instant *2*.

When the axis is displaced downward (to the position *1*) from its mid-point, the force of inertia $F_1$ exerted on the bob is also directed downward. In the other extreme position *2* the force of inertia $F_2$ has an equal magnitude and is directed upward. However, now the torque of the force of inertia in the position *2* is greater than in the position *1* because the *arm* of the force in this position is greater. Therefore on average over a period of the pivot's vibration the force of inertia creates a torque about the axis that tends to turn the pendulum upward, into the vertical inverted position, in which the rod is parallel to the direction of oscillations.

Certainly, if the pendulum makes an acute angle with respect to the downward vertical position, the mean torque of the force of inertia tends to turn the pendulum downward.

Thus, the torque of the force of inertia, averaged over a period of rapid oscillations of the pivot, tends to align the pendulum along the direction of constrained oscillations of the axis. The right-hand panel (b) of Figure 10.4 presents an utterly simple and clear explanation of the origin of this torque. Kapitza [37] called this torque *vibrational*, but we can also call it *inertial*, because its origin is related to the pseudo force of inertia that arises due to the constrained rapid vibrations of the axis. For given values of the driving frequency and amplitude, this torque depends only on the angle of the pendulum's deflection from the direction of the pivot's vibration. This mean inertial torque does not depend on the time explicitly, and its influence on the pendulum can be considered exactly in the same way as the influence of other ordinary external torques, such as the torque of the gravitational force.

This mean torque of the alternating force of inertia gives the desired explanation for the physical reason of existence of the two stable equilibrium positions that correspond to the two preferable orientations of the pendulum's rod along the direction of the pivot's vibration. The principal idea is utterly simple: Although the mean value of the force of inertia $F_{in}(t)$, averaged over the short period of these oscillations, is zero, the value of its *torque* about the axis, averaged over this period, is not zero. This is clearly seen from the right-hand panel of Figure 10.4.

With gravity, the inverted pendulum is stable with respect to small deviations from this position provided the mean torque of the force of inertia is greater in magnitude than the torque of the force of gravity that tends to tip the pendulum down.

Next we show that this stabilization occurs when the following condition is fulfilled: $a^2\omega^2 > 2gl$, or $(a/l)(\omega/\omega_0) > \sqrt{2}$ (see, e.g., [49]). However, this is only an approximate criterion for dynamic stability of the inverted pendulum, which is valid at small amplitudes of forced vibrations of the pivot ($a \ll l$). In Section 10.11 we consider a more rigorous mathematical theory of the phenomenon that allows us to establish a more precise criterion, which is also valid for low frequencies and large amplitudes of constrained oscillations of the pivot.

## 10.6  An Approximate Quantitative Theory of the Inverted Pendulum

Now we can determine the approximate quantitative conditions (valid at $a \ll l$ and $\omega \gg \omega_0$) that provide the dynamic stabilization of the inverted equilibrium position in the presence of the force of gravity. Rapid vertical oscillations of the axis make the inverted position stable, if at small deflections from this position the torque of the force of inertia, averaged over the period of the pivot oscillations (this torque tends to return the pendulum to the inverted position), is greater in magnitude than the torque of the gravitational force that tends to tip the pendulum down.

We can consider the motion of the pendulum whose axis is vibrating with a high frequency as a superposition of two components: A 'slow' component, whose variation during a period of constrained vibrations is small, and a 'fast' (or vibrational) component. Let us imagine an observer who does not notice (or does not want to notice) the vibrational component of this compound motion. If this observer uses, for example, a stroboscopic illumination with the interval between the flashes that equals the period of constrained vibrations of the pendulum's axis, he can see only the slow component of the motion. Our principal interest is to determine this slow component.

When the rod of the pendulum is deflected from the downward vertical position on the average through an angle $\psi$, the instantaneous value $\varphi(t)$ of the deflection angle is subjected to additional rapid sinusoidal oscillation with the frequency $\omega$ about this average value $\psi = \langle \varphi(t) \rangle$ because of the constrained oscillation of the axis. This can be clearly seen from the plots of the angular deflection and velocity (see Figure 10.1). Therefore we can try to search for the instantaneous angle of deflection $\varphi(t)$ as the sum of a slowly varying function $\psi(t)$ and an additional fast term $\delta(t)$ whose mean value is zero. This fast angle $\delta(t)$ oscillates with the high frequency $\omega$ of the pivot vibration. The amplitude of this oscillation is proportional to the sine of the momentary value of the average slow angle $\psi(t)$:

$$\varphi(t) = \psi(t) + \delta(t) = \psi(t) - \frac{z(t)}{l}\sin\psi = \psi(t) - \frac{a}{l}\sin\psi\sin\omega t. \qquad (10.3)$$

Here $a$ is the amplitude of forced vibrations of the axis, $l$ is the length of the pendulum. (When the axis is above its middle position, $z$ is positive and the additional angle $\delta = -(z/l)\sin\psi$ is negative.) Later on we shall find the differential equation for this unknown slow varying function $\psi(t)$ that describes the motion of the pendulum, averaged over the period of rapid oscillations.

The torque of the force of inertia depends on its momentary value $ma\omega^2\sin\omega t$, Eq. (6.3), and on the sine of the angle $\varphi$. The oscillations of the axis cause only small deviations of the momentary deflection angle $\varphi$ from its average value $\psi$ (i.e., $\delta(t) \ll 1$ for all $t$), and so for the sine of the deflection angle we can write the following approximate expression:

$$\sin\varphi = \sin(\psi + \delta) \approx \sin\psi + \delta\cos\psi. \qquad (10.4)$$

With the help of this equation, we can find the approximate value of the gravitational torque about the point of suspension (about the axis of the pendulum), averaged over the period of rapid oscillations of the axis:

$$\langle -mgl\sin\varphi \rangle = -mgl\langle\sin(\psi + \delta)\rangle = -mgl\sin\psi, \qquad (10.5)$$

because the average value of $\delta(t)$ is zero: $\langle\delta(t)\rangle = 0$. We see that the mean torque of the gravitational force is the same as in the case of a pendulum with the immovable suspension point: The oscillating second term in the expansion for the momentary angle, Eq. (10.4), being multiplied to a constant gravitational force, gives no contribution to the mean torque. However, when we take the time average for the torque of the oscillating force of inertia, the first term in the expansion, Eq. (10.4), vanishes, but the oscillating second term gives a nonzero contribution. This occurs by virtue of the sinusoidal dependence on time both of $\delta(t)$ and the force of inertia $F_{in}(t)$ in Eq. (6.3):

$$\langle F_{in}(t)l\sin(\psi + \delta)\rangle = -ma\omega^2 l\, \frac{a}{l}\cos\psi\sin\psi\langle\sin^2\omega t\rangle =$$

$$-\frac{1}{2}ma^2\omega^2\cos\psi\sin\psi. \quad (10.6)$$

The nonzero contribution of this term to the torque of the force of inertia arises because the average value of the sine squared is equal to $1/2$: $\langle\sin^2\omega t\rangle = 1/2$. An explanation of this nonzero average torque on the physical level is given by the right-hand panel of Figure 10.4. For $\psi > \pi/2$ the average value of the torque of the force of inertia is positive: If the pendulum makes an acute angle with the upward vertical direction, this torque tends to turn the pendulum up.

Comparing the right-hand sides of Eqs. (10.5) and (10.6), we see that the torque of the force of inertia can exceed in magnitude the torque of the gravitational force tending to tip the pendulum down, when the following condition is fulfilled:

$$a^2\omega^2 > 2gl. \quad (10.7)$$

Thus, the inverted position of the pendulum is stable if the maximal velocity $wa$ of the vibrating axis is greater than the velocity $\sqrt{2gl}$ attained by a body during a free fall from the height that equals the pendulum length $l$. We can write this criterion of stability in another form, using the expression $\omega_0^2 = g/l$ for the frequency $\omega_0$ of small natural oscillations of the pendulum in the absence of forced vibrations of the axis. Substituting $g = l\omega_0^2$ in Eq. (10.7) we get:

$$\frac{a}{l} \cdot \frac{\omega}{\omega_0} > \sqrt{2}. \quad (10.8)$$

According to this approximate criterion of stability, Eq. (10.8), the product of the dimensionless fractional amplitude of forced oscillations of the axis $a/l$ and the dimensionless fractional frequency of these oscillations $\omega/\omega_0$ must exceed the square root of 2. For instance, for the pendulum whose length $l = 20$ cm and the frequency of forced oscillations of the axis $f = \omega/2\pi = 100$ Hz, the amplitude $a$ must be greater than 3.2 mm. To provide the dynamic stabilization of the inverted pendulum within some finite interval of the angles of deflection from the vertical position, the product of the dimensionless amplitude of forced oscillations of the axis and the dimensionless frequency must be greater than $\sqrt{2}$ by a finite value.

We emphasize that expressions (10.7) and (10.8) give only an approximate criterion of inverted pendulum stability, which is valid for fast enough vertical oscillations of the pivot, whose frequency is much greater than the frequency of natural oscillations of this pendulum in the gravitational field. Later on we will find an improved and extended criterion of dynamic stability that also holds for low frequencies of the pivot.

For a physical pendulum, the condition of dynamic stability in the inverted position is expressed by the same Eq. (10.7) or (10.8) provided we imply by the quantity $l$ the reduced length of the pendulum $I/md$, where $I$ is the moment of inertia with respect to the axis of rotation, $m$ is the mass, and $d$ is the distance between the axis and the center of mass. We note that the criterion (10.7) or (10.8) is independent of friction.

The critical minimum value of the product of the driving amplitude and frequency $a\omega$ found above, Eq. (10.8), agrees with the lower boundary of stability of the inverted pendulum obtained by approximating the exact nonlinear equation of motion by the Mathieu equation, the solutions of which are widely documented in the extensive literature concerning the problem (see, for example, [40], [41]). However, the investigation based on the Mathieu equation and infinite Hill's determinants gives little physical insight into the problem and, more importantly, is restricted to the motion within small angles from the vertical. Conversely, the above explanation shows clearly the physical reason for the dynamic stabilization of the inverted pendulum and is free from the restriction of small angles.

In particular, on the basis of the above developed approach, for given values of the frequency $\omega$ and the amplitude $a$ of forced oscillations of the axis, we can find the maximal admissible mean angle of deflection from the inverted vertical position $\theta_{max} = \pi - \psi_0$ for which the pendulum will return to this position. To do this, we should equate the right-hand sides of Eqs. (10.5) and (10.6) that determine the average values of the torque of the gravitational force, which tends to tip the pendulum down, and of the torque of the force of inertia, which tends to return the pendulum to the inverted position:

$$\cos\theta_{max} = -\cos\psi_0 = \frac{2gl}{a^2\omega^2} = 2\left(\frac{\omega_0}{\omega}\frac{l}{a}\right)^2. \tag{10.9}$$

The greater the product $\omega a$ of the frequency $\omega$ and the amplitude $a$ of constrained vibrations of the axis, the closer the angle $\theta_{max}$ to $\pi/2$. Being deflected from the vertical position by an angle smaller than $\theta_{max}$, the pendulum will execute relatively slow oscillations about this inverted position. This slow motion occurs under the mean torque of both the force of inertia and the force of gravity. Rapid oscillations with the frequency of forced vibrations of the axis superimpose on this slow motion of the pendulum. With friction, the slow motion gradually damps, and the pendulum wobbles up, settling eventually in the inverted position.

Similar behavior of the pendulum can be observed when it is deflected from the lower vertical position. But in this case the frequency of slow oscillations is greater than in the case of the inverted pendulum. Actually, this frequency is greater than the frequency of natural oscillations in the absence of forced vibrations of the axis, because in this case the averaged torque of the force of inertia tends to return the pendulum to the lower vertical position and is added to the torque of the gravitational force. This means that an ordinary clock with a pendulum will be always ahead of time if it is subjected to a fast vertical vibration.

When we take into account the mean torque of the force of inertia, the frequencies $\omega_{\text{up}}$ and $\omega_{\text{down}}$ of small slow oscillations about the inverted position and the lower vertical position are given by the following expressions:

$$\omega_{\text{up}}^2 = \frac{a^2\omega^2}{2l^2} - \omega_0^2, \qquad \omega_{\text{down}}^2 = \frac{a^2\omega^2}{2l^2} + \omega_0^2. \tag{10.10}$$

Substituting $\omega_0 = 0$ into these formulas, we get the expression for the frequency of small slow oscillations of the pendulum with vibrating axis in the absence of the gravitational force. These oscillations can occur about either of the two symmetrical stable equilibrium positions located opposite one another along the direction of forced oscillations of the axis. For vertical oscillations of the axis in the field of gravity, the force of gravity increases the average restoring torque of the force of inertia (and consequently the frequency of slow oscillations) about the lower equilibrium position, and the force of gravity decreases the average restoring torque (and the frequency of slow oscillations) about the upper equilibrium position.

It is worth mentioning that the results regarding the behavior of the pendulum with a vertically vibrating axis are obtained here without the differential equation for the system under consideration. (This equation is discussed in the next section). These results are valid if the amplitude of this constrained vibration of the axis is small compared with the pendulum length, and its frequency is much greater than the frequency of small natural oscillations of the pendulum.

As we mentioned above, at certain intervals of the system parameters (in the intervals of parametric instability) the lower position of the pendulum becomes unstable. However, parametric resonance, as well as the modes of chaotic behavior, occur at such frequencies of constrained oscillations of the pivot that do not satisfy the condition $\omega \gg \omega_0$ of applicability of the approach used in this chapter. (For the principal parametric resonance, $\omega \approx 2\omega_0$.) Therefore the existence of parametric resonance does not disprove our conclusion about the stability of the lower equilibrium position.

## 10.7 Exact Differential Equation for the Pendulum with Oscillating Pivot

The exact differential equation for the pendulum with oscillating pivot includes, besides the torque of the force of gravity, the instantaneous (not averaged over the fast period) value of the torque exerted on the pendulum by the force of inertia that depends explicitly on time $t$:

$$\ddot{\varphi} + 2\gamma\dot{\varphi} + (\omega_0^2 - \frac{a}{l}\omega^2 \sin\omega t)\sin\varphi = 0. \tag{10.11}$$

The second term of Eq. (10.11) takes into account the braking frictional torque, assumed to be proportional to the momentary angular velocity $\dot{\varphi}$ in the mathematical model of the simulated system. The damping constant $\gamma$ is inversely proportional to the quality factor $Q$ commonly used to characterize the viscous friction: $Q = \omega_0/2\gamma$. In the absence of gravity the parametrically driven pendulum is described by Eq. (10.11) with $\omega_0 = 0$. Since in this case the notion of natural frequency loses its sense, it is impossible to use the quality factor defined as $Q = \omega_0/2\gamma$ to characterize friction, but instead we can use another dimensionless quantity $\omega/2\gamma$, where $\omega$ is the driving frequency.

We note that oscillations about the inverted position can be formally described by the same differential equation, Eq. (10.11), with negative values of $\omega_0^2 = g/l$. This is clearly seen if by $\varphi$ in Eq. (10.11) we mean the deviation of the pendulum from the upward vertical. In other words, we can consider $\omega_0^2$ as a control parameter whose variation is physically equivalent to changing the gravitational force exerted on the pendulum. When this control parameter is diminished through zero to negative values, the constant (gravitational) torque in Eq. (10.11) first turns to zero and then changes its sign to the opposite. Such a "gravity" tends to bring the pendulum into the inverted position $\varphi = \pi$, destabilizing the position $\varphi = 0$ of the unforced pendulum: The inverted position with $\omega_0^2 < 0$ in Eq. (10.11) is equivalent to the hanging-down position with the positive value of $\omega_0^2$ of the same magnitude.

Experimental verification of approximate expressions (10.10) for frequencies $\omega_{\text{up}}$ and $\omega_{\text{down}}$ of small slow oscillations about the inverted position and the lower vertical position is given by the graphs in Figure 10.5, obtained in the simulation. The simulation is based on a numerical integration of the exact differential equation, Eq. (10.11), for the momentary angular deflection $\varphi(t)$. To make the verification easier, the pivot frequency $\omega$ was chosen to be $16\,\omega_0$, so that $(a^2/2l^2)\omega^2 = 3.0\,\omega_0^2$. Thus Eq. (10.10) provides the value $\omega_{\text{down}} = 2\omega_0$ for the frequency of slow oscillations about the downward position, which is exactly twice the natural frequency. Then the period of slow oscillations $T_{\text{down}}$ must equal one half of the period $T_0$ of natural oscillations in the absence of pivot vibrations ($T_{\text{down}} = T_0/2$). Figure 10.5 shows that the pendulum executes exactly two cycles of slow oscillations during one period $T_0$, which in this case (at $\omega = 16\,\omega_0$) equals 16 periods $T = 2\pi/\omega$ of pivot vibrations. (The units $T$ are used for the time scale.) For the frequency of slow oscillations about the upward vertical position $\omega_{\text{up}} = \sqrt{2}\omega_0$, so that their period should equal ($T_{\text{up}} = T_0/\sqrt{2}$). This value of the period is also in good agreement with the lower graph in Figure 10.5.

## 10.8   Effective Potential Function for a Pendulum with the Pivot Vibrating at High Frequency

The graphs in Figure 10.5 show clearly that the smooth motion is distorted by the high frequency oscillations most of all near the utmost deflections of the pendulum, and these distortions are relatively small while the pendulum crosses the

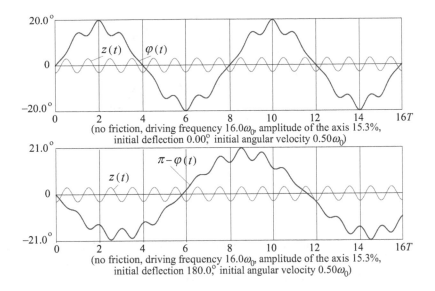

Figure 10.5: The graphs of the momentary angular deflection $\varphi(t)$ for oscillations of the rigid planar pendulum with vibrating axis about the dynamically stabilized lower and upper equilibrium positions respectively, obtained by a numerical integration of the exact differential equation, Eq. (10.11). The sinusoidal graphs of the axis motion $z(t) = -a\cos\omega t$ are also shown.

equilibrium position. This proves that the momentary deflection angle $\varphi(t)$ can be represented approximately as a superposition of the slow varying mean angle $\psi(t)$ and the high frequency term whose angular amplitude is proportional to sine of $\psi(t)$ (see Eq. (10.3)). Indeed, the angular amplitude of the rapid (second) term in Eq. (10.3) is the greatest at the extreme deflections of the pendulum, and this amplitude vanishes when the pendulum in its smooth motion crosses each of the vertical positions.

An observer that doesn't notice the rapid oscillating motion of the pendulum can consider simply that the system moves in an effective potential field $U = U(\psi)$. Such a potential function that governs the smooth motion of the pendulum averaged over the rapid oscillations was first introduced by Landau [4], and derived by several different methods afterwards (see, for example, [47] – [49]). Certainly, some subtle details in the motion of the pendulum revealed by the simulations are lost in the approximate analysis, which refers only to the slow component of the investigated motion. Nevertheless, this analysis allows us to clearly interpret the principal features of the physical system under consideration, and even to evaluate such typically nonlinear properties as the dependence of the period on the amplitude of slow oscillations.

The approximate differential equation for the slow motion of the pendulum can be written under the assumption that the angular acceleration $\ddot\psi(t)$ in this

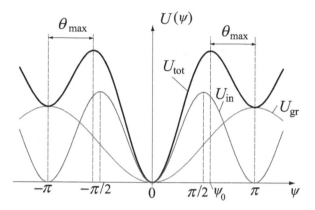

Figure 10.6: Graphs of the gravitational potential energy $U_{gr}$, mean potential energy $U_{in}$ in the field of the force of inertia, and of the total potential energy $U_{tot}(\psi)$ for the pendulum with vertically oscillating pivot.

slow motion is determined by the mean torque $N(\psi)$ exerted on the pendulum in the non-inertial frame of reference associated with its axis:

$$\ddot{\psi} = -\omega_0^2 \sin\psi - \frac{1}{2}\frac{a^2}{l^2}\omega^2 \cos\psi \sin\psi. \tag{10.12}$$

The mean torque in the right-hand side of Eq. (10.12) is calculated approximately under the assumption that the slowly varying angular coordinate $\psi(t)$ is 'frozen.' To facilitate interpretation of the slow motion described by this nonlinear differential equation, we can introduce a potential function $U(\psi)$ that corresponds to the mean torque $N(\psi)$ exerted on the pendulum. The torque is determined by the derivative of this potential function: $N(\psi) = -dU(\psi)/d\psi$. The above-mentioned observer who doesn't notice the rapid oscillating motion of the pendulum can consider simply that the system moves in an effective potential field $U = U(\psi)$. From the right-hand panel of Eq. (10.12) we conclude that the effective potential consists of two parts $U_{gr}(\psi)$ and $U_{in}(\psi)$ that describe the influence of the force of gravity and the force of inertia, respectively:

$$U(\psi) = U_{gr}(\psi) + U_{in}(\psi) = mgl(1 - \cos\psi) + \frac{1}{4}ma^2\omega^2(1 - \cos 2\psi). \tag{10.13}$$

The graphs of $U_{gr}(\psi)$ and $U_{in}(\psi)$ are shown in Figure 10.6. They both have a sinusoidal shape, but the period of $U_{in}(\psi)$ is just one half of the period of $U_{gr}(\psi)$. Their minima at $\psi = 0$ coincide, thus generating the principal minimum of the total potential function $U(\psi) = U_{tot}(\psi)$. This minimum corresponds to the stable lower equilibrium position of the pendulum. But the next minimum of $U_{in}(\psi)$ is located at $\psi = \pi$, where $U_{gr}(\psi)$ has its maximum corresponding to the inverted position of the pendulum.

If the criterion (10.7) or (10.8) is fulfilled, the amplitude of $U_{in}(\psi)$ is greater than that of $U_{gr}(\psi)$. Then the potential function $U(\psi)$ has (in addition to the absolute minimum at $\psi = 0$, which corresponds to the lower equilibrium position) relative minima at $\psi = \pm\pi$. Both additional minima correspond to the same inverted position of the pendulum. Oscillations of a particle trapped in an additional minimum describe the behavior of the inverted pendulum. Slow small oscillations occurring near the bottom of a potential well are almost harmonic.

Frequencies of these oscillations can be found from the differential equation (10.12) for the slow motion, assuming $\sin\psi \approx \psi$, $\cos\psi \approx 1$ in the vicinity of $\psi = 0$ and $\sin\psi = \sin(\pi - \theta) \approx \theta$, $\cos\psi \approx -1$ near $\psi = \pm\pi$:

$$\ddot{\psi} = -(\omega_0^2 + \frac{1}{2}\frac{a^2}{l^2}\omega^2)\psi, \qquad \ddot{\theta} = -(-\omega_0^2 + \frac{1}{2}\frac{a^2}{l^2}\omega^2)\theta. \qquad (10.14)$$

It follows from (10.14) that frequencies $\omega_{down}$ and $\omega_{up}$ of small slow oscillations about the lower ($\psi = 0$) and upper ($\psi = \pm\pi$) equilibrium positions are given by the same expressions: Eqs. (10.10), p. 257, obtained earlier on the basis of a simple physical approach.

The slopes of the shallow additional potential wells are not as steep as the slopes of the principal well at $\psi = 0$. Therefore the frequency $\omega_{up}$ of slow small oscillations about the inverted position is smaller than the frequency $\omega_{down}$ of small oscillations within the principal well (about the lower vertical position), in accordance with the above expressions (10.10) and with the simulations represented by graphs in Figure 10.5. Certainly, some subtle details in the motion of the pendulum revealed by the simulations are lost in our approximate analysis, which refers only to the slow component of the investigated motion. Nevertheless, this analysis allows us to clearly interpret principal features of the physical system under consideration.

The maxima of the total potential energy $U(\psi)$ are determined by Eq. (10.9). The tops of the potential barrier between the two wells occur at deflections $\pm\psi_0$ ($\psi_0 > \pi/2$) from the lower vertical position and $\pm\theta_{max}$ ($\theta_{max} < \pi/2$) from the upper equilibrium position (Figure 10.6). At these positions of the pendulum the mean torque of gravity is balanced by the mean torque of the force of inertia. However, these equilibrium positions are unstable: The slightest disturbance makes the pendulum to slowly slip down into one of the wells and oscillate there, moving from one slope to the other and back. The pattern of such slow oscillations (averaged over the fast period of constrained vibrations) is far from a sine curve. The pendulum stays for a prolonged time near the summit of the potential barrier at the utmost deflection, and then moves rather fast towards the other utmost deflection to linger there again before the backward fast motion. The simulation of such a motion is shown in Figure 10.7.

The results discussed above are obtained by a decomposition of motion on slow oscillations and rapid vibrations with the driving frequency. Hence these results are approximate and valid when the amplitude of constrained vibration of the axis is small compared to the pendulums length ($a \ll l$) and their frequency is high enough ($\omega \gg \omega_0$). It follows from the graph $U = U(\psi)$ (see Figure 10.6)

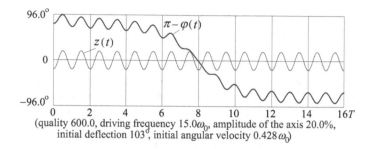

(quality 600.0, driving frequency $15.0\omega_0$, amplitude of the axis 20.0%,
initial deflection $103°$, initial angular velocity $0.428\,\omega_0$)

Figure 10.7: The graphs of oscillations of the pendulum about the inverted position
with maximal possible angular excursion.

that the lower equilibrium position is always stable, and the upper one is stable
if additional minima exist on the curve of potential energy $U = U(\psi)$. These
minima appear when condition (10.7) or (10.8) is fulfilled.

As we already mentioned earlier, for some intervals of the pivot frequency
(intervals of parametric instability) the lower equilibrium position becomes unstable — the phenomenon of parametric resonance occurs at which small initial
oscillations increase progressively. This conclusion does not follow from the investigation based on a decomposition of motion on slow and rapid components.
This is by no means surprising because parametric resonance occurs at such driving frequencies (for the principal parametric resonance $\omega \approx 2\omega_0$) for which this
decomposition is not applicable. In the next chapter we will show that the inverted
(dynamically stabilized) position can also become unstable: At large enough amplitude of the pivot oscillations the pendulum is involved in oscillations about the
inverted position with the period covering two cycles of excitation. This so-called
"flutter" mode of oscillations is closely related to ordinary parametric parametric
resonance of the hanging-down pendulum. We emphasize that parametric resonance, "flutter" mode, and other complicated regimes occur at such frequencies
and amplitudes of the pivot, for which the decomposition of motion on the slow
and rapid components is not applicable.

A more accurate and enhanced criterion of dynamic stabilization of the inverted pendulum, valid in a wider region of system parameters, will be described
in the following chapter.

# 10.9  Subharmonic Resonances of High Orders

In this section we investigate in detail recently discovered kinds of motion of the
pendulum with oscillating pivot — namely the so called *subharmonic resonances*.
The boundaries of the region in the parameter space are determined in which these
resonances can exist, and their relationship with the dynamic stabilization of the
inverted pendulum is discussed. An enhanced and more exact criterion of the in-

verted pendulum stability is determined, which is valid in particular for relatively low frequencies and large amplitudes of the excitation.

The natural slow oscillatory motion in the effective potential well is almost periodic (exactly periodic in the absence of friction). When the driving amplitude and frequency lie within certain ranges, the pendulum, instead of gradually approaching the equilibrium position (either dynamically stabilized inverted position or ordinary downward position) by the process of damped slow oscillations, can be trapped in an $n$-periodic limit cycle locked in phase to the rapid forced vibration of the axis. In such oscillations the phase trajectory repeats itself after $n$ driving periods $T$. Since the motion has period $nT$, and the frequency of its fundamental harmonic equals $\omega/n$ (where $\omega$ is the driving frequency), this phenomenon can be called a *subharmonic resonance* of $n$-th order.

For the inverted pendulum with a vibrating pivot, periodic oscillations of this type were first described by Acheson [51], who called them "multiple-nodding" oscillations. Computer simulations show that the pendulum motion in this regime reminds us of some kind of original dance. We note that these modes of regular periodic oscillations are not specific for the inverted pendulum with a vibrating pivot. Similar 'dancing' oscillations can also be executed (at appropriate values of the driving parameters) about the ordinary (downward-hanging) equilibrium position.

Actually, the origin of these modes is independent of gravity, because such "multiple-nodding" oscillations, synchronized with the pivot, can also occur in the absence of gravity about any of the two equivalent dynamically stabilized equilibrium positions of the pendulum with a vibrating axis (see [52]). Synchronization of these modes with the pivot oscillations creates conditions for supplying the energy to the pendulum needed to compensate for dissipation, and the whole process becomes exactly periodic.

## 10.9.1 Multiple-Nodding Oscillations of the Parametrically Driven Pendulum

An example of multiple-nodding stationary oscillations whose period equals eight periods of the axis is shown in Figure 10.8. The left-hand upper panel of the figure shows the spatial trajectory of the pendulum's bob at these multiple-nodding oscillations of the inverted pendulum.

The left-hand lower panel shows the closed looping trajectory in the phase plane $(\varphi, \dot{\varphi})$. The right-hand panel of Figure 10.8, alongside the graphs of $\varphi(t)$ and $\dot{\varphi}(t)$, also shows their harmonic components and the sinusoidal graphs of the pivot oscillations.

The spectrum of these period-8 oscillations is rich in harmonics. The fundamental harmonic with frequency $\omega/8$ whose period equals eight driving periods dominates the spectrum. We may treat this low-frequency component of the spectrum as a subharmonic (as an 'undertone') of the driving oscillation. This harmonic describes the above-discussed smooth component $\psi(t)$ of the compound period-8 oscillation.

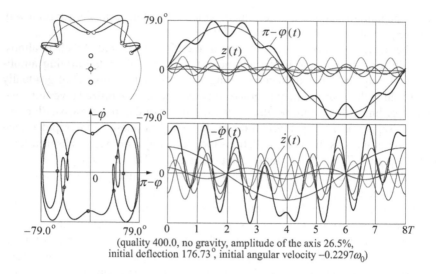

(quality 400.0, no gravity, amplitude of the axis 26.5%, initial deflection 176.73°, initial angular velocity −0.2297$\omega_0$)

Figure 10.8: The spatial path, phase orbit with Poincaré sections, and graphs of large-amplitude stationary period-8 oscillations. The graphs are obtained by a numerical integration of the exact differential equation, Eq. (10.11), with $\omega_0 = 0$, for the momentary angular deflection $\varphi(t)$. Thin lines show separate harmonics. The fundamental harmonic with the frequency $\omega/8$ dominates the spectrum. The seventh and ninth harmonics have nearly equal amplitudes. Graphs of the axis motion $-z(t)$ and $-\dot{z}(t)$ are also shown.

We note that at large swing the third harmonic (frequency $3\omega/8$) is noticeable (see Figure 10.8). This spectral component reflects the non-harmonic character of slow oscillations in the non-parabolic well of the effective potential.

The seventh and ninth harmonics with nearly equal amplitudes give considerable contribution into the spectrum of these period-8 oscillations. Strange as it may seem from the first sight, the 8th harmonic with the driving frequency has zero amplitude, that is, this harmonic is absent in the spectrum. However, this peculiarity also can be easily explained on the basis of the approach developed in this paper.

Indeed, in Eq. (10.3), p. 254, which represents the momentary angular position of the pendulum $\varphi(t)$ as a superposition of slow and fast motions, the rapid component with the driving frequency enters the expression for $\varphi(t)$ being multiplied by the sine of the slow varying coordinate $\psi(t)$. Therefore the rapid component has a varying amplitude, which even changes its sign each time the pendulum crosses the equilibrium position.

Actually, the rapidly oscillating second term in Eq. (10.3) is not a harmonic component in the spectrum of the resulting periodic oscillation, because harmonics of a periodic function are characterized by constant amplitudes.

Next we show that the approximate approach based on the effective potential for the slow motion provides a simple qualitative physical explanation for such an extraordinary (and even counterintuitive at first sight) behavior of the pendulum. Moreover, for subharmonic resonances with $n \gg 1$ this approach yields rather good quantitative results.

The approximate theory developed above allows us to predict conditions at which these $n$-periodic oscillations can occur. For small amplitudes of the slow oscillations, the corresponding minimum of the effective potential can be approximated by a parabolic well in which the smooth motion is almost harmonic.

The natural slow oscillatory motion in the effective potential well is almost periodic (exactly periodic in the absence of friction). A subharmonic resonance of order $n$ can occur if one cycle of this slow motion covers approximately $n$ driving periods, that is, when the driving frequency $\omega$ is close to an integer multiple $n$ of the natural frequency of slow oscillations near either the inverted or the ordinary equilibrium position: $\omega = n\omega_{up}$ or $\omega = n\omega_{down}$. In this case the phase locking can occur, in which one cycle of the slow motion is completed *exactly* during $n$ driving periods. Synchronization of these modes with the oscillations of the pivot creates conditions for systematically supplying the pendulum with the energy needed to compensate for dissipation, and the whole process becomes exactly periodic.

To estimate the frequency of the slow motion (the fundamental frequency), we can use Eq. (10.10), p. 257. As an example, next we consider the pendulum in the absence of gravity, or, which is essentially the same, in the limiting case of very high driving frequencies $\omega \gg \omega_0$ ($\omega/\omega_0 \to \infty$). In this limit both equilibrium positions (ordinary and inverted) are equivalent, and the dimensionless driving amplitude $a/l$ is the only parameter to be predicted as a required condition for the subharmonic resonance of order $n$ (of $n$-periodic oscillations of the pendulum, synchronized with the pivot).

According to Eq. (10.10), for $\omega_0 = 0$ the frequency of slow oscillations is given by $\omega_{slow} = a/(l\sqrt{2})\omega$. For the "quadruple-nodding" mode the slow motion period equals eight periods of the axis, so that $\omega_{slow} = \omega/8$, whence $a/l = \sqrt{2}/8 = 0.177$. This value agrees rather well with the predictions of a more sophisticated quantitative theory of these modes based on the linearized differential equation of the system (see Section 10.9.2 below, Eq. (10.22) with $n = 8$), which gives for such period-8 small oscillations in the absence of gravity the following expression for the driving amplitude: $a_{min} = 63/(32\sqrt{130})\, l = 0.173\, l$. The latter value agrees perfectly with the simulation experiment in the limit of extremely small angular excursions.

Estimating conditions for $n$-periodic oscillations with the help of Eq. (10.10), we assume the slow motion of the pendulum in the effective potential well to be simple harmonic, which is true only if this motion is limited to a small vicinity of the bottom of this well. Therefore we get the lower limit for the driving amplitude at which $n$-periodic oscillations of only infinitely small amplitude can occur. Smooth non-harmonic oscillations of a finite angular excursion that extends over the slanting slopes of the non-parabolic effective potential well are characterized by a greater period than the small-amplitude harmonic oscillations occurring

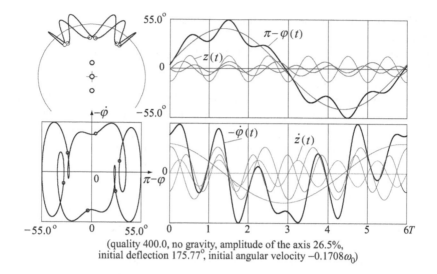

(quality 400.0, no gravity, amplitude of the axis 26.5%,
initial deflection 175.77°, initial angular velocity −0.1708$\omega_0$)

Figure 10.9: The spatial path, phase orbit with Poincaré sections, and graphs of stationary period-6 oscillations. The graphs are obtained by a numerical integration of the exact differential equation, Eq. (10.11) with $\omega_0 = 0$, for the momentary angular deflection $\varphi(t)$. Thin lines show separate harmonics. The fundamental harmonic with the frequency $\omega/6$ dominates the spectrum. The fifth and seventh harmonics have nearly equal amplitudes. Graphs of the axis motion $-z(t)$ and $-\dot{z}(t)$ are also shown.

within the parabolic bottom of this well. Therefore large-amplitude period-8 oscillations shown in Figure 10.8 (their swing equals $\sim 80°$) occur at a considerably greater value of the driving amplitude ($a = 0.265\ l$).

For small angular excursions of the pendulum occurring at driving amplitudes slightly greater than $a_{\min} = 0.173\ l$, the spectrum of period-8 oscillations is formed by the principal harmonic and also by the seventh and ninth harmonics whose frequencies are close to the driving frequency. Their amplitudes equal, respectively, 11.2% and 6.7% of the principal harmonic amplitude. These values observed in the corresponding simulation experiment agree perfectly well with the theoretical values; see Section 10.9.2 below, Eq. (10.24) on p. 271. For the oscillations of a large swing shown in Figure 10.8, the amplitudes of these harmonics slightly differ from the above values, and the contributions of the third, fifth, and eleventh harmonics are also noticeable.

As noted above, in the case of period-8 oscillations of a small swing, the approach based on the effective potential predicts for the driving amplitude $a/l$ a value of $\sqrt{2}/8 = 0.177$, which is rather close to the exact low-amplitude theoretical limit ($a/l = 0.173$). To obtain the slow oscillations of a smaller period (say, of six driving periods; see Figure 10.9), we should increase the driving amplitude. Indeed, when $\omega_{slow} = \omega/6$, Eq. (10.10) yields a greater value $a/l = \sqrt{2}/6 = 0.236$.

However, for such period-6 oscillations this predicted value agrees somewhat worse with the theory based on the linearized differential equation of the system. This theory (see Section 10.9.2 below, Eq. (10.22) with $n = 6$) gives for small period-6 oscillations in the absence of gravity a value of the minimal driving amplitude of $a_{min} = 35/(18\sqrt{74})\, l = 0.226\, l$, which perfectly agrees with the corresponding simulation experiment. Not surprisingly, for the $n$-periodic oscillation with a small $n$ we cannot expect good quantitative predictions from the effective potential approach because in such cases the period of a "smooth" motion contains only a few driving periods. The "rapid" component of the motion here is not rapid enough for good averaging.

With gravity, these complex $n$-periodic multiple-nodding modes exist both for the inverted and non-inverted pendulum. Assuming $\omega_{down,\, up} = \omega/n$ ($n$ driving cycles during one cycle of the slow oscillation), we find for the minimal normalized driving amplitudes (for the boundaries of the subharmonic resonances in the presence of gravity) the values

$$m_{min} = \sqrt{2(1/n^2 \mp k)}, \qquad (10.15)$$

where we have introduced a notation $k = (\omega_0/\omega)^2$. This dimensionless parameter $k$ (inverse normalized drive frequency squared), being physically less meaningful than $\omega/\omega_0$, is nevertheless more convenient for further investigation, because the improved criterion acquires a simpler form in terms of $k$. As we already indicated above (see Section 10.7, p. 258), negative values of this parameter $k$ correspond to negative $g$ values (negative $\omega_0^2$ values) in the exact differential equation, Eq. (10.11), and can be treated as referring to the inverted pendulum. Then the boundaries of subharmonic resonances can be expressed both for the hanging-down and inverted pendulums by the same formula: $m_{min} = \sqrt{2(1/n^2 - k)}$. The limit of this expression at $n \to \infty$ gives the above-mentioned approximate condition of stability of the inverted pendulum, Eq. (10.8), p. 255, now expressed in terms of parameter $k$:

$$m_{min} = \sqrt{-2k}. \qquad (10.16)$$

(In this expression $k < 0$, because it is applicable to the inverted pendulum.) Indeed, the lower limit of stability of the inverted pendulum, Eq. (10.8), p. 255, can be regarded as the condition of subharmonic resonance of an infinite order in the inverted position — on the edge of stability the frequency of slow oscillation about the inverted position tends to zero.

## 10.9.2 Spectrum of Small-Amplitude $n$-Periodic Oscillations

As we have seen from the above described simulations, the spectrum of stationary $n$-periodic oscillations of a small angular excursion consists primarily of the fundamental harmonic $A \sin(\omega t/n)$ with the frequency $\omega/n$, and two high harmonics of the orders $n - 1$ and $n + 1$. This spectrum composition is consistent

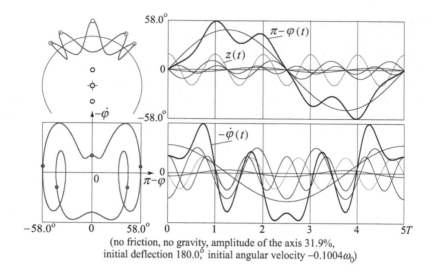

(no friction, no gravity, amplitude of the axis 31.9%,
initial deflection 180.0°, initial angular velocity −0.1004$\omega_0$)

Figure 10.10: The spatial path, phase orbit with Poincaré sections, and time-dependent graphs of the subharmonic resonance of the fifth order. The graphs are obtained by a numerical integration of the exact differential equation, Eq. (10.11), for the momentary angular deflection of the pendulum $\varphi(t)$. Separate harmonics are shown by thin lines. The fundamental harmonic (frequency $\omega/5$) dominates the spectrum. Next the fourth and sixth harmonics (frequencies $4\omega/5$ and $6\omega/5$) contribute to a considerable extent. At large swing the second harmonic (frequency $2\omega/5$) is also noticeable.

with the above-considered representation of the pendulum motion as a superposition of slow and rapid motions given by Eq. (10.3), p. 254. Indeed, according to this equation with $\sin \psi \approx \psi$, in this approximation

$$\varphi(t) = \psi(t) - m \sin \psi \cos \omega t \approx \psi(t) - m\psi \cos \omega t =$$
$$= A \sin(\frac{\omega}{n} t) - mA \sin(\frac{\omega}{n} t) \cos \omega t = \tag{10.17}$$
$$= A \sin(\frac{\omega}{n} t) + \frac{mA}{2} \sin(\frac{n-1}{n} \omega t) - \frac{mA}{2} \sin(\frac{n+1}{n} \omega t).$$

This kind of spectral composition is clearly seen from the plots in Figures 10.8 and 10.9, as well as from Figure 10.10 for the resonance of the fifth order. While the pendulum crosses the equilibrium position, both high harmonics add in the opposite phases and thus almost don't distort the smooth motion (described by the principal harmonic). Near the utmost deflections the phases of high harmonics coincide, and thus here their sum causes the most serious distortions of the smooth motion.

According to Eq. (10.17), both high harmonics must have equal amplitudes $(m/2)A$. However, we see from the above-mentioned plots that these amplitudes are slightly different. Therefore we can try to improve the approximate solution for $\varphi(t)$, Eq. (10.17), as well as the theoretical values for the lower boundaries of subharmonic resonances, Eq. (10.15), by assuming for the possible solution a similar spectrum but with unequal amplitudes, $A_{n-1}$ and $A_{n+1}$, of the two high harmonics (for $n > 2$, the case of $n = 2$ will be considered separately):

$$\varphi(t) = A_1 \sin(\omega t/n) + A_{n-1} \sin[(n-1)\omega t/n] + A_{n+1} \sin[(n+1)\omega t/n]. \quad (10.18)$$

Since oscillations at the boundaries have infinitely small amplitudes, we can use instead of Eq. (10.11) the following linearized (Mathieu) equation:

$$\ddot{\varphi} + 2\gamma\dot{\varphi} + (\omega_0^2 - m\omega^2 \sin \omega t)\varphi = 0. \quad (10.19)$$

Substituting $\varphi(t)$, Eq. (10.18), into this equation (with $\gamma = 0$) and expanding the products of trigonometric functions, we obtain a system of approximate equations for the coefficients $A_1$, $A_{n-1}$ and $A_{n+1}$:

$$\begin{aligned}
2(kn^2 - 1)A_1 + mn^2 A_{n-1} - mn^2 A_{n+1} &= 0, \\
mn^2 A_1 + 2[n^2(k-1) + 2n - 1]A_{n-1} &= 0, \quad (10.20) \\
-mn^2 A_1 + 2[n^2(k-1) - 2n - 1]A_{n+1} &= 0.
\end{aligned}$$

The homogeneous system has a nontrivial solution if its determinant equals zero. This condition yields an equation for the corresponding critical (minimal) driving amplitude $m_{\min}$ at which $n$-period mode $\varphi(t)$, Eq. (10.18), can exist. Solving the equation, we find:

$$m_{\min}^2 = \frac{2}{n^4} \frac{[n^6 k(k-1)^2 - n^4(3k^2 + 1) + n^2(3k+2) - 1]}{[n^2(1-k) + 1]}. \quad (10.21)$$

Then, for this critical driving amplitude $m_{\min}$, the fractional amplitudes $A_{n-1}/A_1$ and $A_{n+1}/A_1$ of high harmonics for a given order $n$ of the subharmonic resonance can easily be found as the solutions to the homogeneous system of equations, Eqs. (10.20).

### 10.9.3 Lower Boundary of the Dynamic Stabilization

As we already mentioned, it is possible to identify the lower boundary of the dynamic stabilization of the inverted pendulum with the condition of subharmonic resonance of an infinite order. Therefore the limit of $m_{\min}$, Eq. (10.21), at $n \to \infty$ gives an improved formula for the lower boundary of the dynamic stabilization of the inverted pendulum instead of the commonly known approximate criterion $m_{\min} = \sqrt{-2k}$, which is valid for $k \ll 1$ (see Eq. (10.16), p. 267):

$$m_{\min} = a_{\min}/l = \sqrt{-2k(1-k)} \quad (k < 0). \quad (10.22)$$

The minimal amplitude $m_{\min}$ that provides the dynamic stabilization is shown as a function of parameter $k = (\omega_0/\omega)^2$ (inverse normalized driving frequency

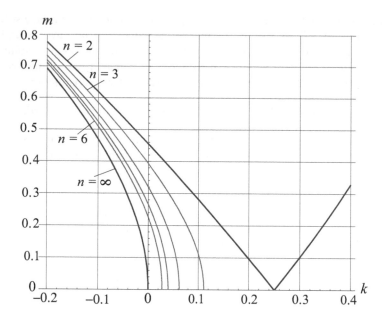

Figure 10.11: The normalized driving amplitude $m = a/l$ versus $k = (\omega_0/\omega)^2$ (inverse normalized driving frequency squared) at the boundaries of the dynamic stabilization of the inverted pendulum (the left curve marked as $n \to \infty$), and at subharmonic resonances of several orders $n$ (see text for details).

squared) by the left curve ($n \to \infty$) in Figure 10.11. The other curves to the right from this boundary show the dependence on $k$ of minimal driving amplitudes for which the subharmonic resonances of several orders can exist (the first curve for $n = 6$ and the others for $n$ values diminishing down to $n = 2$ from left to right). At positive values of $k$ these curves correspond to the subharmonic resonances of the parametrically excited hanging-down pendulum.

Subharmonic oscillations of a given order $n$ (for $n > 2$, case $n = 2$ will be considered separately) are possible to the left of $k = 1/n^2$, that is, for the driving frequency $\omega > n\omega_0$.

The curves in Figure 10.11 show that when the driving frequency $\omega$ is increased beyond the value $n\omega_0$ (i.e., as parameter $k$ is decreased from the critical value $1/n^2$ toward zero), the threshold driving amplitude (over which $n$-order subharmonic oscillations are possible) rapidly increases. The limit of very high driving frequency ($\omega/\omega_0 \to \infty$), in which the gravitational force is insignificant compared with the force of inertia (or, which is essentially the same, the limit of zero gravity $\omega_0/\omega \to 0$), corresponds to $k = 0$, that is, to the points of intersection of the curves in Figure 10.11 with the $m$-axis.

The continuations of these curves further to negative $k$ values describe the transition through zero gravity to the "gravity" directed upward, which is equivalent to the case of an inverted pendulum in the ordinary (directed downward) gravitational field. Therefore these curves at negative $k$ values give the threshold driving amplitudes for subharmonic resonances of the inverted pendulum.[1]

The curve $n = 2$ in Figure 10.11 corresponds to the upper boundary of dynamic stabilization for the inverted pendulum: As we show below, when this boundary is exceeded, the inverted pendulum destabilizes and is trapped into the limit cycle of stationary oscillations with period $2T$ (so-called "flutter mode"). Curves that lie between the lower ($n \to \infty$) and upper ($n = 2$) boundaries of dynamic stabilization correspond to subharmonic oscillations of different orders. These oscillations do not disprove the criterion of the inverted pendulum's stability because the pendulum is trapped in an $n$-periodic limit cycle only if the initial state belongs to the basin of attraction that corresponds to this limit cycle. Otherwise the pendulum eventually comes to rest in the inverted position.

A complete investigation of the parametrically excited pendulum is complicated by the extensive set of parameters that characterize the system ($\omega_0$, $\omega$, $a$, $\gamma$). A considerable simplification is achieved by eliminating one of the parameters, namely, the natural frequency $\omega_0 = \sqrt{g/l}$, when we turn to studying the pendulum in the absence of gravity. This simplified model is also useful for qualitative understanding of the pendulum's behavior in the presence of gravity in cases of high driving frequency and/or large driving amplitude, when the gravitational force plays the role of a small addition to the force of inertia. Many of the above-mentioned complicated counterintuitive modes are not related to the force of gravity, and can be studied in their purest form when they are observed in the simple device with the oscillating pivot in the absence of gravity, which is described by Eq. (10.11) with $\omega_0 = 0$.

The points of intersection of the curves in Figure 10.11 with the $m$-axis, corresponding to the threshold conditions at zero gravity ($k = 0$), give, according to Eq. (10.21), the following values of the normalized driving amplitudes:

$$m_{\min} = \frac{\sqrt{2}(n^2 - 1)}{n^2\sqrt{n^2 + 1}}. \tag{10.23}$$

The fractional amplitudes $A_{n-1}/A_1$ and $A_{n+1}/A_1$ of the most important high harmonics of $\varphi(t)$ [expressed approximately by Eq. (10.18)] for the case of zero gravity ($k = 0$) are given by the following formulas:

$$\frac{A_{n-1}}{A_1} = \frac{n+1}{\sqrt{2}\sqrt{n^2 + 1}(n - 1)}, \quad \frac{A_{n+1}}{A_1} = \frac{n-1}{\sqrt{2}\sqrt{n^2 + 1}(n + 1)}. \tag{10.24}$$

For $n = 8$ (quadruple-nodding oscillations), Eq. (10.21) yields $m_{\min} = a_{\min}/l = 63/(32\sqrt{130}) = 0.173$. This critical value of the driving amplitude was already mentioned in Section 10.9.1, p. 263, and it agrees exactly with the simulation experiment for period-8 small oscillations. The above Eqs. (10.24) also

---

[1]Actually the curves in Figure 10.11 are plotted not according to Eq. (10.21), but rather with the help of a somewhat more complicated formula (not cited here), which is obtained by holding one more high order harmonic component in the trial function $\varphi(t)$.

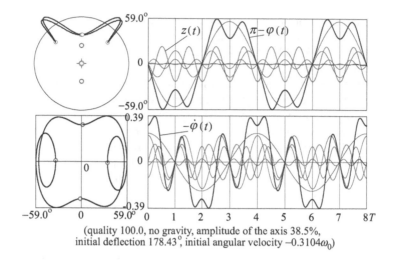

(quality 100.0, no gravity, amplitude of the axis 38.5%,
initial deflection 178.43°, initial angular velocity $-0.3104\omega_0$)

Figure 10.12: The spatial path, phase orbit, and graphs of period-4 oscillations. This example shows 'double-nodding' oscillations about one of the dynamically stabilized equilibrium positions in the absence of gravity.

yield the fractional contributions of the seventh and ninth harmonics: $A_7/A_1 = 9/(7\sqrt{130}) = 0.113$, $A_9/A_1 = 7/(9\sqrt{130}) = 0.068$ — the values that also agree perfectly well with the simulations based on numerical integration of the exact differential equation.

For $n = 6$ Eq. (10.23) yields $m_{min} = a_{min}/l = 35/(18\sqrt{74}) = 0.226$, and for fractional contributions of the fifth and seventh harmonics Eq. (10.24) gives, respectively, $A_5/A_1 = 7/(5\sqrt{74}) = 0.163$, and $A_7/A_1 = 5/(7\sqrt{74}) = 0.083$. These theoretical values agree quite well with the simulations (see Section 10.9.1).

For $n = 4$ (double-nodding oscillations) Eq. (10.23) and Eq. (10.24) yield $m_{min} = a_{min}/l = 15/(8\sqrt{34}) = 0.321$, $A_3/A_1 = 5/(3\sqrt{34}) = 0.286$, $A_5/A_1 = 3/(5\sqrt{34}) = 0.103$. If in the approximate solution we also take into account the seventh harmonic, for zero gravity and zero friction we find more accurate values of the critical driving amplitude: $m_{min} = a_{min}/l = 0.320$, and of the fractional contributions of high harmonics: $A_3/A_1 = 0.288$, $A_5/A_1 = 0.102$, $A_7/A_1 = 0.015$. We can compare these values with results of the simulation experiment: $a_{min}/l = 0.320$, $A_3/A_1 = 0.287$, $A_5/A_1 = 0.101$, $A_7/A_1 = 0.016$.

Critical values for the driving amplitudes that provide small steady-state parametric oscillations with odd $n$ values ($n = 3, 5, \ldots$) calculated on the basis of a linearized theory also show good agreement with the simulations described in Section 10.9.1 (p. 263). Thus, the boundaries $m_{min}$ and amplitudes of high harmonics $A_{n-1}/A_1$ and $A_{n+1}/A_1$ for subharmonic resonances of different orders $n$ calculated above with the help of an approximate theory agree perfectly well with the simulation experiments based on numerical integration of the exact differen-

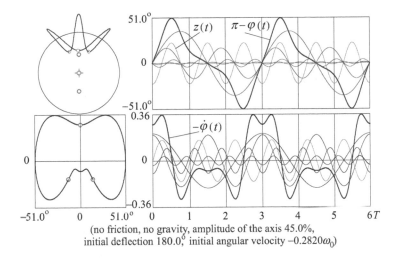

(no friction, no gravity, amplitude of the axis 45.0%,
initial deflection 180.0°, initial angular velocity $-0.2820\omega_0$)

Figure 10.13: The spatial path, phase orbit, and graphs of period-3 steady-state oscillations without gravity in the absence of friction.

tial equation, Eq. (10.11). Figures 10.8, 10.9, and 10.10 show the graphs of these subharmonic oscillations for $n = 8$, $n = 6$, and $n = 5$, respectively. Figures 10.12 and 10.13 show the graphs for $n = 4$ and $n = 3$, obtained in the simulations.

Friction introduces a phase shift between forced oscillations of the pivot and harmonics of the steady-state $n$-periodic motion of the pendulum. By virtue of this phase shift the pendulum is supplied with energy needed to compensate for frictional losses. With friction, the direct and backward spatial paths of the pendulum do not coincide, and the symmetry of the phase trajectory with respect to the ordinate axis is destroyed. This is clearly seen from a comparison of Figures 10.8, 10.9 and 10.12 for subharmonic resonances in the presence of weak friction with Figures 10.10 and 10.13, which refer to an idealized case in which friction is absent.

## 10.9.4 Subharmonic Resonances of Fractional Orders

In this section we discuss several new exotic modes of regular behavior of the parametrically driven pendulum, kindred to the above-described subharmonic resonances. These modes were discovered recently in the simulation experiments.

Figure 10.14 shows a regular period-8 motion of the pendulum, which can be characterized as a subharmonic resonance of a fractional order, specifically, of the order 8/3 in this example. Here the amplitude of the fundamental harmonic (whose frequency equals $\omega/8$) is much smaller than the amplitude of the third harmonic (frequency $3\omega/8$). This third harmonic dominates the spectrum, and can be regarded as the principal one, while the fundamental harmonic can be regarded

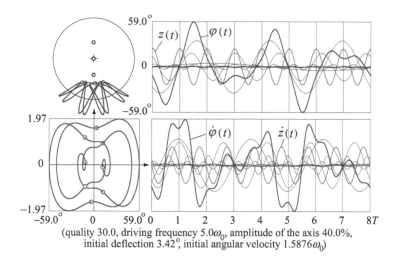

(quality 30.0, driving frequency $5.0\omega_0$, amplitude of the axis 40.0%,
initial deflection 3.42°, initial angular velocity $1.5876\omega_0$)

Figure 10.14: The spatial path, phase orbit, and graphs of stationary oscillations that can be treated as a subharmonic resonance of a fractional order 8/3. Graphs of separate harmonics are shown by thin lines. The third harmonic (frequency $3\omega/8$) dominates the spectrum.

as its third subharmonic (that is, as an 'undertone'). Considerable contributions to the spectrum are given also by the fifth and eleventh harmonics of the fundamental frequency.

Approximate boundary conditions for small-amplitude stationary oscillations of this type ($n/3$-order subresonance) can be found analytically from the linearized differential equation by a method similar to that used above for $n$-order subresonance: We can try as $\varphi(t)$ a solution consisting of spectral components with frequencies $3\omega/n$, $(n-3)\omega/n$, and $(n+3)\omega/n$:

$$\varphi(t) = A_3 \sin(3\omega t/n) + A_{n-3} \sin[(n-3)\omega t/n] + A_{n+3} \sin[(n+3)\omega t/n].$$
$$(10.25)$$

Substituting this trial function $\varphi(t)$ into Eq. (10.19) (with $\gamma = 0$) and expanding the products of trigonometric functions, we obtain a system of equations for the coefficients $A_3$, $A_{n-3}$ and $A_{n+3}$. The condition of existence of a non-trivial solution to the system yields the following expression for the minimal driving amplitude at which the corresponding subresonance of $n/3$-order can occur:

$$m_{\min} = \frac{3\sqrt{2}(n^2 - 3^2)}{n^2\sqrt{n^2 + 3^2}}.$$
$$(10.26)$$

(Compare Eq. (10.26) with a similar expression, Eq. (10.23), for the critical driving amplitude of the integer-order subharmonic resonances.)

For the critical amplitude $a_{\min}$ of the (8/3) fractional order subharmonic resonance, Eq. (10.26) yields (for the case of the absence of gravity) the value

$a_{\min}/l = 165/(32\sqrt{146}) = 0.427$. (In the presence of gravity this resonance is illustrated by Figure 10.14.) For the fractional amplitudes $A_{n-3}$ and $A_{n+3}$ of harmonics in the trial function, Eq. (10.25) with $n = 8$, this approximate linearized theory gives the values $A_5/A_3 = 33/(5\sqrt{146}) = 0.546$, $A_{11}/A_3 = 15/(11\sqrt{146}) = 0.113$. More precise values (which agree well with the simulations) are obtained by also including the thirteenth harmonic in the trial function, Eq. (10.25): $a_{\min}/l = 0.419$, $A_5/A_3 = 0.560$, $A_{11}/A_3 = 0.111$, $A_{13}/A_3 = 0.044$.

The analytical results of calculations for $n \geq 8$ agree well with the simulations, especially if one more high harmonic is included in the trial function $\varphi(t)$. If the driving amplitude exceeds the critical value, the angular excursion of the pendulum at these modes increases, and additional harmonics appear in the spectra of such oscillations.

Similar (though more complicated) calculations of the critical driving amplitudes and spectrum on the basis of a linearized differential equation are also possible for various modes of the parametrically driven pendulum in the presence of gravity and friction.

## 10.9.5 Coexistence of Subharmonic Resonances of Different Orders $n$

As we have shown above, for subharmonic resonances of high orders in the case of zero gravity ($n \gg 1$), Eq. (10.23) yields the approximate value $m_{\min} \approx \sqrt{2}/n$ obtained with the help of the simple approach which treats the condition of $n$-order subharmonic resonance as the coincidence of $n$ driving periods with one period of the slow motion of the pendulum near the bottom of the effective potential well. The fractional amplitudes of both high harmonics $A_{n-1}/A_1$ and $A_{n+1}/A_1$, given by Eq. (10.24), are almost equal at $n \gg 1$ and approach the common value $1/(\sqrt{2}n) = m_{\min}/2$, in accordance with Eq. (10.17), which describes the $n$-period subharmonic oscillations as a superposition of the slow and rapid motions.

The approach based on the effective potential provides us not only with a qualitative understanding of these complex periodic modes, but also, being applicable for large-amplitude motions, explains the coexistence of several $n$-periodic modes with different $n$ values at identical values of the system parameters.

For the oscillations of a large swing shown in Figure 10.8, the contribution of the third harmonic to the spectrum is also noticeable. In our approximate approach, the appearance of this spectral component is explained by deviations in the shape of the effective potential well (in which the slow oscillation is executed) from a parabolic well, that is, by the non-harmonic character of the slow oscillation with a large angular excursion.

For a large angular excursion, the smooth motion of the pendulum occurs in the non-parabolic effective potential well with a 'soft' restoring force, in which the period becomes longer if we increase the amplitude. By virtue of this dependence of the periods of non-harmonic smooth motions on the swing, different modes (modes with different values of $n$) can coexist at the same amplitude of the pivot.

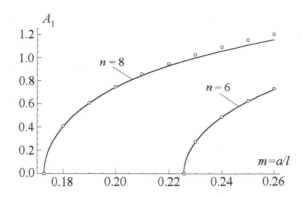

Figure 10.15: The principal harmonic amplitudes for $n = 8$ and $n = 6$ modes versus the driving amplitude $m = a/l$ given by an approximate theory (see text for details) and by the simulation experiment.

Indeed, the period of a large-amplitude slow oscillation can be equal to, say, six driving periods, while the period of oscillation with a somewhat greater amplitude in the same non-parabolic potential well can be equal to eight driving periods.

Figures 10.8 and 10.9 show the simulations of such coexisting period-8 and period-6 modes, respectively, obtained at the identical parameters of the system (zero gravity, $a/l = 0.265$). That is, both smooth motions occur in the same potential well. In which of these competing modes the pendulum eventually becomes trapped in a certain simulation depends on the starting conditions. The set of initial conditions that leads, after an interval in which transients decay, to a given dynamic equilibrium (to the same steady-state periodic motion, or attractor) in the limit of large time, constitutes the basin of attraction of this attractor. The coexisting periodic motions in Figures 10.8 and 10.9 represent competing attractors and are characterized by different domains of attraction.

Figure 10.15 shows the dependence on the driving amplitude $m = a/l$ of the fundamental harmonic amplitudes $A_1$ for both $n = 8$ and $n = 6$ modes.

To estimate how the swing of oscillations executed at the subharmonic resonance of a given order $n$ depends on the excess $(a - a_{\min})$ of the driving amplitude $a$ over the critical (threshold) value $a_{\min}$, and how the fractional amplitude of the third harmonic depends on the swing, we can expand $\sin \psi$ and $\sin 2\psi$ in the differential equation that describes the slow motion, Eq. (10.12), p. 260, in a power series, preserving the two first terms:

$$\ddot{\psi} + \omega_0^2 (\psi - \frac{1}{6}\psi^3) + \frac{1}{2}m^2\omega^2(\psi - \frac{2}{3}\psi^3) = 0. \qquad (10.27)$$

Here we again use the notation $m = a/l$ for the normalized driving amplitude. We can try to search for the solution of Eq. (10.27) in the form of a superposition of the fundamental and third harmonics:

$$\psi = A_1 \sin \omega_1 t + A_3 \sin 3\omega_1 t. \qquad (10.28)$$

Substituting $\psi$ from Eq. (10.28) into Eq. (10.27), and equating to zero the coefficient of $\sin \omega_1 t$, we find how the frequencies of slow oscillations depend on the amplitude $A_1$:

$$\omega_{\text{down, up}}^2 = \frac{1}{2} m^2 \omega^2 (1 - \frac{1}{2} A_1^2) \pm \omega_0^2 (1 - \frac{1}{8} A_1^2). \tag{10.29}$$

This expression reduces to Eq. (10.10) if $A_1 \to 0$. Equating the frequencies $\omega_{\text{down, up}}$ to the fundamental harmonic frequency $\omega/n$, we obtain an approximate dependence of the fundamental harmonic amplitude $A_1$ on the excess of the normalized driving amplitude over its critical value $(m - m_{\text{min}})$. For the case $\omega_0 = 0$ (absence of gravity) we find:

$$A_1 = \sqrt{2} \sqrt{1 - m_{\text{min}}^2/m^2} \approx 2\sqrt{1 - m_{\text{min}}/m}. \tag{10.30}$$

The latter approximate expression is valid if the driving amplitude only moderately exceeds the critical value (if $(m - m_{\text{min}}) \ll m_{\text{min}}$). For $n = 8$ and $n = 6$ the dependencies of $A_1$ on $m$ are plotted by solid curves in Figure 10.15 together with experimental values of $A_1$ obtained by numerical simulations. If the driving amplitude $m$ is greater than $m_{\text{min}} = 0.226$ for $n = 6$, each of the subharmonic oscillations with $n = 8$ and $n = 6$ can exist at the same values of the driving amplitude.

The amplitude $A_3$ of the third harmonic in Eq. (10.28) can be estimated similarly by equating to zero the coefficient of $\cos 3\omega_1 t$, when $\psi$ from Eq. (10.28) is substituted into Eq. (10.27). It is convenient to express $A_3$ as a function of the amplitude $A_1$ of the slow motion:

$$A_3 = \frac{A_1^3}{3(16 - 7A_1^2)}. \tag{10.31}$$

The corresponding graph is shown by a solid line in Figure 10.16. The points refer to the simulation of the subharmonic oscillations with $n = 8$.

## 10.10 The Upper Boundary of Dynamic Stabilization and the Principal Parametric Resonance

When the amplitude $a$ of the pivot vibrations is increased beyond a certain critical value $a_{\text{max}}$, the dynamically stabilized inverted position of the pendulum loses its stability. After a disturbance the pendulum does not come to rest in the up position, no matter how small the release angle, but instead eventually settles into a limit cycle, executing finite amplitude steady-state oscillation (about the inverted vertical position). The period of such oscillation is twice the driving period, and its swing grows as the excess of the drive amplitude over the threshold $a_{\text{max}}$ is increased.

This loss of stability of the inverted pendulum was first described in 1992 by Blackburn et al. [44] and demonstrated experimentally in [45]. The authors [44]

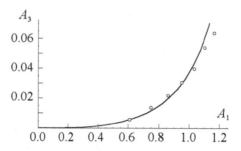

Figure 10.16: The third harmonic amplitude for $n = 8$ mode versus the amplitude of the principal harmonic given by the approximate theory (see text) and by the simulation experiment.

called these limit-cycle oscillations the "flutter" mode. The graphs and the double-lobed phase trajectory of such oscillations are shown in Figure 10.17.

## 10.10.1   The "Flutter" Mode and Ordinary Parametric Resonance

Obviously, the steady-state "flutter" oscillations can be regarded as a special case of subharmonic resonances, specifically, the case with $n = 2$. As we already mentioned, for small values of $n$ it is impossible to correctly represent the pendulum motion as consisting of the slow and rapid components. The driving amplitude $a_{max}$ is not small compared with the length $l$ of the pendulum. Consequently, this case occurs beyond the limits of applicability of the approach based on the effective potential. This approach cannot explain the destabilization of the inverted pendulum, as well as the loss of stability of the hanging-down pendulum at conditions of ordinary parametric resonance. (In the latter case the driving amplitude can be small, but the necessary driving frequency is not high enough for the separation of rapid and slow motions.)

However, the simulation shows (see Figure 10.17) a very simple spectral composition of period-2 steady oscillations occurring over the upper boundary of dynamic stability. Namely, the spectrum consists of the fundamental harmonic whose frequency equals $\omega/2$ (half the driving frequency $\omega$) with a small addition of the third harmonic with the frequency $3\omega/2$. We note that large-amplitude oscillations of the non-inverted pendulum in conditions of the principal parametric resonance are characterized by a similar spectrum (see Figure 10.18). This similarity of the spectra is by no means occasional: Both the ordinary parametric resonance and the period-2 "flutter" mode that destroys the dynamic stability of the inverted state belong essentially to the same branch of possible steady-state period-2 oscillations of the parametrically excited pendulum. We can treat this branch as a subharmonic resonance of order $n = 2$.

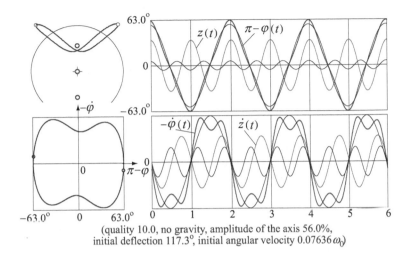

(quality 10.0, no gravity, amplitude of the axis 56.0%,
initial deflection 117.3°, initial angular velocity $0.07636\,\omega_0$)

Figure 10.17: Stationary period-2 oscillations occurring over the upper boundary of dynamic stability (the "flutter" mode). The spectrum consists of the fundamental harmonic (frequency $\omega/2$) and the third harmonic (frequency $3\omega/2$).

Therefore the upper boundary of dynamic stability for the inverted pendulum can be found directly from the linearized differential equation of the system, Eq. (10.11), by the same method that is commonly used for determination of conditions that lead to the loss of stability of the non-inverted pendulum through excitation of ordinary parametric resonance (the ranges of parametric instability; see, for example, [4]). We can apply the linearized Eq. (10.19) to this problem because at the boundary of dynamic stability the amplitude of oscillations is infinitely small. The periodic solution to Eq. (10.19), which corresponds to the boundary of instability, can be represented as a superposition of the fundamental harmonic whose frequency $\omega/2$ equals half the driving frequency, and the third harmonic with the frequency $3\omega/2$:

$$\varphi(t) = A_1 \sin(\omega t/2) + A_3 \sin(3\omega t/2). \qquad (10.32)$$

The phases of harmonics in (10.32) correspond to pivot oscillations in the form $z(t) = a \cos \omega t$. Substituting $\varphi(t)$ from Eq. (10.32) into the linearized differential equation, Eq. (10.19), with $\gamma = 0$, and expanding the products of trigonometric functions, we obtain an expression in which we should equate to zero the coefficients of $\sin(\omega t/2)$ and $\sin(3\omega t/2)$. Thus we get a system of homogeneous equations for the coefficients $A_1$ and $A_3$, which has a nontrivial solution when its determinant equals zero. This requirement yields a quadratic equation for the desired normalized critical driving amplitude $a_{max}/l = m_{max}$. The relevant root of this equation (in the case $\omega_0 = 0$, which corresponds to the absence of gravity or to the high frequency limit of the pivot oscillations with gravity) is

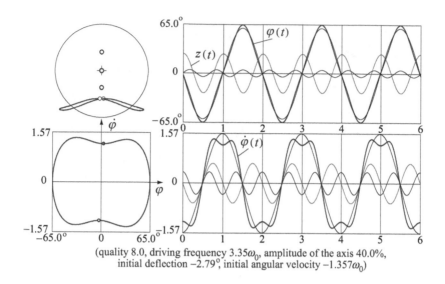

(quality 8.0, driving frequency $3.35\omega_0$, amplitude of the axis 40.0%,
initial deflection $-2.79°$, initial angular velocity $-1.357\omega_0$)

Figure 10.18: The spatial path, phase orbit, and graphs of stationary oscillations with period $2T$ (ordinary parametric resonance). The spectrum consists of the fundamental harmonic (frequency $\omega/2$) and its third harmonic (frequency $3\omega/2$), whose graphs are shown by thin lines.

$m_{max} = 3(\sqrt{13} - 3)/4 = 0.454$. The corresponding ratio of amplitudes of the third harmonic to the fundamental one equals $A_3/A_1 = (\sqrt{13} - 3)/6 = 0.101$.

A somewhat more complicated calculation in which the higher harmonics in $\varphi(t)$ (up to the seventh) are taken into account yields for $m_{max}$ and $A_3/A_1$ the values that coincide (within the assumed accuracy) with those cited above. These values agree well with the simulation experiment in conditions of the absence of gravity ($\omega_0 = 0$) and very small angular excursion of the pendulum. When the normalized amplitude of the pivot $m = a/l$ exceeds the critical value $m_{max} = 0.454$, the swing of the period-2 "flutter" oscillation (amplitude $A_1$ of the fundamental harmonic) increases in proportion to the square root of this excess: $A_1 \sim \sqrt{a - a_{max}}$. This dependence follows from the nonlinear differential equation of the pendulum, Eq. (10.11), if $\sin \varphi$ in it is approximated as $\varphi - \varphi^3/6$, and also agrees well with the simulation experiment for amplitudes up to $45°$ (see Figure 10.19).

As the amplitude $a$ of the pivot is increased over the value $0.555\,l$, bifurcation of the symmetry-breaking occurs: The angular excursions of the pendulum to one side and to the other become different, destroying the spatial symmetry of the oscillation and hence the symmetry of the phase orbit.

As the pivot amplitude is increased further, after $a = 0.565\,l$ the system undergoes a sequence of period-doubling bifurcations, and finally, at $a = 0.56622\,l$ (for $Q = \omega/2\gamma = 20$), the oscillatory motion of the pendulum becomes replaced,

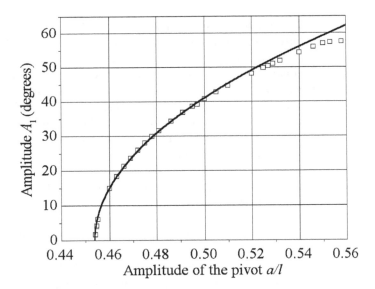

Figure 10.19: Amplitude $A_1$ of the fundamental harmonic in the "flutter" mode over the upper boundary $a_{max}$ of dynamic stabilization. The solid curve corresponds to the theoretical dependence $A_1 \sim \sqrt{a - a_{max}}$.

at the end of a very long chaotic transient, by a regular unidirectional period-1 rotation.

## 10.10.2 Boundaries of the Second-Order Subharmonic Resonance

Similar (though more complicated) theoretical investigation of the boundary conditions for period-2 stationary oscillations in the presence of gravity allows us to obtain the dependence of the critical (destabilizing) amplitude $m$ of the pivot on the driving frequency $\omega$. In terms of parameter $k = g/(l\omega^2)$ this dependence has the following form:

$$m_{max} = \frac{1}{4}(\sqrt{117 - 232k + 80k^2} - 9 + 4k), \qquad k = \frac{g}{l\omega^2}. \qquad (10.33)$$

The graph of this boundary as function of $k$ is shown in Figure 10.11 (p. 270) by the curve marked as $n = 2$. The critical driving amplitude tends to zero at $k \to 1/4$ (at $\omega \to 2\omega_0$). This condition corresponds to ordinary parametric resonance of the hanging-down pendulum: At small driving amplitudes this resonance is excited if the driving frequency equals the doubled natural frequency. If the driving frequency exceeds $2\omega_0$ (that is, if $k < 0.25$), a finite driving amplitude is required for infinitely small steady parametric oscillations even in the absence of friction.

The curve $n = 2$ intersects the ordinate axis at $m = 3(\sqrt{13} - 3)/4 = 0.454$. This case ($k = 0$) corresponds to the above-mentioned limit of a very high driving frequency ($\omega/\omega_0 \to \infty$) or zero gravity ($\omega_0 = 0$), so that $m = 0.454$ gives the upper limit of stability for each of the two dynamically stabilized equivalent equilibrium positions: If $m > 0.454$ at $g = 0$, the "flutter" mode is excited. The continuation of this curve to the region of negative values of $k = g/(l\omega^2)$ corresponds to the transition from ordinary downward gravity through zero to "negative," or upward "gravity," or, equivalently, to the case of the inverted pendulum in an ordinary (directed down) gravitational field.

Thus, the same formula, Eq. (10.33), gives the driving amplitude (as a function of the driving frequency) at which both the equilibrium position of the hanging-down pendulum is destabilized due to excitation of ordinary parametric oscillations, and the dynamically stabilized inverted equilibrium position is destabilized due to excitation of period-2 "flutter" oscillations. We can treat this as an indication that both phenomena belong to the same branch of the pendulum's regular behavior: They are closely related and have a common physical nature. All the curves that correspond to subharmonic resonances of higher orders ($n > 2$) lie between this curve and the lower boundary of dynamic stabilization of the inverted pendulum (curve $n \to \infty$ in Figure 10.11).

Actually, Eq. (10.33) in the vicinity of $k = 1/4$ ($\omega = 2\omega_0$) gives both boundaries of the instability interval that surrounds the principal parametric resonance, which occurs at $k = 1/4$, that is, at $\omega = 2\omega_0$. For $k > 1/4$ ($\omega < 2\omega_0$) Eq. (10.33) yields negative $m$ whose absolute value $|m|$ corresponds to stationary oscillations at the other boundary (to the right of $k = 0.25$, see Figure 10.11). Such oscillations are also represented by two harmonic components with frequencies $\omega/2$ and $3\omega/2$, but their phases differ from those in Eq. (10.32) — these harmonics are of cosine type (for $m > 0$). For negative $m$, which physically means simply the opposite phase of the pivot oscillations, harmonics of oscillations at the second boundary will be of a sine type, just as in (10.32).

In Figure 10.20 the boundaries of the principal interval of parametric instability form the "tongue" shown by curves 1 and 2 as functions of normalized driving frequency $\omega/\omega_0$ (instead of the more convenient but physically less meaningful quantity $k = (\omega_0/\omega)^2$ used in Figure 10.11).

For the hanging-down pendulum, in the absence of friction the critical amplitude tends to zero as the frequency of the pivot approaches $2\omega_0$ from either side. Curve 3 shows in the parameters plane ($\omega/\omega_0$, $a/l$) the tongue-shaped region of principal parametric resonance in the presence of friction (for $Q = \omega_0/2\gamma = 5.0$). The non-inverted vertical position of the pendulum whose pivot is vibrating at frequency $2\omega_0$ loses stability when the normalized amplitude of this vibration exceeds the threshold value of $1/2Q$. This curve almost merges with curves 1 and 2 as the frequency $\omega$ deviates from the resonant value $2\omega_0$. (For a detailed discussion of the role of friction see Section 10.10.3.) In the high-frequency limit, for which the role of gravity is negligible, the normalized critical pivot amplitude $a/l$ tends to the value 0.454 that corresponds to destabilization of the two symmetric equilibrium positions in the absence of gravity.

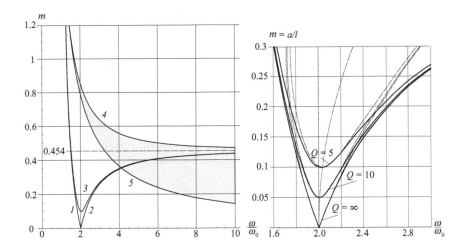

Figure 10.20: The boundaries of parametric instability (driving amplitude versus normalized driving frequency). Curves *1* and *2* correspond to frequency boundaries of the principal interval of parametric instability ($\omega \approx 2\omega_0$) for the non-inverted pendulum in the absence of friction, curve *3* — the same with friction ($Q = 5.0$). Curves *4* and *5* correspond to the upper and lower boundaries of dynamic stability for the inverted pendulum. Curve *3* is also shown in an increased scale (right).

Curve *4* of this diagram corresponds to destabilization of the inverted pendulum by excitation of the "flutter" oscillations. The smaller the frequency of the pivot, the greater the critical amplitude at which the inverted position becomes unstable. Actually curve *4* for the boundary of the "flutter" mode is the continuation (through infinite values of the driving frequency) of curve *2* (or curve *3* in the presence of friction). The latter is the boundary of ordinary parametric resonance of the non-inverted pendulum. This relationship between the two phenomena becomes especially obvious if we compare curve *4* with its equivalent in Figure 10.11, which is the curve marked as $n = 2$ at negative $k$ values.

Curve *5* in Figure 10.20 shows in the parameter plane the lower boundary of dynamic stabilization of the inverted pendulum, which is defined by criterion (10.22). Thus, the region of stability of the inverted pendulum occupies the shaded panel of the parameter plane between curves *5* and *4*.

### 10.10.3 The Influence of Friction

For small (and moderate) driving amplitudes, the principal parametric resonance occurs at a driving frequency (frequency of the pivot) whose value is approximately twice the natural frequency of the pendulum: $\omega \approx 2\omega_0$. We can calculate the threshold of parametric excitation of the hanging-down pendulum by vertical

oscillations of the pivot in the condition of the principal resonance by equating the
work done by the force of inertia during a cycle of a steady-state motion of the
pendulum to the energy dissipated due to friction.

For the calculation of the threshold, it is convenient to consider that the pivot's
motion is described by a sine function $z(t) = -a \sin \omega t$. This specific phase of
the pivot oscillation can be provided by an appropriate choice of the time origin.
In this case the small steady-state oscillations at the threshold are approximately
described by a cosine function: $\varphi(t) = C_1 \cos(\omega t/2)$.

The torque of the force of inertia is $F_{in} l \sin \varphi$, and the elementary work $dW$
done by this torque during an infinitesimal time interval $dt$ is

$$F_{in} l \sin \varphi d\varphi = F_{in} l \sin \varphi \, \dot\varphi dt =$$
$$-I \frac{a}{l} \omega^2 \sin \omega t \sin \varphi \, \dot\varphi dt \approx -I \frac{a}{l} \omega^2 \sin \omega t \, \varphi \dot\varphi dt. \qquad (10.34)$$

Here $I$ is the pendulum's moment of inertia. Integrating this expression over the
period of the pendulum motion $T = 2\pi/\omega_0 = 4\pi/\omega$, we find the total work
done during $T$, that is, the increment $\Delta E$ in the total energy $E$ during two driving
periods due to the parameter variation:

$$\Delta E = I\omega^2 C_1^2 (a/l)\omega/2\pi. \qquad (10.35)$$

The work of the frictional force determines the dissipation of mechanical energy. The elementary (negative) work $dW$ done by the torque of this force during
$dt$ is $-2I\gamma(\dot\varphi)^2 dt$. Integrating this work over the period of oscillation, we find
$-I\gamma c_1^2 \omega^2 \pi$. We note that both the frictional losses and the energy supplied by
oscillations of the pivot are proportional to the square of the amplitude $C_1$. This
means that over the threshold, friction cannot restrict the growth of the amplitude. Equating the absolute values of the work done by the force of inertia and by
the frictional force yields $\omega(a/l) = 2\gamma$. Since at resonance $\omega \approx 2\omega_0$, we obtain
the following approximate expression for the threshold value of the normalized
amplitude of the pivot:

$$m_{\text{thres}} = \frac{a_{\text{thres}}}{l} = \frac{\gamma}{\omega_0} = \frac{1}{2Q}. \qquad (10.36)$$

If this threshold value is exceeded, parametric resonance occurs in some interval of
driving frequencies extending on both sides of the resonant frequency $\omega_{\text{res}} = 2\omega_0$.
For a given value of the driving amplitude, the wider the interval, the smaller the
friction. To find the boundaries of parametric instability in the presence of friction,
we should include the damping term $2\gamma\dot\varphi$ into the linearized differential equation
of the pendulum, Eq. (10.19). With friction, the solution to this equation includes,
in contrast to Eq. (10.32), both sine and cosine terms:

$$\varphi(t) = A_1 \sin(\omega t/2) + A_3 \sin(3\omega t/2) + B_1 \cos(\omega t/2) + B_3 \cos(3\omega t/2). \quad (10.37)$$

Substituting $\varphi(t)$ given by Eq. (10.37) into Eq. (10.19), we obtain the homogeneous system of approximate equations for $A_1$, $A_3$, and $B_1$, $B_3$. Desired

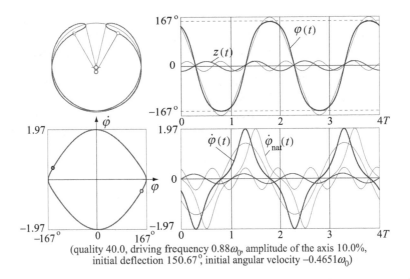

(quality 40.0, driving frequency $0.88\omega_0$, amplitude of the axis 10.0%,
initial deflection $150.67°$, initial angular velocity $-0.4651\omega_0$)

Figure 10.21: The spatial path, phase orbit, and graphs of stationary oscillations with large swing at $\omega < 2\omega_0$. The spectrum consists of the fundamental harmonic (frequency $\omega/2$) and its third harmonic (frequency $3\omega/2$), whose graphs are shown by thin lines.

boundaries of parametric instability are found from the condition of existence of a non-trivial solution to this system. Expressions for the boundaries that follow from this calculation are rather complicated and not cited here. The corresponding graph (for $Q = 5$) is shown by curve 3 in Figure 10.20.

It occurs that friction influences noticeably the boundaries of parametric instability only in the vicinity of $\omega = 2\omega_0$, that is, at small amplitudes of the pivot amplitude $m = a/l$. These boundaries are shown in detail by three thick curves in the right panel of Figure 10.20 for $Q = 5$, $Q = 10$, and for the absence of friction ($Q \to \infty$). Thin curves are plotted according to the following approximate expression (10.38), which is valid for small driving amplitudes ($m = a/l \ll 1$) in the vicinity of $\omega = 2\omega_0$:

$$\omega_{1,2} = (2 \pm 2\sqrt{m^2 - 1/(2Q)^2} + 7m^2/2)\omega_0. \tag{10.38}$$

If the driving parameters lie in the region inside these "tongues," the hanging equilibrium position is unstable, and the pendulum leaves it after the slightest perturbation. The growth of the amplitude is restricted by nonlinear effects (by dependence of the natural frequency on the amplitude). The growth of the natural period with the amplitude at large swing causes a violation of conditions that are favorable for parametric resonance (two excitation cycles during one natural period). As a result, an oscillation of a finite swing is established after fading of the transient beats.

However, by virtue of the same nonlinear properties of the pendulum, stationary parametric oscillations with an amplitude approaching $180°$ are possible in the system with small friction even at small values of the driving amplitude. An example of such parametric oscillations is shown in Figure 10.21. Due to the growth of the natural period of the pendulum at large swing, synchronization of its motion with the pivot oscillations (phase locking) can occur at $\omega < 2\omega_0$. If the pendulum swing equals $167°$, the driving frequency that provides the phase locking should be only $0.88\,\omega_0$ (compare with condition $\omega = 2\omega_0$ of parametric resonance at small swing).

Parametric oscillations shown in Figure 10.21 are very close in their properties to the nonlinear natural non-damping oscillations of the same swing. To illustrate this similarity, the thin line in the lower panel of Figure 10.21 shows for comparison the graph of angular velocity $\dot{\varphi}_{\mathrm{nat}}(t)$ for natural oscillations of the same swing in the absence of friction. As we can clearly see from the graphs, the period of these natural oscillations is the same as the period of corresponding parametric oscillations in the system with friction, and the shape of graphs $\dot{\varphi}_{\mathrm{nat}}(t)$ and $\dot{\varphi}(t)$ is exactly coinciding.

The role of the pivot oscillations in this case reduces to supplying the pendulum with energy needed to compensate for frictional losses. As a whole, such parametrically excited oscillations have much in common with the so-called 'bell-ringer mode,' which can be observed at direct forced excitation of a rigid pendulum by a sinusoidal external torque (see Chapter 9).

# 10.11  An Enhanced Criterion for Kapitza's Pendulum Stability

In this section an enhanced and more precise criterion for dynamic stabilization of the parametrically driven inverted pendulum is obtained: The boundaries of stability are determined with greater precision and are valid in a wider region of the system parameters than the above-discussed approximate results. As we already mentioned, the lower boundary of stability is associated with the phenomenon of subharmonic resonances in this system, namely, with the subharmonic resonance of an infinitely large order. The upper boundary of dynamic stabilization of the inverted pendulum is related to the ordinary parametric resonance (i.e., to the parametric destabilization of the lower equilibrium position). These relationships allow us to determine the boundaries of dynamic stabilization in a wider region of the system parameters (including relatively low frequencies and large amplitudes of excitation), and with a greater precision compared to previous results. Computer simulation of the physical system aids the analytical investigation and proves the theoretical results.

Among new discoveries regarding the inverted pendulum, the most important for finding the stability boundaries are the destabilization of the (dynamically stabilized) inverted position at large driving amplitudes through excitation of period-2 ("flutter") oscillations [44]–[45], and the existence of $n$-periodic

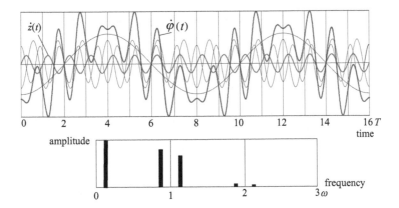

Figure 10.22: Angular velocity $\dot{\varphi}(t)$ time dependence (with the graphs of separate harmonics) and its spectrum for subharmonic resonance of the eighth order.

multiple-nodding regular oscillations [51]. The relationship of "flutter" oscillations in the inverted pendulum with ordinary parametric resonance of the hanging pendulum is discussed in [49]. A physical interpretation of periodic multiple-nodding oscillations as subharmonic parametric resonances is given above in Section 10.9.1 (see also [52]).

## 10.11.1 Subharmonic Resonances at Arbitrary Frequencies and Amplitudes of the Pivot Vibration

In the above discussion in Section 10.9.2 of conditions at which subharmonic resonances can occur, we assumed that steady-state oscillations whose period equals $n$ periods of the pivot vibration consist of the principal harmonic with the frequency $\omega/n$, and of two harmonic components with frequencies $(1 - 1/n)\omega$ and $(1 + 1/n)\omega$; see Eq. (10.18), p. 269, for the trial function $\varphi(t)$. This assumption is valid if the pivot is forced to oscillate at sufficiently small amplitudes and high enough frequencies. The assumption is also supported by an example shown in Figure 10.22 of such stationary subharmonic oscillations whose period equals 8 cycles of the pivot's motion. The graphs show time dependence of the angular velocity $\dot{\varphi}(t)$ together with its harmonics, and the spectrum of the velocity.

The fundamental harmonic component whose period equals 8 driving periods dominates the spectrum. We may treat it as a subharmonic (as an 'undertone') of the driving oscillation. This principal harmonic of the frequency $\omega/n$ describes the smooth component $\psi(t)$ of the compound period-8 oscillation: $\psi(t) = A\sin(\omega t/n)$.

However, Figure 10.22, as well as Figure 10.8 (see p. 264), obtained by numeric integration of the exact differential equation for the investigated system, show the presence of small contributions in $\varphi(t)$ of harmonics with frequencies

$(2n-1)\omega/n$ and $(2n+1)\omega/n$ (the graphs of these harmonics are small and hence not shown in the above-mentioned figures).

We may also include such harmonic components in the trial function in our attempt to improve the desired criterion and to make this criterion applicable in a wider region of the system parameters:

$$\varphi(t) = A_1 \sin(\frac{\omega}{n}t) + A_{n-1}\sin(\frac{n-1}{n}\omega t) + A_{n+1}\sin(\frac{n+1}{n}\omega t) +$$
$$+ A_{2n-1}\sin(\frac{2n-1}{n}\omega t) + A_{2n+1}\sin(\frac{2n+1}{n}\omega t). \quad (10.39)$$

Since oscillations at the threshold of the subharmonic resonance have infinitely small amplitudes, we can use instead of the exact differential equation, Eq. (10.11), the linearized (Mathieu) equation, Eq. (10.19), p. 269. Substituting $\varphi(t)$ from Eq. (10.39) into this equation (with $\gamma = 0$) and expanding the products of trigonometric functions, we obtain the following system of approximate equations for the coefficients $A_1$, $A_{n-1}$ and $A_{n+1}$, $A_{2n-1}$ and $A_{2n+1}$:

$$2(kn^2 - 1)A_1 + mn^2 A_{n-1} - mn^2 A_{n+1} = 0,$$
$$mn^2 A_1 + 2[n^2(k-1) + 2n - 1]A_{n-1} - mn^2 A_{2n-1} = 0,$$
$$-mn^2 A_1 + 2[n^2(k-1) - 2n - 1]A_{n+1} + mn^2 A_{2n+1} = 0, \quad (10.40)$$
$$mn^2 A_{n-1} + 2[n^2(k-4) + 4n - 1]A_{2n-1} = 0,$$
$$mn^2 A_{n+1} + 2[n^2(k-4) - 4n - 1]A_{2n+1} = 0.$$

Here $k$ is the parameter that enters into Eq. (10.19): $k = g/(l\omega^2)$. The homogeneous system (10.40) has a nontrivial solution if its determinant equals zero. This condition yields an equation (not cited here) for the corresponding threshold (minimal) normalized driving amplitude $m_{min} = a_{min}/l$ at which $n$-periodic mode $\varphi(t)$ given by expression (10.39) can exist.

The equation for the threshold driving amplitude can be solved numerically with the help of, say, the *Mathematica* package by Wolfram Research, Inc.). Then, after substituting this critical driving amplitude $m_{min}$ into the system (10.40), fractional amplitudes $A_{n-1}/A_1$, $A_{n+1}/A_1$, $A_{2n-1}/A_1$ and $A_{2n+1}/A_1$ of high harmonics for a given order $n$ can be found as the solutions to the homogeneous system of Eqs. (10.40).

If we ignore the contribution of harmonics with frequencies $(2n-1)\omega/n$ and $(2n+1)\omega/n$ in $\varphi(t)$, that is, assume $A_{2n-1}$ and $A_{2n+1}$ to be zero, system (10.40) simplifies considerably. The corresponding approximate solution can be found above in Section 10.9.2, Eq. (10.21).

For the full system (10.40) the final expressions for $m_{min}$ and for the amplitudes of harmonics are too bulky to be cited here. We have used them in Figure 10.23 for plotting the curves of $m_{min}$ as functions of parameter $k = (\omega_0/\omega)^2$ (inverse normalized driving frequency squared). The curves in Figure 10.23 correspond to subharmonic oscillations of different orders $n$ (thin curves).

To verify our analytical results for subharmonic oscillations in a computer simulation, we choose a value $k = -0.3$, corresponding to the drive frequency

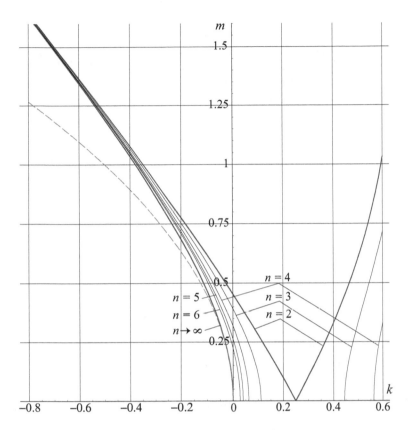

Figure 10.23: The normalized driving amplitude $m = a/l$ versus $k = (\omega_0/\omega)^2$ (inverse normalized driving frequency squared) at the lower boundary of the dynamic stabilization of the inverted pendulum (the left curve marked as $n \to \infty$), and at subharmonic resonances of several orders $n$ (see text for details).

$\omega = 1.826\,\omega_0$, for which the approach based on separation of slow and rapid motions of the pendulum is obviously inapplicable.

The above-described calculation applied to the subharmonic oscillation of order $n = 8$ predicts for the threshold normalized drive amplitude $m_{\min} = a_{\min}/l$ a value 87.73% of the pendulum's length.

Results of the simulation (based on numerical integration of the exact differential equation of the system) are presented in Figure 10.24. This simulation perfectly confirms the theoretical prediction.

The set of Poincaré sections in the phase plane consists of eight fixed points, and the phase orbit becomes closed after eight cycles of the pivot oscillations. The fractional amplitudes of harmonics obtained in the simulation agree perfectly well with the theoretical prediction.

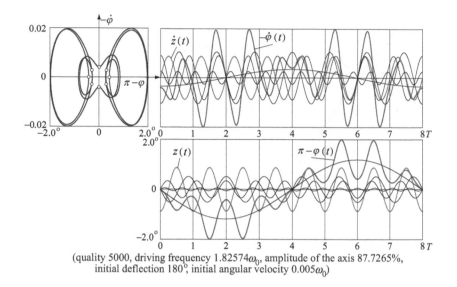

(quality 5000, driving frequency $1.82574\omega_0$, amplitude of the axis 87.7265%, initial deflection 180°, initial angular velocity $0.005\omega_0$)

Figure 10.24: Phase trajectory with Poincaré sections, angular velocity $-\dot\varphi(t)$ and angle $\pi - \varphi(t)$ time dependencies (with the graphs of separate harmonics) for subharmonic resonance of the eighth order. For convenience of presentation, the angle is measured from the inverted position.

## 10.11.2   Improved Lower Boundary of Dynamic Stabilization

As we already noted, the criterion of stability for the inverted pendulum can be related to the condition of the subharmonic resonance of an infinitely large order $n$, which occurs in the vicinity of the inverted position $\varphi = \pm\pi$. Hence the limit of $m_{\min}$ at $n \to \infty$ gives an improved formula for the lower boundary of dynamic stabilization of the inverted pendulum.

If we use the approximate solution of (10.40) in which the contribution of higher harmonics with frequencies $(2n-1)\omega/n$ and $(2n+1)\omega/n$ is ignored, the limit $n \to \infty$ gives for the lower boundary the value $m_{\min} = \sqrt{-2k(1-k)}$ in the region $k < 0$ (see Eq. (10.22), p. 269). For the solution of the full system (10.40) with higher harmonics $A_{2n-1}$ and $A_{2n+1}$ included, the limit of $m_{\min}$ at $n \to \infty$ yields the following expression:

$$m_{\min} = 2\sqrt{\frac{k(k-1)(k-4)}{3k-8}},  \tag{10.41}$$

which should be used instead of the commonly known approximate criterion $m_{\min} = \sqrt{-2k}$, given by Eq. (10.7), Eq. (10.8), or Eq. (10.16). The minimal amplitude $m_{\min} = a_{\min}/l$ that corresponds to the improved criterion (10.41) of dynamic stabilization is shown as a function of $k = (\omega_0/\omega)^2$ by the thick left curve marked as $n \to \infty$ in Figure 10.23. This curve is localized wholly in the region of negative $k$ values.

To compare the improved criterion (10.41) with the commonly known approximate criterion of the inverted pendulum stability given by Eq. (10.16), the latter is also shown in Figure 10.23 by the dashed thin left curve (red in the electronic version). We note how these two curves diverge dramatically at low frequencies and large amplitudes of the pivot oscillations.

The other curves to the right from this boundary show the dependence on $k$ of minimal driving amplitudes for which the subharmonic resonances of several orders can exist (the first curve for $n = 6$, and the others for $n$ values diminishing down to $n = 2$ from left to right). At negative $k$ values these curves give the threshold drive amplitudes for subharmonic oscillations about the inverted position. Case $k = 0$ corresponds to zero gravity (or infinitely high frequency of the pivot vibration). Points of intersection of the curves with the ordinate axis on this diagram give minimal drive amplitudes for which in the absence of gravity subharmonic oscillations of certain order $n$ can exist about any of the two dynamically stabilized positions (Figure 10.8, p. 264 shows an example of such period-8 oscillations) in the absence of gravity.

Continuations of the curves to positive $k$ values correspond to subharmonic parametric resonances (multiple-nodding oscillations) about the downward equilibrium position. The curve for $n = 2$ corresponds to ordinary parametric resonance, in which two cycles of excitation take place during one full oscillation of the pendulum. In the absence of friction the threshold drive amplitude for this resonance tends to zero at $\omega \to \omega_0/2$, that is, at $k \to 1/4$. From Figure 10.23 we see clearly that the curve, corresponding to $n = 2$ subharmonic oscillations of the inverted pendulum (the "flutter" mode), and the principal parametric resonance of ordinary (hanging) pendulum belong to essentially the same branch of period-2 regular behavior. Indeed, in the $k > 0$ region this branch gives the boundaries of the ordinary parametric resonance of the hanging-down pendulum, while in the $k < 0$ region this branch gives the upper boundary of dynamic stabilization for the inverted pendulum (see Section 10.10 and Section 10.10.2).

We note that the existence of subharmonic oscillations in the same region of the $k$—$m$ plane does not disprove criterion (10.41) of the inverted pendulum stability. Indeed, the pendulum is trapped into an $n$-periodic subharmonic limit cycle (with $n > 2$) only if the initial state belongs to a certain small basin of attraction that corresponds to this limit cycle. Otherwise the pendulum eventually comes to rest in the inverted position (or to unidirectional rotation, if the pendulum is released beyond a certain critical initial deviation).

For experimental verification of the improved criterion (10.22) we again choose relatively low drive frequency $\omega = 1.826\,\omega_0$ ($k = -0.3$), for which distinctions between the conventional and improved criteria are especially noticeable. At this frequency Eq. (10.41) gives for the lower boundary of stability the drive amplitude $a_{\min} = 0.868\,l$. The upper boundary (see Section 10.10) at $k = -0.3$ equals $a_{\max} = 0.929\,l$. Computer simulations in Figure 10.25 show how within this narrow region of stability, at $a = 0.875\,l$ (just over the lower boundary, Fig-

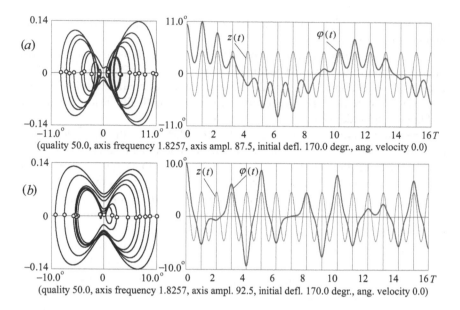

(quality 50.0, axis frequency 1.8257, axis ampl. 87.5, initial defl. 170.0 degr., ang. velocity 0.0)

(quality 50.0, axis frequency 1.8257, axis ampl. 92.5, initial defl. 170.0 degr., ang. velocity 0.0)

Figure 10.25: Phase trajectories and time histories of gradually damping oscillations about the inverted position just over the lower ($a$) and just below the upper ($b$) boundaries of dynamic stabilization.

ure 10.25$a$) and at $a = 0.925\,l$ (just below the upper boundary, Figure 10.25$b$), the pendulum, being initially deviated through 10° and released with zero velocity, in both cases shown in Figure 10.25 returns gradually to the inverted position.

We note the peculiarities of the transients shown in Figure 10.25 that lead to the state of rest in the inverted position. In both cases Poincaré sections, corresponding to time instants $t_q = qT$ ($q = 0, 1, 2\ldots$, $T = 2\pi/\omega$ – period of the pivot oscillations), are located near the $\varphi$-axis and gradually condense, approaching the origin of the phase plane. This origin corresponds to the state of rest in the inverted position.

Just over the lower boundary of stability the graph of time history in Figure 10.25$a$ resembles, during a limited time interval, the corresponding graph of a subharmonic resonance (of a multiple-nodding oscillation) of some high order (compare with Figure 10.24): The pendulum executes many 'nods' on one side of the inverted position, then on the other side with a somewhat smaller amplitude, and so on, gradually approaching the upper vertical. These intermittent damping 'nods' are described by one-sided shrinking loops of the phase orbit that pass from one side of the phase plane to the other each time the pendulum crosses the vertical line. The greater the number of such 'nods,' the closer to the upper boundary: We remember that this boundary corresponds to the subharmonic resonance of infinite number in which the pendulum makes an infinite number of 'nods' before crossing the upper vertical.

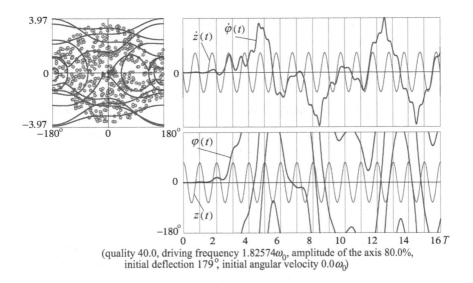

(quality 40.0, driving frequency $1.82574\omega_0$, amplitude of the axis 80.0%,
initial deflection $179°$, initial angular velocity $0.0\omega_0$)

Figure 10.26: Chaotic oscillations below the boundary of stability.

Conversely, just below the upper boundary (Figure 10.25$b$), the character of gradually damping oscillations about the inverted position reminds us of the "flutter" mode (see Section 10.10) with its double-lobed phase curve (Figure 10.17): One cycle of the pendulum oscillations covers approximately two drive periods and is represented by a double-lobed phase orbit. This orbit shrinks gradually around the origin of the phase plane. We note that near the upper boundary the graph of time history and the phase trajectory (Figure 10.25$b$) resemble those of a "flutter" oscillation. Over this upper boundary the pendulum eventually occurs trapped into the period-2 steady-state oscillation instead of returning to the vertical position.

If the initial deflection exceeds some critical value, at first the pendulum goes slowly further from the vertical, then executes random revolutions to one and the other side, and eventually (after a long 'tumbling' chaotic transient) becomes trapped into period-1 unidirectional rotation. The smaller this critical deviation, the closer the drive parameters are to the boundary of stability. For $k = -0.3$, $m = 0.875$ and $Q = 50$ the initial deviation from the inverted position should not exceed $14°$. Friction reduces the basin of attraction of the equilibrium in the inverted state: At $Q = 20$ the initial deviation should not exceed $10°$.

At smaller than $0.868\,l$ values of the drive amplitude the inverted pendulum is unstable. Figure 10.26 shows how at $k = -0.3$ and $a = 0.800\,l$ the pendulum, being released at only $1°$ deviation from the inverted position, occurs eventually in a chaotic regime ('tumbling' chaos). The graphs in Figure 10.26 show the initial stage of the time history. The set of Poincaré sections in the phase plane gives an impression of the further random behavior, characterized by a strange attractor. We note that the inverted pendulum at these drive parameters should be stable according to conventional criterion, Eq. (10.22), which at $k = -0.3$ gives a considerably smaller value $a_{min} = 0.775\,l$ (compare with improved value $a_{min} = 0.868\,l$) for the lower boundary of stability.

### 10.11.3　Improved Upper Boundary of Dynamic Stabilization

The curve $n = 2$ in Figure 10.23 (its part in $k \leq 0$ region) corresponds to the upper boundary of dynamic stabilization for the inverted pendulum: After a disturbance the pendulum does not come to rest in the up position, no matter how small the release angle, but instead eventually settles into a limit cycle, executing finite amplitude steady-state oscillation (about the inverted vertical position). The period of such an oscillation is twice the driving period, and its swing grows as the excess of the drive amplitude over the threshold $a_{max}$ is increased.

As we already mentioned, this loss of stability of the inverted pendulum has been first described in 1992 by Blackburn et al. [44] and demonstrated experimentally in [45]. The authors [44] called these limit-cycle oscillations the "flutter" mode. Above we have shown (see Section 10.10.1, p. 278) that the "flutter" mode and the principal parametric resonance belong to the same branch of the period-2 stationary regime (this unambiguously follows from Figure 10.23). Hence the same analytical method can be used to calculate conditions of their excitation. Simulations show a very simple spectral composition for both, namely a superposition of the fundamental harmonic whose frequency $\omega/2$ equals half the driving frequency, the third harmonic with the frequency $3\omega/2$, and maybe a tiny admixture of the fifth harmonic:

$$\varphi(t) = A_1 \cos(\omega t/2) + A_3 \cos(3\omega t/2) + A_5 \cos(5\omega t/2). \qquad (10.42)$$

The phases of harmonics in (10.42) correspond to pivot oscillations in the form $z(t) = a \cos \omega t$. Substituting $\varphi(t)$ into the linearized differential equation, Eq. (10.19), p. 269, with $\gamma = 0$ and expanding the products of trigonometric functions, we obtain an expression in which we should equate the coefficients of $\cos(\omega t/2)$, $\cos(3\omega t/2)$, and $\cos(5\omega t/2)$ to zero. Thus we get a system of homogeneous equations for the coefficients $A_1$, $A_3$, and $A_5$ of harmonics in the trial function (10.42):

$$\begin{aligned}
(4k - 2m - 1)A_1 - 2mA_3 &= 0, \\
-2A_1 + (4k - 9)A_3 - 2mA_5 &= 0, \\
-2mA_3 + (4k - 25)A_5 &= 0.
\end{aligned} \qquad (10.43)$$

This system has a nontrivial solution when its determinant equals zero. If we neglect the contribution of the fifth harmonic in $\varphi(t)$, Eq. (10.42), that is, if we let $A_5 = 0$, we get the following approximate expression for the upper boundary of stability (see also Eq. (10.33), p. 281):

$$m_{max} = \frac{1}{4}\left[\sqrt{(4k - 9)(20k - 13)} + 4k - 9\right]. \qquad (10.44)$$

If the fifth harmonic is included, the requirement for a non-trivial solution to the system (10.43) yields a cubic equation for the desired normalized critical driving amplitude $a_{max}/l = m_{max}$. The relevant root of this equation (too cumbersome to be shown here) is used for plotting the curve $n = 2$ in Figure 10.23. However, for the interval of $k$ values under consideration ($-0.8 - 0.6$) the approximate expression (10.44) gives a curve that is visually indistinguishable from the curve $n = 2$ in Figure 10.23.

The curve $n = 2$ intersects the ordinate axis at $m \approx 3(\sqrt{13} - 3)/4 = 0.454$. This case ($k = 0$) corresponds to the above-mentioned limit of a very high driving frequency ($\omega/\omega_0 \to \infty$) or zero gravity ($\omega_0 = 0$), so that $m = 0.454$ gives the upper limit of stability for each of the two dynamically stabilized equivalent equilibrium positions: If $m > 0.454$ at $g = 0$, the "flutter" mode is excited.

The lower and upper boundaries of the dynamic stability gradually converge while the drive frequency is reduced: Figure 10.23 shows that the interval between $m_{min}$ and $m_{max}$ shrinks to the left, when $|k|$ is increased. Both boundaries merge at $k \approx -1.41$ ($\omega \approx 0.8423\omega_0$) and $m \approx 2.451$. The diminishing island of dynamic stability of the pendulum in the inverted state vanishes in the surrounding sea of chaotic motions.

The improved theoretical values for the lower and upper boundaries of stability are obtained here for the frictionless system ($\gamma = 0$). Computer simulations (based on numerical integration of exact differential equation, Eq. (7.15), show that relatively weak friction ($Q \geq 15 - 30$) does not noticeably influence these boundaries. This can be easily explained physically if we take into account that in conditions of dynamic stabilization (sufficiently large frequency or amplitude of parametric excitation) the role of inertial forces is much more important.

Conversely, the basin of attraction for equilibrium of the pendulum in the inverted position is sensitive to friction: The interval of initial deviations within which the pendulum returns eventually to the inverted position becomes smaller as friction is increased.

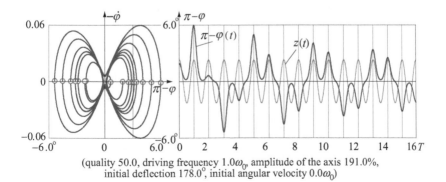

(quality 50.0, driving frequency $1.0\omega_0$, amplitude of the axis 191.0%, initial deflection $178.0°$, initial angular velocity $0.0\omega_0$)

Figure 10.27: Dynamic stability of the inverted pendulum at low frequency and large amplitude of the pivot oscillations.

For further experimental verification of the improved values for both upper and lower boundaries of the inverted pendulum's stability, and for comparison of the improved criterion, Eq. (10.41), with the conventional one, Eq. (10.22), we now choose $k = -0.5$ (drive frequency $\omega = 1.4142\,\omega_0$). At this $k$ value the improved criterion, Eq. (10.41), yields for the low boundary of stability the value $m_{\min} = 1.1920$ (the amplitude of the pivot oscillation $a_{\min}$ equals 119.20% of the pendulum length $l$), while the conventional one, Eq. (10.22), gives only $m_{\min} = 1.00$ (100% of the pendulum length $l$: $a_{\min} = l$). This value of the pivot amplitude is far below the real boundary of dynamic stability.

The improved theoretical value for the upper boundary at $k = -0.5$ is $m_{\max} = 1.2226$ (122.26%), while the approximate theory (in which a harmonic with the frequency $5\omega/2$ is not taken into account, see Section 10.10) gives a slightly greater value: According to Eq. (10.44) $m_{\max} = 1.2265$ (122.65%). Simulations show that below $m = 115.68\%$ the motion is chaotic ('tumbling' chaos); in the interval $m = 115.69\% - 119.19\%$ the pendulum, after a long chaotic transient, is trapped in period-1 non-uniform unidirectional rotation (in contradiction with the conventional criterion, Eq. (10.22), which predicts stability of the inverted position in this interval), and only in the interval $m = 119.20\% - 122.27\%$ the pendulum, being released at a small deviation from the inverted position, eventually comes to rest, in perfect accordance with the improved criterion, Eq. (10.41).

Then, over the upper boundary of stability, within the interval $m = 122.28\% - 123.02\%$, the pendulum occurs in a "flutter" mode; at $m = 123.03\% - 147.01\%$

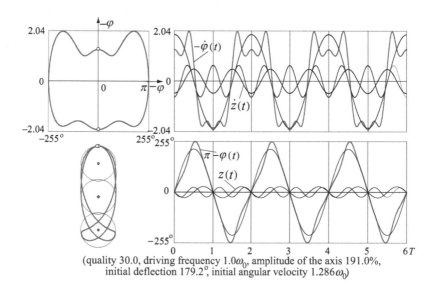

(quality 30.0, driving frequency $1.0\omega_0$, amplitude of the axis 191.0%, initial deflection 179.2°, initial angular velocity $1.286\omega_0$)

Figure 10.28: Period-2 oscillation of a large amplitude about the inverted position: The phase orbit, spatial trajectory, and graphs of angular velocity $\dot{\varphi}(t)$ and angle $\varphi(t)$ time dependencies (with the graphs of separate harmonics).

executes unidirectional rotation; at $m = 147.02\% - 150.6\%$ the pendulum, after a long chaotic transient, comes to period-2 oscillation about the inverted position with an amplitude of approximately 260° (similar to the oscillation shown in Figure 10.28); at $m \geq 150.7\%$ the pendulum eventually settles into the unidirectional rotation.

Further simulations shown in Figures 10.27–10.29 refer to the case of especially low frequency of the pivot oscillations: $\omega = \omega_0$ ($k = -1$). The theoretical values for the lower and upper boundaries of stability, according to the improved criteria, at this frequency are $m_{\min} = 1.9069$ (190.69% of the pendulum length) and $m_{\max} = 1.9138$ (191.38%), respectively. Figure 10.27 shows how in this narrow interval the pendulum, being released at 178°, first goes further from the vertical for about 6° maximum angular excursion and then steadily approaches the inverted position by the process of gradually damping "flutter"-like oscillations.

The basin of attraction for equilibrium in the inverted position is rather small: At slightly different initial conditions the pendulum, after a long transient, occurs in a steady-state large-amplitude (approximately 255°) period-2 oscillation about the inverted position. In this extraordinary motion, the angular excursion of the pendulum from one extreme position to the other takes one period of excitation and is greater than a full circle (about 510°). Therefore, otherwise we can treat this regime as alternating clockwise and counterclockwise revolutions over the

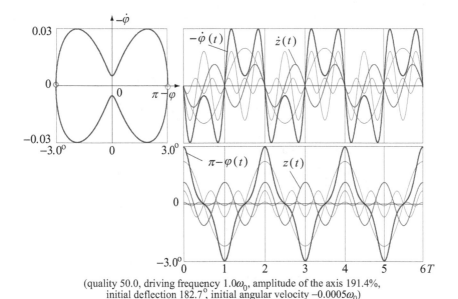

(quality 50.0, driving frequency $1.0\omega_0$, amplitude of the axis 191.4%,
initial deflection 182.7°, initial angular velocity $-0.0005\omega_0$)

Figure 10.29: The phase orbit and graphs of $\varphi(t)$ and $\dot{\varphi}(t)$ for "flutter" oscillations of the inverted pendulum just over the upper boundary of stability.

top. The phase orbit (with two fixed points of Poincaré sections) and the spatial trajectory of the pendulum bob in Figure 10.28 give an impression of such an extraordinary pendulum motion, which coexists with the state of rest in the inverted position. Just over the upper boundary of stability (191.38%), the pendulum eventually settles into the "flutter" mode (Figure 10.29).

## 10.12   Complicated Regular Motions of the Parametrically Driven Pendulum

Behavior of the planar rigid pendulum excited parametrically by vertical oscillations of its pivot is much richer in various modes than we can expect for such a simple physical system relying on our intuition. Its nonlinear large-amplitude motions can hardly be called 'simple.' The simulations show that variations of the parameter set (dimensionless driving amplitude $a/l$, normalized driving frequency $\omega/\omega_0$, and quality factor $Q$) result in different regular and chaotic types of dynamical behavior. Variations in the initial conditions can result in different regimes that coexist at the same values of the system parameters. This property of nonlinear systems is usually referred to as *multistability*.

One more type of a counterintuitive steady-state regular behavior is shown in Figure 10.30. This mode can be characterized as resulting from a multiplication of the period of a subharmonic resonance, specifically, as tripling of the sixth-order subharmonic resonance described in Section 10.9.1. Comparing Figure 10.30 with

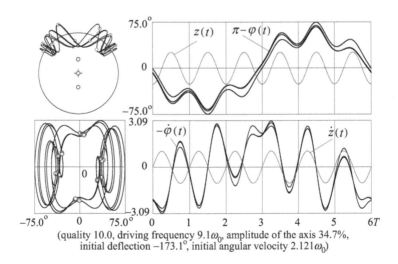

(quality 10.0, driving frequency $9.1\omega_0$, amplitude of the axis 34.7%, initial deflection $-173.1°$, initial angular velocity $2.121\omega_0$)

Figure 10.30: The spatial path, phase orbit, and time-dependent graphs of $\varphi(t)$ and $\dot{\varphi}(t)$ for stationary period-18 regular oscillations. The graphs show three consecutive cycles of six driving periods each.

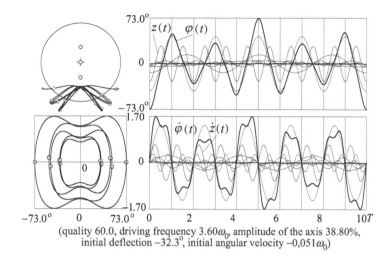

(quality 60.0, driving frequency $3.60\omega_0$, amplitude of the axis 38.80%,
initial deflection $-32.3°$, initial angular velocity $-0,051\omega_0$)

Figure 10.31: The spatial path of the pendulum's bob, phase orbit, and time-dependent graphs of $\varphi(t)$ and $\dot{\varphi}(t)$ with graphs of their harmonic components for stationary period-10 oscillations.

Figure 10.9, p. 266, we see that in both cases the motion is quite similar during any cycle consisting of six consecutive driving periods, but in Figure 10.30 the motion during each subsequent cycle of six periods is slightly different from the preceding cycle.

After three cycles of six driving periods the phase orbit becomes closed and then repeats itself, so that the period of this stationary motion equals 18 driving periods. However, the harmonic component whose period equals six driving periods dominates the spectrum (just like in the spectrum of period-6 oscillations in Figure 10.9, p. 266), while the fundamental harmonic (frequency $\omega/18$) of a small amplitude is responsible only for tiny divergences between the adjoining cycles consisting of six driving periods each. Such multiplications of the period are characteristic of large amplitude oscillations at subharmonic resonances both for the inverted and hanging-down pendulum.

Figure 10.31 shows an example of stationary oscillation about the lower equilibrium position with a period that equals ten driving periods. This large amplitude motion can be treated as originating from a period-2 oscillation (that is, from ordinary principal parametric resonance) by a five-fold multiplication of the period. The harmonic component with half the driving frequency ($\omega/2$) dominates the spectrum. But in contrast to the preceding example, the divergences between adjoining cycles consisting of two driving periods each are generated by the contribution of a harmonic component with the frequency $3\omega/10$ rather than of the fundamental harmonic (frequency $\omega/10$) whose amplitude is much smaller.

Figure 10.32 shows one more example of complicated steady-state oscilla-

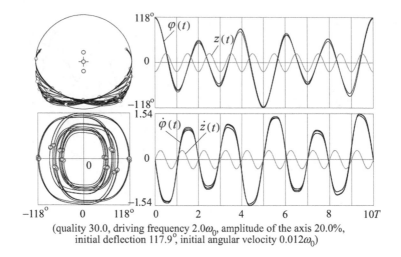

(quality 30.0, driving frequency $2.0\omega_0$, amplitude of the axis 20.0%,
initial deflection $117.9°$, initial angular velocity $0.012\omega_0$)

Figure 10.32: The spatial path, phase orbit, and graphs of stationary period-30 oscillations. The graphs show three consecutive cycles of ten driving periods each. One full period consists of three almost identical cycles, each of which covers 10 drive periods.

tion of the parametrically driven pendulum. This period-30 motion can be treated as generated from the period-2 principal parametric resonance first by five-fold multiplication of the period (resulting in period-10 oscillation), and next by multiplication (tripling) of the period. Such large-period stationary regimes are characterized by small domains of attraction consisting of several disjoint islands in the phase plane of initial conditions.

Other numerous modes of regular, periodic behavior are formed by unidirectional period-2 or period-4 (or even period-8) rotation of the pendulum or by oscillations alternating with revolutions to one or to both sides in turn. Such modes have periods constituting several driving periods.

## 10.13   Chaotic Motions of the Parametrically Driven Pendulum

At large enough driving amplitudes the pendulum whose pivot is forced to oscillate in the vertical direction exhibits different chaotic regimes. Chaotic behavior of various nonlinear systems has been a subject of intense interest during recent decades, and the parametrically forced pendulum serves as an excellent physical model for studying general laws of the dynamical chaos.

Next we describe several different kinds of chaotic regimes, which for the time being have not been extensively investigated in the literature. Poincaré mapping,

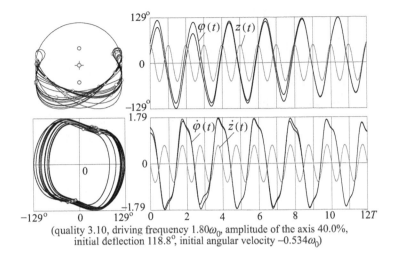

(quality 3.10, driving frequency $1.80\omega_0$, amplitude of the axis 40.0%,
initial deflection $118.8°$, initial angular velocity $-0.534\omega_0$)

Figure 10.33: Chaotic attractor whose Poincaré sections form a two-band set of points.

that is, a stroboscopic picture of the phase plane for the pendulum taken once during each driving cycle after initial transients have died away, gives an obvious and convenient means to distinguish between regular periodic behavior and persisting chaos. A steady-state subharmonic motion of order $n$ would be seen in the Poincaré map as a systematic jumping of the representative point among $n$ fixed mapping points. When the pendulum exhibits a chaotic motion, the points of Poincaré sections wander randomly, never repeating some pattern exactly. Their behavior in the phase plane gives an impression of the strange attractor for the motion in question.

Figure 10.33 shows an example of a purely oscillatory two-band chaotic attractor for which the set of Poincaré sections consists of two disjoint islands. This attractor is characterized by a fairly large domain of attraction in the phase plane. The two islands of the Poincaré map are visited regularly (strictly in turn) by the representing point, but within each island the point wanders irregularly from cycle to cycle. This means that for this kind of motion the flow in the phase plane is chaotic, but the distance between any two initially close phase points within this attractor remains limited in the progress of time: The greatest distance in the phase plane is determined by the size of these islands of the Poincaré map.

Figure 10.34 shows the chaotic attractor that corresponds to a slightly reduced friction compared to the case shown in Figure 10.33, while all other parameters are unchanged. Gradual reduction of friction causes the islands of Poincaré sections to grow and coalesce, and finally to form a strip-shaped set occupying considerable region of the phase plane. As in the preceding example, each cycle of these oscillations (consisting of two driving periods) slightly but randomly varies from

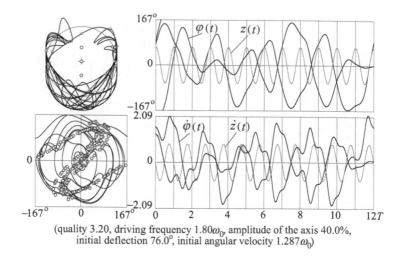

(quality 3.20, driving frequency $1.80\omega_0$, amplitude of the axis 40.0%,
initial deflection 76.0°, initial angular velocity $1.287\omega_0$)

Figure 10.34: Chaotic attractor with a strip-like set of Poincaré sections.

the preceding one.

However, in this case the large and almost constant amplitude of oscillations occasionally (after a large but unpredictable number of cycles) considerably reduces or, vice versa, increases (sometimes so that the pendulum makes a full revolution over the top). These decrements and increments occasionally result in switching the phase of oscillations: Motion of the pendulum, say, to the right side that occurred during even driving cycles is replaced by the motion in the opposite direction. During long intervals between these rare events the motion of the pendulum is purely oscillatory with only slightly (and randomly) varying amplitude of the oscillation.

This kind of intermittent irregular behavior differs from the well-known so-called 'tumbling' chaotic attractor (see Figure 10.26, p. 293) that exists over a relatively broad range of parameter space. The tumbling attractor is characterized by random oscillations (whose amplitude varies strongly from cycle to cycle), often alternated with full revolutions to one or the other side.

Figure 10.35 illustrates one more kind of a strange attractor. In this example the motion is always purely oscillatory, and nearly repeats itself after each six driving periods. The six bands of Poincaré sections make two groups of three isolated islands each. The representing point visits these groups in alternation. Moreover, the representing point visits the islands of each group in a quite definite order, but within each island the points continue to bounce randomly from one place to another without any apparent order.

The six-band attractor has a rather extended (and very complicated in shape) domain of attraction. Nevertheless, at these values of the control parameters the system exhibits multiple asymptotic states: The chaotic attractor coexists with sev-

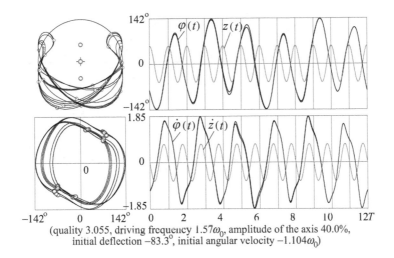

(quality 3.055, driving frequency $1.57\omega_0$, amplitude of the axis 40.0%,
initial deflection $-83.3°$, initial angular velocity $-1.104\omega_0$)

Figure 10.35: An oscillatory six-band chaotic attractor.

eral periodic regimes. One of these periodic regimes is shown in Figure 10.36. The
period of regular asymmetric oscillations in this example equals four driving periods. Such asymmetric regimes exist in pairs, whose phase orbits are mirror images
of one another.

Chaotic regimes exist also for purely rotational motions. Poincaré sections for
such rotational chaotic attractors can make several isolated islands in the phase
plane. A possible scenario of transition to such chaotic modes from unidirectional
regular rotation lies through an infinite sequence of period-doubling bifurcations
occurring when a control parameter (the driving amplitude or frequency or the
braking frictional torque) is slowly varied without interrupting the motion of the
pendulum. However, there is no unique route to chaos for the more complicated
chaotic regimes described above.

## 10.14  Concluding Remarks

In this chapter we have shown that at sufficiently high frequency and small amplitude of the pivot oscillations, many remarkable peculiarities in behavior of the
pendulum can be clearly explained on the basis of the method of averaging in
which rapid and slow motions of the pendulum are separated. In particular, this
approach and the related concept of the effective potential for slow motion are
very useful for understanding the dynamic stabilization of the inverted pendulum,
as well as for explanation of the origin of recently discovered subharmonic resonances of high orders. Coexistence of resonances of different orders at identical
parameters of the system also can be easily explained on this basis. Corresponding
approximate quantitative theory of parametric excitation allows us to calculate the

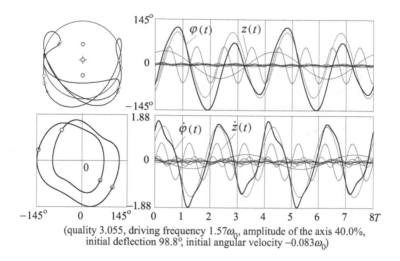

(quality 3.055, driving frequency $1.57\omega_0$, amplitude of the axis 40.0%,
initial deflection 98.8°, initial angular velocity $-0.083\omega_0$)

Figure 10.36: Regular period-4 asymmetric oscillations that coexist with the chaotic oscillations shown in Figure 10.35.

spectrum of subharmonic resonances and to find their boundaries in the parameter space of the system.

The relationship between subharmonic resonances and the phenomenon of dynamic stabilization of the inverted pendulum reveals a way of obtaining an improved criterion of dynamic stabilization, Eq. (10.22), which is valid in a wider region of frequencies and amplitudes of the pivot oscillations, including values of these parameters for which the method of averaging (of separation of rapid and slow motions) is inapplicable.

Relying on the common nature of the second-order subharmonic resonance and ordinary parametric resonance, we have presented a physical explanation for the recently discovered phenomenon of destabilization of the (dynamically stabilized) inverted position: At sufficiently large amplitudes of the pivot oscillations the inverted pendulum becomes trapped into the limit cycle of "flutter" oscillations with period $2T$.

This phenomenon of "flutter" oscillations is completely analogous to destabilization of the hanging-down pendulum when conditions of ordinary parametric resonance are fulfilled. Ordinary parametric resonance and the "flutter" mode belong to the same branch of 2-periodic stationary oscillations, so that the criterion of destabilization for the inverted position (10.33) can be obtained by the same method as for destabilization of the downward position at ordinary parametric resonance. Results of corresponding analytical calculations for the boundaries of the stability region agree perfectly well with computer simulations based on a numeric integration of the exact differential equation that describes the rigid planar pendulum with vertically oscillating pivot.

In this chapter we have touched only a small part of existing stationary states, regular and chaotic motions of the parametrically driven pendulum. The pendulum's dynamics exhibits a great variety of other asymptotic rotational, oscillatory, and combined (both rotational and oscillatory) multiple-periodic stationary states (attractors), whose basins of attraction are characterized by a surprisingly complex (fractal) structure. Computer simulations also reveal intricate sequences of bifurcations, leading to numerous intriguing chaotic regimes. Most of them remain beyond the scope of this chapter, and those mentioned here are still awaiting a plausible physical explanation. With good reason, we can suppose that this seemingly simple physical system is really inexhaustible.

# Chapter 11

# Torsion Pendulum with Dry and Viscous Damping

**Annotation.** Free and forced oscillations of a torsion spring pendulum damped by viscous and dry friction are investigated analytically and with the help of numerical simulations. A simplified mathematical model is assumed (Coulomb law), which nevertheless can explain many peculiarities in behavior of various oscillatory systems with dry friction. The amplitude of free oscillations diminishes under dry friction linearly, and the motion stops after a final number of cycles. The amplitude of a sinusoidally driven pendulum with dry friction grows at resonance without limit if the threshold is exceeded. At strong enough non-resonant sinusoidal forcing, dry friction causes transients that typically lead to definite limit cycles — periodic steady-state regimes of symmetric non-sticking forced oscillations that are independent of initial conditions. However, at the subharmonic sinusoidal forcing, interesting peculiarities of the steady-state response are revealed, such as multiple coexisting regimes of asymmetric oscillations that depend on initial conditions. Under certain conditions, simple dry friction pendulum shows complicated stick-slip motions and chaos.

## 11.1   Basics of the Theory

### 11.1.1   Introduction

Various mechanical vibration systems with combined viscous and dry (Coulomb) friction are of considerable importance in numerous applications of dynamics in engineering.

When friction is viscous, the spring oscillatory systems are described by linear differential equations. This case allows an exhaustive explicit analytical solution that is usually studied in undergraduate courses at universities and can be found in most textbooks on general physics. However, the influence of dry friction on oscillatory systems remains as a rule beyond the scope of the academic literature and traditional physics courses.

Dry friction results in a nonlinearity. With dry friction, the system acquires a non-smooth, discontinuous nonlinear character. If the coefficient of dry friction is sufficiently small, the oscillating body slides under harmonic forcing and its velocity is zero only for the instants at which the direction of motion reverses. This kind of motion of a dry friction oscillator with no stick phase is usually referred to as a pure *slip motion*, or *non-sticking motion*. At strong enough dry friction, sticking may occur: The body remains at rest for a finite time during the driving cycle after its velocity reaches zero. A detailed historical review of dry friction and stick-slip phenomena can be found in [53].

Dry friction as a nonlinearity is the current focus of research activities. Even the simplest dry friction model, the Coulomb friction, can explain the principal peculiarities in the motion of a dry friction oscillator. Damping of free oscillations under dry friction is very clearly described in the textbook of Pippard [3] (see also [54]). Different approaches to the problem are discussed in [55], [56]. Den Hartog [57] was the first to solve, in 1930, the periodic sliding response of a harmonically forced oscillator with both viscous and dry-friction damping. Later on, the analytical solutions of non-sticking responses were widely discussed in the contemporary scientific literature (see [59]–[66] and references therein). The problem was treated by using a number of various analytical and numerical techniques. In recent years, there has been an increasing interest in periodic and chaotic motions of discontinuous dynamical systems because of their important role in engineering (see, for example, [67]).

In the literature the analytical solution to the problem of oscillations in a system with dry friction is usually obtained by a simple method of stage-by-stage integration of the differential equations that describe the system. These equations are linear for the time intervals occurring between consecutive turning points, if the simplest (Coulomb) model is assumed for dry friction. The intervals are bounded by the instants at which the velocity is zero. The complete solution is obtained by fitting these half-cycle solutions to one another for adjoining time intervals. By virtue of the piecewise linear nature of the relevant differential equations, explicit solutions can be found for the time intervals between the successive turnarounds.

In our approach to the problem we try to rely primarily on the physics underlying the investigated phenomena. In this chapter we are concerned with free oscillations of a torsion spring pendulum, and with forced oscillations of the pendulum kinematically driven by an external sinusoidal force, including cases of damping caused by dry (Coulomb) friction, and both by viscous and dry friction. Mathematically, the pendulum driven by an external force is equivalent to the spring-mass system with the body residing on the horizontally oscillating base. The simple formulae of analytical solutions are confirmed by graphs obtained in computer simulations. New results cover quantitative description of the resonant growth of oscillations under sinusoidal forcing, and closed-form analytical solutions at sub-resonant frequencies. These solutions correspond to multiple asymmetric steady-state regimes coexisting at the same values of the system parameters. Characteristics of such regimes depend on the initial conditions. Our analytical and numerical solutions are illustrated by a simplified version of the relevant simulation program (Java applet) available on the web [58].

## 11.1.2 The Physical System

The rotating component of the torsion spring oscillator investigated in the paper is a balanced flywheel whose center of mass lies on the axis of rotation (Figure 11.1), similar to devices used in mechanical watches.

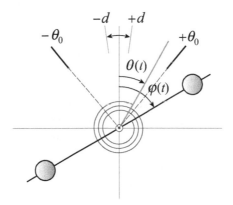

Figure 11.1: Schematic diagram of the driven torsion oscillator with dry friction.

A spiral spring with one end attached to the flywheel flexes when the flywheel is turned. The other end of the spring is attached to the exciter — a driving rod, which can be turned by an external force about the axis common with the flywheel axis. The spring provides a restoring torque whose magnitude is proportional to the angular displacement of the flywheel relative to the driving rod. In other words, we assume that the flywheel is in equilibrium (the spring is unstrained) when the rod of the flywheel is parallel to the driving rod.

In the case of unforced (free, or natural) oscillations in an isolated system, the motion is initiated by an external influence which occurs before a particular instant. This influence determines the initial mechanical state of the system, that is, the displacement and the velocity of the oscillator at the initial instant. These initial conditions determine the amplitude and phase of subsequent free oscillations. The frequency and damping rate of free oscillations are determined solely by the physical properties of the system, and do not depend on the initial conditions.

Oscillations are called forced if an oscillator is subjected to an external periodic influence whose effect on the system can be expressed by a separate term, a periodic function of the time, in the differential equation of motion. We are interested in the response of the system to the periodic external force. The behavior of oscillatory systems under periodic external forces is one of the most important issues in the theory of oscillations. A noteworthy distinctive characteristic of forced oscillations is the phenomenon of resonance, in which a small periodic disturbing force can produce an extraordinarily large response in the oscillator. Resonance is found everywhere in physics and so a basic understanding of this fundamental problem has wide and various applications.

The phenomenon of resonance depends upon the whole functional form of the driving force and occurs over an extended interval of time rather than at some particular instant. In this paper we draw attention to peculiarities of resonance in an oscillator with dry friction.

In our model of an oscillatory system, free oscillations of the flywheel occur when the driving rod is immovable ($\theta = 0$). Forced oscillations are excited when the driving rod rotates back and forth sinusoidally about its middle position $\theta = 0$ between the angles $-\theta_0$ and $\theta_0$ (see Figure 11.1): $\theta(t) = \theta_0 \sin \omega t$. This mode differs from the dynamical mode usually considered in textbooks, according to which oscillations are excited by a given external force exerted on the system. Our mode can be called kinematical, because in this mode oscillations are excited by forcing one part of the system (the driving rod) to execute a given motion (in our case a simple harmonic motion). This kinematical mode is especially convenient for observation, because the motion of the exciter can be seen simultaneously with oscillations of the flywheel.

In the Coulomb model of dry friction, as long as the system is moving, the magnitude of dry friction is assumed to be constant, and its direction is opposite that of the velocity, that is, its direction changes each time the direction of the velocity changes. When the system is at rest, the force of static dry friction takes on any value from some interval $-F_{\max}$ to $F_{\max}$. The actual value of static frictional force can be found from the requirement of balancing the other forces exerted on the system. In other words, the force of static friction adjusts itself to make equilibrium with other external forces acting on the body. The magnitude of the force of kinetic dry friction is assumed in this model to be equal to the limiting force $F_{\max}$ of static friction.

In real physical systems dry friction is characterized by more complicated dependencies on the relative velocity (see, for example, [68], [69]). The limiting force of static friction is usually greater than the force of kinetic friction. When the speed of a system increases from zero, kinetic friction at first decreases, reaches a minimum at some speed, and then gradually increases with a further increase in speed. These peculiarities are ignored in the idealized z-characteristic of dry friction. Nevertheless, this idealization allows us to understand many important features of oscillations in real physical systems.

In our model of a torsion oscillator some amount of dry friction can exist in the bearings of the flywheel axis. Because the magnitude of static frictional torque can assume any value up to $N_{\max}$, there is a range of values of angular displacement called the *stagnation interval* or *dead zone* in which static friction can balance the restoring elastic torque of the strained spring. At any point within this interval the system can be at rest in a state of neutral equilibrium, in contrast to a single position of stable equilibrium provided by the spring in the case of an oscillator with viscous friction.

The stagnation interval extends equally to either side of the point at which the spring is unstrained. The stronger the dry friction in the system, the more extended the stagnation interval. The boundaries of the interval $\pm d$ are determined by the limiting torque $N_{max}$ of static friction. In Figure 11.1 these boundaries $-b$ and $+b$ are shown for the case in which the driving rod is in its middle position $\theta = 0$.

### 11.1.3 The Differential Equation of the Oscillator

The rotating flywheel of the torsion oscillator is simultaneously subjected to the restoring torque $-D(\varphi - \theta)$ produced by the spring, the torque $-B\dot{\varphi}$ of viscous friction that is proportional to the angular velocity, and the torque $N_{fr}$ of kinetic dry friction. When the exciter is forced to move periodically according to $\theta(t) = \theta_0 \sin \omega t$, the differential equation describing the rotational motion of the flywheel with the moment of inertia $J$ is thus

$$J\ddot{\varphi} = -D(\varphi - \theta_0 \sin \omega t) - B\dot{\varphi} + N_{fr}. \tag{11.1}$$

The torque $N_{fr}$ is directed oppositely to angular velocity $\dot{\varphi}$, and is constant in magnitude while the flywheel is moving, but may have any value in the interval from $-N_{max}$ up to $N_{max}$ while the flywheel is at rest:

$$N_{fr}(\dot{\varphi}) = -N_{max} \operatorname{sign} \dot{\varphi} = \begin{cases} -N_{max} & \text{for} \quad \dot{\varphi} > 0, \\ N_{max} & \text{for} \quad \dot{\varphi} < 0. \end{cases} \tag{11.2}$$

Here $N_{max}$ is the limiting value of the static frictional torque. It is convenient to express the value $N_{max}$ in terms of the maximal possible deflection angle $d$ of the flywheel at rest, when the driving rod (see Figure 11.1) is immovable at its middle position $\theta = 0$: $N_{max} = Dd$. The angle $d$ corresponds to the boundary of the stagnation interval. Dividing all terms of Eq. (11.1) by $J$, we get

$$\ddot{\varphi} + 2\gamma\dot{\varphi} + \omega_0^2 d \operatorname{sign} \dot{\varphi} + \omega_0^2 \varphi = \omega_0^2 \theta_0 \sin \omega t. \tag{11.3}$$

The damping constant $\gamma$ is a measure of the intensity of viscous friction. It is introduced here by the relation $2\gamma = B/J$. The frequency $\omega_0 = \sqrt{D/J}$ characterizes undamped natural oscillations. The sign $\dot{\varphi}$ function is meant to take the undetermined values between 1 and $-1$ at zero argument, which corresponds to stick phase. The actual value of the static dry friction torque is such that the system is in equilibrium. The differential equation for an oscillator with dry friction, Eq. (11.1), as well as Eq. (11.3), is nonlinear because the torque $N_{fr}(\dot{\varphi})$ abruptly changes when the sign of $\dot{\varphi}$ changes at the extreme points of oscillation. This is the so-called Filippov system [70]. In the idealized case of the z-characteristic this is a piecewise smooth system, and we may consider the following two linear equations instead of Eq. (11.3):

$$\ddot{\varphi} + 2\gamma\dot{\varphi} + \omega_0^2(\varphi + d) = \omega_0^2 \theta_0 \sin \omega t \qquad \text{for} \quad \dot{\varphi} > 0, \tag{11.4}$$

$$\ddot{\varphi} + 2\gamma\dot{\varphi} + \omega_0^2(\varphi - d) = \omega_0^2 \theta_0 \sin \omega t \qquad \text{for} \quad \dot{\varphi} < 0. \tag{11.5}$$

Whenever the sign of the angular velocity $\dot{\varphi}$ changes, the pertinent equation of motion also changes. The nonlinear character of the problem reveals itself in alternate transitions from one of the linear Eqs. (11.4)–(11.5) to the other.

## 11.1.4    Damping of Free Oscillations under Dry Friction

For the case of free (unforced) oscillations the right-hand side of Eqs. (11.4)–(11.5) is zero. The case in which the dry friction is absent (the dead zone vanishes: $d = 0$) and damping of free oscillations occurs solely due to viscous friction, is discussed in Chapter 1. For this idealized case the differential equation of motion becomes linear. It has a well known analytical solution, according to which the amplitude of free oscillations under viscous friction decreases exponentially with time. That is, the consecutive maximal deflections of the oscillator from its equilibrium position form a diminishing geometric progression because their ratio is constant.

In an idealized linear system such oscillations continue indefinitely, their amplitude asymptotically approaching zero. The duration of exponential damping can be characterized by a conventional decay time $\tau = 1/\gamma$. The exponential character of damping caused by viscous friction follows from the proportionality of friction to velocity. Some other relationship between friction and velocity produces damping with different characteristics.

The role of dry friction in the damping of free oscillations is considered in detail in Chapter 2. The solution to Eqs. (11.4)–(11.5) for non-zero dry friction ($d \neq 0$) can be found by using the method of the stage-by-stage integration of each of the linear equations for the half-cycle during which the direction of motion is unchanged. These solutions are then joined together at the instants of transition from one equation to the other in such a way that the displacement at the end point of one half-cycle becomes the initial displacement at the beginning of the next half-cycle. This array of solutions continues until the end point of a half-cycle lies within the dead zone.

An important feature of free oscillations damped by dry friction is that the motion completely ceases after a finite number of cycles. As the system oscillates, each subsequent change of its velocity occurs at a smaller displacement from the mid-point of the stagnation interval. Eventually the turning point of the motion occurs within the stagnation interval, where static friction can balance the restoring torque of the spring, and so the motion abruptly stops. At which point of the interval this event occurs depends on the initial conditions, which may vary from one situation to the next.

In systems with both dry and viscous friction the damping of oscillations can also be investigated by the stage-by-stage solving of the equations of motion and by using the mechanical state at the end of the previous half-cycle as the initial conditions for the next in turn half-cycle. The phase trajectory consists in this case of the shrinking alternating halves of spiral loops that are characteristic of a linear damped oscillator. The focal points of these spirals alternate between the boundaries of the stagnation interval. The loops of the phase trajectory are no longer

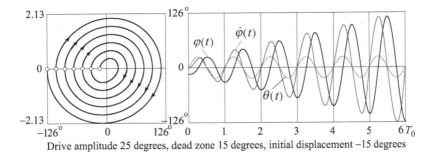

Drive amplitude 25 degrees, dead zone 15 degrees, initial displacement −15 degrees

Figure 11.2: Phase trajectory with Poincaré sections (left) and graphs of $\varphi(t)$ and $\dot{\varphi}(t)$ (right) for oscillations at resonance with dry friction.

equidistant. Nevertheless their shrinking does not last indefinitely: The phase trajectory in this case also terminates after some finite number of turns around the origin when it reaches the stagnation interval on the $\varphi$-axis.

## 11.2 Sinusoidally Driven Oscillator with Dry Friction

### 11.2.1 Resonance in the Oscillator with Dry Friction under Sinusoidal Excitation

In this section we analyze forced oscillations of the torsion spring pendulum in conditions of resonance, that is, when the frequency of excitation $\omega$ equals natural frequency $\omega_0$ of the oscillator ($T = T_0 = 2\pi/\omega_0$). Generally at large enough dry friction sticking may occur: The flywheel remains at rest for a finite time after the velocity reaches zero. However, if the amplitude of excitation $\theta_0$ in Eqs. (11.4)–(11.5) exceeds some threshold value, the motion of the flywheel is purely sliding (non-sticking), and in the absence of viscous friction the amplitude of oscillations grows indefinitely. An example of such resonant oscillations is shown in Figure 11.2. The phase trajectory and the graphs of $\varphi(t)$ and $\dot{\varphi}(t)$ are obtained by computer simulation that is based on numeric integration of Eqs. (11.4)–(11.5). We note the linear growth of the amplitude: The succession of maximal deflections of the flywheel forms an arithmetic progression. Next we find analytically the threshold for excitation of such growing oscillations, and calculate the increment of the amplitude after each driving cycle.

We choose for simplicity the initial deflection $\varphi(0)$ of the flywheel coinciding with the left boundary of the dead zone, that is, $\varphi(0) = -d$, and initial angular velocity zero: $\dot{\varphi}(0) = 0$. Such initial conditions provide the sliding (non-sticking) motion from the very beginning with two turnarounds during each cycle of excitation. During the first half of the excitation period ($0 < t < T_0/2$) the angular velocity is positive ($\dot{\varphi}(t) > 0$), and we should use Eq. (11.4). The solution to this equation (with $\gamma = 0$), satisfying the above indicated initial conditions, can be written as follows:

$$\varphi(t) = -\frac{1}{2}\theta_0(\omega_0 t \cos\omega_0 t - \sin\omega_0 t) - d, \quad \dot{\varphi}(t) = \frac{1}{2}\theta_0\omega_0^2 t \sin\omega_0 t, \quad 0 < t < T_0/2.$$
(11.6)

According to (11.6), the next maximal elongation to the right side occurs at $t = T_0/2$ and equals $\frac{1}{2}\pi\theta_0 - d$. This elongation is greater in magnitude than the preceding (initial) elongation $d$ to the left side by $\frac{1}{2}\pi\theta_0 - 2d$.

To find the increment in the amplitude during the second half-cycle of excitation, when the flywheel rotates in the opposite direction, we should use Eq. (11.5). An analytical solution to this equation is given below in Section 11.2.2. It occurs that the increment in amplitude during the second half-cycle is the same as during the first half-cycle. Therefore during the whole cycle the increment in amplitude equals $\pi\theta_0 - 4d$. Specifically, for $\theta_0 = 25°$ and $d = 15°$ (the values corresponding to the simulation shown in Figure 11.2) the amplitude should increase during each cycle by $18.54°$. The simulation in Figure 11.2 shows that during the first six cycles the amplitude increased by $126° - 15° = 111°$, which gives for increment during one cycle the value $18.5°$, in good agreement with the theoretical prediction.

## 11.2.2  Analytical Solution for the Second Half-Cycle of the Resonant Excitation

For the first half-cycle of excitation at resonance ($\omega = \omega_0$) the motion of the flywheel is given by Eq. (11.6), if initially the flywheel is at rest ($\dot{\varphi}(0) = 0$) exactly at the left side of the stagnation zone: $\varphi(0) = -d$. If we take some arbitrary initial deflection $\varphi(0) = \varphi_0$ to the left side from the equilibrium ($\varphi_0 < 0$), which lies beyond the dead zone ($|\varphi_0| > d$) and initial velocity $\dot{\varphi}(0) = 0$, the motion of the flywheel will also be non-sticking from the very beginning, and during the time interval $0 < t < T_0/2$ will be described by the following expression:

$$\varphi(t) = (\varphi_0 + d)\cos\omega_0 t - \frac{1}{2}\theta_0(\omega_0 t \cos\omega_0 t - \sin\omega_0 t) - d, \quad 0 < t < T_0/2.$$
(11.7)

At the end of the first half-cycle (at $t = T_0/2$) the angular velocity of the flywheel becomes zero, while its deflection to the right side reaches $\varphi_1 = -\varphi_0 + \frac{1}{2}\theta_0\pi - 2d$. These values of $\varphi$ and $\dot{\varphi}$ should be used as the initial conditions at $t = T_0/2$ for the differential Eq. (2.5) that describes (with $\gamma = 0$) the second half-period $T_0/2 < t < T_0$ of the forced motion, during which $\dot{\varphi} < 0$. To solve this equation, it is convenient to move the time origin $t = 0$ to $T_0/2$. In these new notations Eq. (2.5) takes the following form:

$$\ddot{\varphi} + \omega_0^2(\varphi + d) = -\omega_0^2\theta_0\sin\omega_0 t.$$
(11.8)

The solution to Eq. (11.8), satisfying initial conditions $\varphi(0) = \varphi_1$ and $\dot{\varphi}(0) = 0$, can be written as follows:

$$\varphi(t) = (\varphi_1 - d)\cos\omega_0 t + \frac{1}{2}\theta_0(\omega_0 t \cos\omega_0 t - \sin\omega_0 t) + d. \tag{11.9}$$

To find the angular position $\varphi(T_0)$ and the angular velocity $\dot{\varphi}(T_0)$ of the flywheel at the end of the first cycle of excitation, we should substitute $t = T_0/2$ into Eq. (11.9):

$$\varphi(T_0) = -\varphi_1 + d - \frac{1}{2}\theta_0\pi + d = \varphi_0 - \theta_0\pi + 4d. \tag{11.10}$$

Hence the magnitude of angular elongation to the left increased during the first cycle of excitation by the value $|\varphi(T_0) - \varphi_0| = \pi\theta_0 - 4d$. This increment is independent of the initial deflection $\varphi_0$. The succession of maximal deflections at resonance in the oscillator with dry friction forms an increasing arithmetic progression.

In case $d = 0$ (zero width of the dead zone, that is, absence of dry friction) the solution given by Eq. (11.7) takes the following form:

$$\varphi(t) = \varphi_0 \cos\omega_0 t - \frac{1}{2}\theta_0(\omega_0 t \cos\omega_0 t - \sin\omega_0 t). \tag{11.11}$$

Obviously, for initial conditions $\varphi(0) = \varphi_0$, $\dot{\varphi}(0) = 0$ this solution is valid for any $t$ value, not only for the first half-cycle of excitation $0 < t < T_0/2$. According to Eq. (11.11), in the absence of any friction (dry and viscous), the amplitude of resonant forced oscillations changes in magnitude during one cycle of excitation by the same amount, $\pi\theta_0$. If the oscillator is excited from the state of rest in the equilibrium position, its amplitude grows linearly from the very beginning. This growth continues indefinitely. From the energy considerations, this can be easily explained by certain phase relations between rotary oscillations of the flywheel and the sinusoidally varying torque exerted on the flywheel by the spring: This torque always acts in the direction of rotation, thus increasing the energy of the flywheel.

However, if the initial displacement of the flywheel is positive ($\varphi(0) > 0$), the external torque at the initial stage is directed against the angular velocity, and the amplitude of oscillations diminishes, in spite of the exact tuning to resonance, through value $\pi\theta_0$ during each cycle. The energy is transferred from the oscillator to the exciter. This situation is illustrated in Figure 11.3. After the amplitude reduces to zero, the phase relations between the exciting rod and the flywheel become favorable for the transfer of energy to the oscillator, and the amplitude starts to grow indefinitely. In the absence of dry friction, the initial linear reduction and further growth of the amplitude occur equally fast, in contrast to the case with dry friction (see Figure 11.6, p. 319), in which friction speeds up the reduction and slows down the growth of the amplitude.

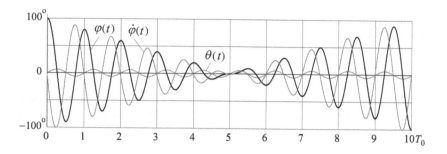

Figure 11.3: Oscillations at resonance without friction (with initial displacement).

### 11.2.3   The Threshold of the Resonant Growth

The growth of oscillations amplitude occurs if the value of increment $\pi\theta_0 - 4d$ during a cycle of excitation is positive. Hence the threshold of resonance $(\theta_0)_{min}$ for the oscillator with dry friction is given by the following condition:

$$\theta_0 > \frac{4}{\pi} d, \qquad (\theta_0)_{min} = \frac{4}{\pi} d. \qquad (11.12)$$

For given width $d$ of the dead zone (for given dry friction), Eq. (11.12) defines the critical (minimal) value $(\theta_0)_{min}$ of the drive amplitude, which provides non-sticking forced oscillations of the flywheel after a rather short transient.

However, during the transient, depending on the initial conditions, sticking is possible. For $\theta_0 > (\theta_0)_{min}$, after the transient is over, at the initial moment $t_n = nT = nT_0$ of each cycle of excitation in-turn, angular velocity $\dot{\varphi}(t_n)$ of the flywheel is zero: $\dot{\varphi}(nT) = 0$. This means that Poincaré sections in the phase plane (corresponding to time moments $t_n = nT$) approach the abscissa axis during the transient and remain on its negative side further on. Since the increment in the elongation is the same for each cycle, the points of Poincaré sections on the axis are equidistant (see Figure 11.2).

Stationary periodic oscillations at the threshold conditions are shown in Figure 11.4. At arbitrary initial values of $\varphi$ and $\dot{\varphi}$ the phase trajectory eventually approaches a limit cycle similar to the cycle shown in the left-hand side of Figure 11.4. The amplitude of steady-state forced oscillations at the threshold depends on initial conditions. If initial velocity is zero ($\dot{\varphi}(0) = 0$), steady-state oscillations occur from the very beginning, without any transient, in the case where initial displacement $\varphi(0)$ is negative and lies beyond the dead zone, that is, if $\varphi(0) < 0$, $|\varphi(0)| \geq d$. The amplitude of these oscillations equals $|\varphi(0)|$. This mode of oscillations is unstable with respect to variations in parameters $\theta_0$ and $d$: A slight increment of the drive amplitude or decrement in the dead zone width causes an indefinite growth of the amplitude.

If the amplitude of the exciter $\theta_0$ is smaller than the critical value $(\theta_0)_{min}$ given by Eq. (11.12), but greater than the dead zone width $d$, a steady-state regime

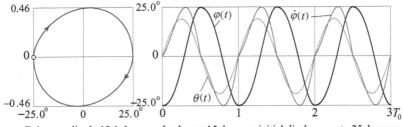

Drive amplitude 19.1 degrees, dead zone 15 degrees, initial displacement −25 degrees

Figure 11.4: Oscillations at the threshold conditions at resonance with dry friction.

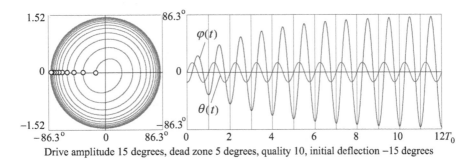

Drive amplitude 15 degrees, dead zone 5 degrees, quality 10, initial deflection −15 degrees

Figure 11.5: Oscillations at resonance with dry and viscous friction.

with two sliding phases and two sticking phases establishes after the transient is over. For $\theta_0$ smaller than the dead zone width $d$, the flywheel, depending on the initial conditions, either remains immovable from the very beginning, or makes several movements with sticking and then finally stops at some point of the dead zone.

### 11.2.4 Resonance in the Presence of Dry and Viscous Friction

The resonant growth of amplitude over the threshold is restricted if some amount of viscous friction is present in the system. In a dual-damped system steady-state oscillations with a constant amplitude eventually establish for arbitrary initial conditions. An example of resonant oscillations in the system with both dry and viscous friction is shown in Figure 11.5. Eqs. (11.4)–(11.5) allow us to calculate the amplitude $a$ of such resonant symmetric steady-state oscillations.

We choose the time origin $t = 0$ at the beginning of the next-in-turn drive cycle. At this moment the flywheel occurs at the extreme displacement to the left side ($\varphi(0) = -a$) and has the angular velocity zero ($\dot{\varphi}(0) = 0$). During the first half-cycle of the drive ($0 < t < T_0/2$) it moves to the right, so that $\dot{\varphi}$ is positive during this interval. Therefore we should use Eq. (11.4) with $\omega = \omega_0$ in its right-hand part. It is convenient to introduce instead of $\varphi(t)$ a new unknown function $\psi(t) = \varphi(t) + d$, which, according to (11.4), satisfies the following equation:

$$\ddot{\psi} + 2\gamma\dot{\psi} + \omega_0^2\psi = \omega_0^2\theta_0 \sin \omega_0 t. \qquad (11.13)$$

We can search for its periodic partial solution in the form $\psi(t) = A \cos \omega_0 t$. This function satisfies Eq. (11.13), if $A = -(\omega_0/2\gamma)\theta_0 = -Q\theta_0$. Next we add to this partial solution the general solution of the corresponding homogeneous equation:

$$\psi(t) = -Q\theta_0 \cos \omega_0 t + (C \cos \omega_0 t + S \sin \omega_0 t) \exp(-\gamma t). \qquad (11.14)$$

It follows from the initial condition $\dot{\psi}(0) = 0$ that in (11.14) $S = (\gamma/\omega_0)C$. To find $C$, we require that in the steady-state symmetric regime elongations to both sides should be equal: $\varphi(0) = -\varphi(T_0/2)$. From this condition we get

$$C = \frac{2d}{1 - \exp(-\gamma T_0/2)} = \frac{2d}{1 - \exp(-\pi/2Q)}. \qquad (11.15)$$

Substituting these $C$ and $S$ values in Eq. (11.14), we obtain the time dependence of the angular displacement $\varphi(t) = \psi(t) - d$ for the first half-cycle of excitation. The desired amplitude $a$ of this steady-state resonant oscillation is given by $-\varphi(0)$:

$$a = Q\theta_0 - d \left( \frac{2}{1 - \exp(-\pi/2Q)} - 1 \right) \approx Q \left( \theta_0 - \frac{4d}{\pi} \right). \qquad (11.16)$$

The latter approximate expression is valid in the case of rather weak viscous friction, when $Q \gg 1$. In the absence of dry friction (at $d = 0$) the growth of amplitude at resonance is restricted due to viscous friction by the value $Q\theta_0$, which is $Q$ times greater than the amplitude of the driving rod $\theta_0$, in accordance with the first term in Eq. (11.16). With dry friction, the steady-state amplitude is approximately $Q$ times greater than the excess of the drive amplitude $\theta_0$ over the threshold $4d/\pi$. We emphasize that dry friction alone is unable to restrict the growth of amplitude over the threshold at $\omega = \omega_0$. Nevertheless, Eq. (11.16) shows that when dry friction is added to the system with viscous friction, the steady-state amplitude at resonance is smaller than $Q\theta_0$. From the numerical simulation (Figure 11.5) we see that with $\theta_0 = 15°$ and $Q = 10$ the resonant amplitude equals only 86.3°, if the dead zone $d$ equals 5° (compare with $Q\theta_0 = 150°$ at $d = 0$). This experimental value 86.3° is in good agreement with the theoretical result expressed by (11.16), according to which the steady-state amplitude should be 86.2°.

In conditions of exact tuning to resonance (at $\omega = \omega_0$) the energy is transferred to the oscillator from the external source (from the exciter) with maximal efficiency, if at the beginning of each excitation cycle the flywheel occurs at an extreme elongation to the left-hand side. Indeed, in this case the sinusoidally varying external torque exerted on the flywheel by the exciter acts during the whole cycle in the direction of the flywheel rotation, and over the threshold (at $\pi\theta_0 > 4d$) overcomes the torque of dry friction: The amplitude grows linearly (see Figure 11.2) increasing during a cycle by $\pi\theta_0 - 4d$. Conversely, if at the beginning of the excitation cycle the flywheel occurs at an extreme elongation to the right-hand side,

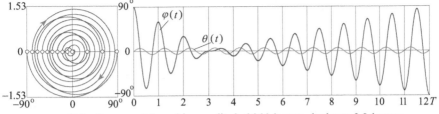

Drive frequency $1.0\omega_0$, drive amplitude 6.366 degrees, dead zone 2.5 degrees,
initial angle 90 degrees, initial angular velocity $0.0\omega_0$

Figure 11.6: Phase diagram with Poincaré sections and graph of $\varphi(t)$ of oscillations with dry friction at resonance with initial deflection $\varphi(0) = +90°$.

the external torque of the spring during the whole cycle is directed against the flywheel's angular velocity together with the frictional torque. In this case the amplitude reduces during each cycle by the amount $\pi\theta_0 + 4d$. After the amplitude reduces to zero, the phase relations between the flywheel and exciter change to the opposite and become favorable for the transfer of energy to the oscillator: The amplitude begins to grow.

An example of such behavior is shown in Figure 11.6. At the drive amplitude $\theta_0 = 6.366°$ and the dead zone $2.5°$, the amplitude linearly reduces during each cycle of the initial stage of the process by $\pi\theta_0 + 4d = 30°$. After 3 full driving cycles the amplitude diminishes from initial $90°$ to zero. During the further resonant growth the amplitude linearly increases during each cycle by $\pi\theta_0 - 4d = 10°$, and after next 9 cycles becomes $90°$.

## 11.2.5  Non-Resonant Forced Oscillations

In the case of exact tuning to resonance, in contrast to the oscillator with viscous damping, dry friction alone is unable to restrict the growth of the amplitude of forced oscillations over the threshold. In non-resonant cases ($\omega \neq \omega_0$) of harmonic excitation, after a transient of a finite duration, steady-state oscillations of constant amplitude can establish due to dry friction even in the absence of viscous friction. Non-resonant forced oscillations in the oscillator with dry friction received significant attention in the literature. Since the pioneer's work of Den Hartog [57] in 1930, several researchers [59]–[63] have investigated the system analytically and numerically, and obtained exact solutions, describing the steady-state non-sticking motion with two turnarounds per cycle for a harmonically excited dry friction oscillator.

An example of such non-resonant oscillations in the system with a considerable amount of dry friction is shown in Figure 11.7. The periodic motion consists of two non-sticking phases of equal duration $T/2$. The angular velocity is negative in one phase and positive in the other. Unfortunately, it is impossible to express

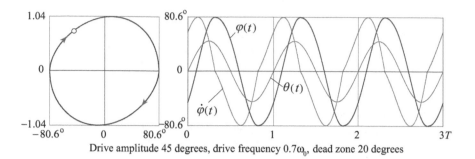

Drive amplitude 45 degrees, drive frequency $0.7\omega_0$, dead zone 20 degrees

Figure 11.7: Phase diagram and graphs of $\varphi(t)$ and $\dot{\varphi}(t)$ of non-sticking steady-state oscillations with dry friction at $\omega = 0.7\omega_0$.

$\varphi(t)$ in a closed analytical form, because the turnaround points dividing the two phases are determined by a transcendental equation. To find the amplitude $a(\omega)$ of this symmetric oscillation, it is sufficient to consider only one phase between successive turnarounds, which is described by differential Eq. (2.5). The calculations are similar to those described above for the resonant case (though more complicated). Using periodicity and symmetry of the desired solution, we find the following dependence of the steady-state amplitude on the driving frequency $\omega$ and amplitude $\theta_0 = 45°$ of the excitation:

$$a(\omega) = \theta_0 \sqrt{\frac{1}{(1 - \omega^2/\omega_0^2)} - \frac{d^2(\omega_0/\omega)^2 \sin^2 \pi(\omega_0/\omega)}{\theta_0^2(\cos \pi(\omega_0/\omega) + 1)^2}}. \qquad (11.17)$$

For the frequency of excitation $\omega = 0.7\omega_0$, drive amplitude $\theta_0 = 45°$, and dead zone $d = 20°$ we get from (11.17) for the steady-state amplitude the value $80.63°$, which is in perfect agreement with the numerical simulation illustrated by Figure 11.7.

Expression (11.17) for the steady-state amplitude of non-sticking oscillations coincides (in somewhat different notations) with results published earlier in the literature [59], [63]. Frequency-response resonant curves (amplitude-frequency characteristics) given by (11.17) for the oscillator with dry friction are shown in Figure 11.8 for several values of relative width $d/\theta_0$ of the dead zone (in the frequency region $\omega > 0.5\omega_0$).

We emphasize that expression (11.17) is valid only for sliding (non-sticking) symmetric motions of the oscillator. Such motions are possible if the following simple implicit condition (see [59]) on the parameters is fulfilled:

$$a(\omega, \theta_0, d) \geq \frac{d}{\theta_0} \left(\frac{\omega_0}{\omega}\right)^2. \qquad (11.18)$$

Solving Eq. (11.18) numerically for the unknown $d$ at $\omega = 0.7\omega_0$ and $\theta_0 = 45°$ (these values were used for the simulation shown in Figure 11.7), we find that

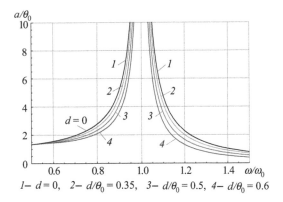

1– $d = 0$,   2– $d/\theta_0 = 0.35$,   3– $d/\theta_0 = 0.5$,   4– $d/\theta_0 = 0.6$

Figure 11.8: Amplitude–frequency characteristics of sinusoidally driven spring oscillator with dry friction damping.

the maximal width $d_{\text{max}}$ of the dead zone for which steady-state non-sticking symmetric motions are possible equals 32.5°. A closed-form formula for the domain of steady-state symmetric non-sticking oscillatory responses was obtained in [60]. It provides the minimum driving torque amplitude $(\theta_0)_{\text{min}}$ required to prevent sticking for given width $d$ of the dead zone and given drive frequency $\omega$:

$$(\theta_0)_{\text{min}} = d\sqrt{\left(\frac{\omega_0^2}{\omega^2} - 1\right)^2 \left[1 + \frac{(\omega/\omega_0)^2 \sin^2(\pi\omega_0/\omega)}{(1 + \cos(\pi\omega_0/\omega)^2}\right]}. \qquad (11.19)$$

Certainly, this equation can also be used to find the maximal width $d_{\text{max}}$ of the dead zone for which steady-state non-sticking symmetric motions are possible at given frequency $\omega$ and amplitude $\theta_0$ of the driving torque. Substituting $\omega = 0.7\omega_0$ and $\theta_0 = 45°$ in (11.19), we get $d_{\text{max}} = 32.5°$, in accordance with the above estimate (11.18).

The upper part of Figure 11.9 illustrates oscillations occurring on this edge of such a non-sticking regime (dead zone 32.5°). For initial conditions $\varphi(0) = 0$, $\dot{\varphi}(0) = 0$, sticking occurs several times during a short transient, which ends with non-sticking symmetric steady-state oscillations. According to (11.17), their amplitude must equal 66.3°, in good agreement with the simulation.

For comparison, the lower part of Figure 11.9 shows the steady-state oscillations at the same values of the frequency and amplitude of the exciter ($\omega = 0.7\omega_0$ and $\theta_0 = 45°$), but for a somewhat greater amount of dry friction (dead zone 37°). In this case sticking occurs twice during each cycle of excitation.

Not surprisingly, the amplitude of steady-state symmetric forced oscillations with sticking observed in the simulation is smaller than the theoretical value that Eq. (11.17) predicts (55° as opposed to 62°).

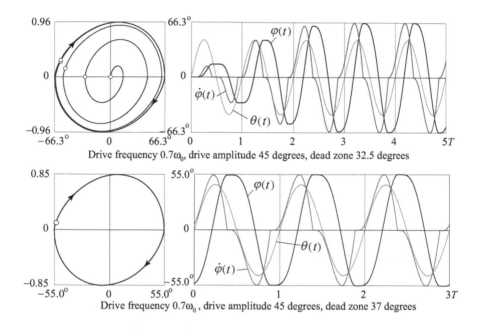

Drive frequency 0.7$\omega_0$, drive amplitude 45 degrees, dead zone 32.5 degrees

Drive frequency 0.7$\omega_0$, drive amplitude 45 degrees, dead zone 37 degrees

Figure 11.9: Phase diagram and graphs of $\varphi(t)$ and $\dot{\varphi}(t)$ of the transient that leads to non-sticking steady-state oscillations at $\omega = 0.7\omega_0$ and critical width $d = 32.5°$ of the dead zone (upper part), and steady-state symmetric oscillations with sticking at $d = 37°$ (lower part).

## 11.3    Harmonic Excitation at Sub-Resonant Frequencies

Generally characteristics of steady-state behavior of the periodically forced oscillator with dry friction, as well as of the oscillator with viscous friction, are uniquely defined by the system parameters (natural frequency and the quality factor), and by the frequency and amplitude of the excitation. Certain exceptions are revealed if the frequency $\omega$ of sinusoidal excitation coincides with one of subharmonics of the natural frequency: $\omega = \omega_0/n$, where $n$ is an integer number.

Analytical steady-state solutions at sub-harmonic excitation were considered for the first time in [64]. Here we suggest a simpler and physically more transparent approach to the problem, and discuss peculiarities of such oscillations in more detail. Simple closed-form solutions are illustrated by time-dependent graphs and phase orbits obtained with the help of computer simulations.

According to Eq. (11.17), at frequencies of excitation $\omega = \omega_0/2$, $\omega = \omega_0/4$,

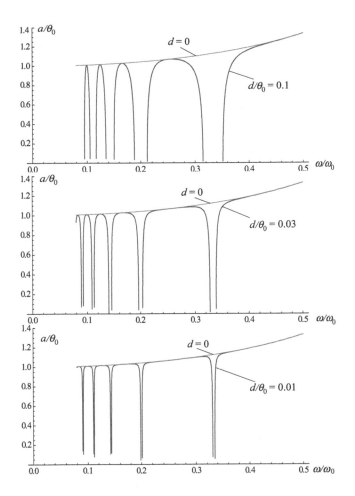

Figure 11.10: Frequency–response curves for the oscillator with dry friction given by Eq. (11.17) at excitation frequencies $\omega < \omega_0/2$ for several values of the dead zone width ($d/\theta_0 = 0.1$, $d/\theta_0 = 0.03$, $d/\theta_0 = 0.01$).

...the amplitude of steady-state non-sticking symmetric oscillation, independently of the dead zone width $d$, should be equal to the amplitude of forced steady-state oscillation in the absence of friction (that is, at $d = 0$): $a(\omega) = \theta_0/(1 - \omega^2/\omega_0^2)$. (Certainly, this arbitrary value of $d$ should satisfy the condition (11.19) for non-sticking motions, which at $\omega = \omega_0/n$ gives $d \leq \theta_0/3$.)

Figure 11.10 shows that frequency-response curves for different $d$ values at $\omega = \omega_0/2$ graze the curve for $d = 0$. Computer simulations testify that in these cases steady-state oscillations are generally asymmetric: The angular excursion to one side is greater than to the other. This means that at $\omega = \omega_0/n$ occurrence of special analytical solutions can be expected. Below we show that in contrast with

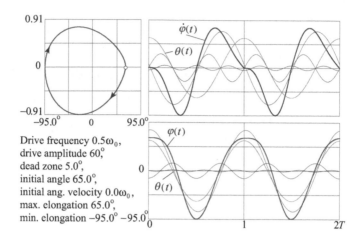

Figure 11.11: Phase diagram and time-dependent graphs of $\dot{\varphi}(t)$ and $\varphi(t)$ with the graphs of their harmonics for non-sticking asymmetric steady-state oscillations at $\omega = \omega_0/2$ and small width of the dead zone ($d = 5.0°$).

the general case of forced oscillations, for which the steady-state regime is described by the unique solution that is independent of the initial conditions, a continuum of asymmetric non-sticking solutions exists at $\omega = \omega_0/2n$. Each solution gives an asymmetric limit cycle (attractor) that corresponds to initial conditions from a certain basin of attraction.

To study the excitation of the oscillator at $\omega = \omega_0/n$, it is more convenient to choose further on for the time origin $t = 0$ the moment at which the exciter reaches its maximal deflection $\theta_0$, that is, to assume for $\theta(t)$ the following time dependence: $\theta(t) = \theta_0 \cos \omega t$. With this choice, as we will see later, the turnarounds in the steady-state motion of the oscillator occur approximately at $t = 0$ and $t = T/2$. This simplifies the form of analytical solutions.

For definiteness we restrict further discussion to the case $n = 2$. If $\omega = \omega_0/2$, the spectrum of steady-state asymmetric oscillations at sufficiently small dry friction (narrow dead zone) consists primarily of the principal harmonic with the frequency of excitation $\omega$ and its second harmonic, whose frequency $2\omega$ equals the natural frequency $\omega_0$ of the oscillator (see Figure 11.11). A small admixture of the third harmonic is also noticeable.

Mathematically, the principal harmonic corresponds to the forced periodic partial solution of the nonhomogeneous differential equation of motion (11.3) with $\gamma = 0$ and with the sinusoidal forcing term $\omega_0^2 \theta_0 \cos \omega t$ whose frequency $\omega$ equals $\omega_0/2$. This partial solution is $\frac{4}{3}\theta_0 \cos \omega t$. The torque of dry friction moves the mid-point of this oscillation to $-d$ (left boundary of the dead zone) if $\dot{\varphi} > 0$ and to $d$ if $\dot{\varphi} < 0$. This periodic (square-wave) displacement of the mid-point caused by dry friction explains the appearance of the third harmonic in the solution shown in Figure 11.11. The second harmonic with the frequency $2\omega = \omega_0$

is the general solution of the homogeneous equation that corresponds to (11.3). This general solution describes natural oscillations with the frequency $\omega_0 = 2\omega$, and can be represented (at $\gamma = 0$) as $A \cos 2\omega t + B \sin 2\omega t$, where $A$ and $B$ are arbitrary constants. In contrast to a system with viscous friction, now this general solution — oscillation with the natural frequency — does not damp out in the course of time during the transient. The simulation shows (in accordance with the requirement $\dot{\varphi}(0) = \dot{\varphi}(T/2) = 0$) that in the steady-state regime the phase of this second harmonic is such that $B = 0$ (see Figure 11.11). Hence the asymmetric steady-state motion at $\omega = \omega_0/2$ can be approximately described by the following equations:

$$\varphi_+(t) = \tfrac{4}{3}\theta_0 \cos \omega t + A_+ \cos 2\omega t - d, \qquad \dot{\varphi} > 0, \qquad (11.20)$$
$$\varphi_-(t) = \tfrac{4}{3}\theta_0 \cos \omega t + A_- \cos 2\omega t + d, \qquad \dot{\varphi} < 0, \qquad (11.21)$$

and

$$\dot{\varphi}_+(t) = -\tfrac{4}{3}\omega\theta_0 \sin \omega t - 2\omega A_+ \sin 2\omega t, \qquad \dot{\varphi} > 0, \qquad (11.22)$$
$$\dot{\varphi}_-(t) = -\tfrac{4}{3}\omega\theta_0 \sin \omega t - 2\omega A_- \sin 2\omega t, \qquad \dot{\varphi} < 0. \qquad (11.23)$$

One condition on constants $A_+$ and $A_-$ follows from the requirement of continuity of $\varphi(t)$ at the turnaround points, when the sign of velocity reverses. These are the moments $t = 0$ and $t = T/2$ (see Figure 11.11). From $\varphi_+(0) = \varphi_-(0)$ we get $A_+ - A_- = 2d$, or $A_+ = A_- + 2d$ (condition $\varphi_-(T/2) = \varphi_+(T/2)$ yields the same relation between $A_+$ and $A_-$). Therefore only one of these constants remains arbitrary.

The steady-state regime described by Eqs. (11.20)–(11.23) occurs from the very beginning (that is, without any transient) if the initial conditions are chosen properly. At $t = 0$ we get from (11.22) or (11.23) that the required initial angular velocity equals zero: $\dot{\varphi}(0) = 0$. This value is independent of the dead zone width $d$ and the drive amplitude $\theta_0$. Since during the first half-cycle $\dot{\varphi}(0)$ is negative, for the required initial displacement we should use Eq. (11.21), which yields $\varphi_0 = \varphi(0) = \tfrac{4}{3}\theta_0 + A_- + d$. We see that the arbitrary constants $A_+$ and $A_-$, which determine the contribution of the second harmonic into the steady-state motion, depend on an arbitrary initial displacement $\varphi_0$:

$$A_+ = \varphi_0 - \frac{4}{3}\theta_0 + d, \qquad A_- = \varphi_0 - \frac{4}{3}\theta_0 - d. \qquad (11.24)$$

This means that in the system with dry friction, in contrast to the oscillator with viscous friction, different initial displacements generally lead to different regimes of steady-state oscillations (to different limit cycles). Substituting these values of $A_+$ and $A_-$ in (11.20)–(11.23), we get the closed-form analytical solutions for asymmetric steady-state subresonant regimes of the dry-friction oscillator at $\omega = \omega_0/2$. Below we show that such steady-state regimes occur from the very beginning if the value of $\varphi_0$ belongs to a certain interval.

Now we can derive some interesting properties of the discussed steady-state solutions. Extreme elongations correspond to the turnaround points and hence occur at $t \approx 0$ and at $t \approx T/2$. Maximum displacement to the right-hand side occurs

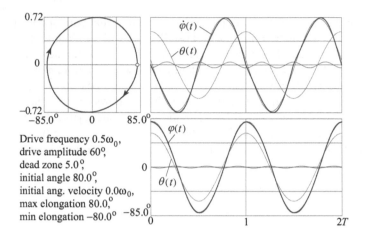

Figure 11.12: Phase diagram and time-dependent graphs of $\dot{\varphi}(t)$ and $\varphi(t)$ with the graphs of their harmonics for non-sticking symmetric steady-state oscillations at $\omega = \omega_0/2$ and small width of the dead zone ($d = 5.0°$).

at $t \approx 0$ and, according to Eq. (11.20) or (11.21), equals $\varphi_{\max} = \frac{4}{3}\theta_0 + A_- + d$ (or, equivalently, $\varphi_{\max} = \frac{4}{3}\theta_0 + A_+ - d$). Extreme elongation to the left-hand side occurs $t \approx T/2$ and equals $|\varphi_{\min}| = \frac{4}{3}\theta_0 - A_- - d = \frac{4}{3}\theta_0 - A_+ + d$.

We get that the total angular excursion $\varphi_{\max} + |\varphi_{\min}|$ between the extreme points for all possible solutions (11.20)–(11.21) equals $\frac{8}{3}\theta_0$:

$$\varphi_{\max} + |\varphi_{\min}| = \varphi_{\max} - \varphi_{\min} = \frac{8}{3}\theta_0. \tag{11.25}$$

It depends solely on the drive amplitude $\theta_0$, and does not depend on the intensity of dry friction (on the width $d$ of the dead zone).

The difference between the extreme elongations characterizes the asymmetry of this steady-state regime:

$$\varphi_{\max} - |\varphi_{\min}| = 2(A_+ - d) = 2(A_- + d) = 2(\varphi_0 - \frac{4}{3}\theta_0). \tag{11.26}$$

The extreme elongations to both sides are equal to one another if $\varphi_0 - \frac{4}{3}\theta_0 = 0$. In this case $A_+ = d$ and $A_- = -d$, and the second harmonic in the oscillation described by Eqs. (11.20)–(11.21) vanishes. Such symmetric steady-state oscillation with the amplitude $\frac{4}{3}\theta_0$ occurs only if the initial displacement $\varphi_0$ equals $\frac{4}{3}\theta_0$ and the initial velocity zero.

The phase trajectory and time-dependent graphs of $\dot{\varphi}(t)$ and $\varphi(t)$ (together with the graphs of their harmonics) for such symmetric oscillation are shown in Figure 11.12. (These graphs at $d \ll \theta_0$ almost merge with the graphs of their principal harmonics.)

The initial conditions that lead to the greatest asymmetry of the limit cycle can be found as follows. We are interested in the unsticking regime with two turnarounds per one excitation cycle. These turnarounds occur near $t = 0$ and $t = T/2$, when the angular velocity $\dot{\varphi}$ changes sign. We can rely on physical considerations in finding the condition for a turnaround occurring without sticking for a finite time interval. Indeed, to avoid sticking at this point, the restoring torque of the spring exerted on the flywheel should be greater or at least equal to the greatest possible torque of dry friction. The desired condition corresponds to the equality of these two torques. When the torque exerted by the spring equals the torque of static friction in magnitude, the angular acceleration $\ddot{\varphi}$ of the flywheel equals zero. Next we consider this condition for each of the turnarounds, occurring at $t = 0$ and $t = T/2$.

- For the first turnaround occurring at $t = 0$ we should require $\ddot{\varphi}(0) = 0$ using Eq. (11.21) that corresponds to negative angular velocity $\dot{\varphi}(t) < 0$ (see Figure 11.11). From this requirement we find immediately $A_- = -\frac{1}{3}\theta_0$. Substituting this value to Eq. (11.24), we get the first (lower) boundary $\varphi_{0(\text{lower})}$ of admissible initial deflections:

$$\varphi_{0(\text{lower})} = \theta_0 + d. \tag{11.27}$$

With this initial displacement one of the two possible most asymmetric steady-state oscillations occurs, in which the extreme deflections are:

$$\varphi_{\max} = \theta_0 + d, \qquad \varphi_{\min} = -\frac{5}{3}\theta_0 + d. \tag{11.28}$$

- To find the other boundary of the admissible initial deflections $\varphi_{0(\text{upper})}$, we should require $\ddot{\varphi}(T/2) = 0$ using Eq. (11.20). This yields $A_+ = \frac{1}{3}\theta_0$, and for the second (upper) boundary of admissible initial deflections we get:

$$\varphi_{0(\text{upper})} = \frac{5}{3}\theta_0 - d. \tag{11.29}$$

With this initial displacement the other of the two possible most asymmetric steady-state oscillations occurs, in which the extreme deflections are:

$$\varphi_{\max} = \frac{5}{3}\theta_0 - d, \qquad \varphi_{\min} = -\theta_0 - d. \tag{11.30}$$

To verify these theoretical predictions with the help of numerical simulations, we choose the drive amplitude $\theta_0 = 60°$, which means that in the steady-state regime the total angular excursion between the extreme points $\varphi_{\max} + |\varphi_{\min}|$ should equal $\frac{8}{3}\theta_0 = 160°$ independently of the intensity of dry friction. Width $d$ of the dead zone in this simulation equals $5°$. One of the two possible steady-state oscillations with the greatest asymmetry (see Figure 11.11) occurs from the very beginning (without any transient), according to (11.27), at the initial conditions

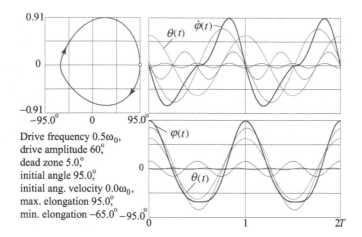

Drive frequency 0.5$\omega_0$,
drive amplitude 60°,
dead zone 5.0°
initial angle 95.0°
initial ang. velocity 0.0$\omega_0$,
max. elongation 95.0°
min. elongation −65.0° −95.0°

Figure 11.13: Phase diagram and time-dependent graphs of $\dot{\varphi}(t)$ and $\varphi(t)$ with their harmonics for non-sticking asymmetric steady-state oscillations at $\omega = \omega_0/2$ and small width of the dead zone ($d = 5.0°$).

$\varphi(0) = \theta_0 + d = 65°$, $\dot{\varphi}(0) = 0$. The extreme elongations should be $\varphi_{max} = \theta_0 + d = 65°$ and $\varphi_{min} = -\frac{5}{3}\theta_0 + d = -95°$. These theoretical predictions agree perfectly well with the computer simulation (see Figure 11.11).

Steady-state oscillation with equal elongations to both sides ($\varphi_{max} = |\varphi_{min}| = 80°$) at the same values of the system parameters ($\theta_0 = 60°$, $d = 5°$) occurs if the initial displacement $\varphi_0 = \frac{4}{3}\theta_0 = 80°$. This case is illustrated by the simulation shown in Figure 11.12.

The second case of the greatest asymmetry ($\dot{\varphi}(T/2) = 0$) occurs, according to (11.29), at the initial conditions $\frac{5}{3}\theta_0 - d = 95°$, $\dot{\varphi}(0) = 0$. Extreme elongations in this case, according to (11.30), are $\varphi_{max} = \frac{5}{3}\theta_0 - d = 95°$ and $\varphi_{min} = -\theta_0 - d = -65°$. These asymmetric oscillations are illustrated by the simulation shown in Figure 11.13.

Three different limit cycles of non-sticking oscillations that are shown in Figures 11.11–11.13 correspond to the same values of the system parameters $\omega = \omega_0/2$, $\theta_0 = 60.0°$, and $d = 5.0°$. Actually, at $\omega = \omega_0/2$ there exists a continuum of different steady-state non-sticking motions with the same total angular excursion $\frac{8}{3}\theta_0$, proportional to the drive amplitude $\theta_0$. This is a manifestation of multistability — a typical feature of nonlinear systems. If the initial angular velocity $\dot{\varphi}(0)$ equals zero, and the initial angular displacement $\varphi_0$ lies in the interval from $\varphi_{0(lower)} = \theta_0 + d$ to $\varphi_{0(upper)} = \frac{5}{3}\theta_0 - d$, the steady-state motion starts without a transient. The character of these steady-state oscillations varies in this interval of initial displacements from one of the most asymmetric cases at $\varphi_{0(lower)} = \theta_0 + d$ (see Figure 11.11) through the symmetric case occurring at $\varphi_0 = \frac{4}{3}\theta_0$ (Figure 11.12) to the other most asymmetric case at $\varphi_{0(upper)} = \frac{5}{3}\theta_0 - d$ (Figure 11.13). If the initial displacement lies beyond this interval, or the initial

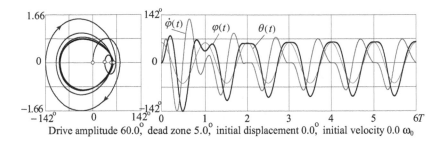

Figure 11.14: Phase diagram and time-dependent graphs of $\varphi(t)$ and $\dot{\varphi}(t)$ for a transient at $\omega = \omega_0/2$, $\theta_0 = 60.0°$, $d = 5.0°$, and at the initial conditions $\varphi(0) = 0$, $\dot{\varphi}(0) = 0$.

velocity is not equal to zero, one of the limit cycles from the same continuum is eventually established after a transient process, during which oscillations with sticking for finite time intervals take place. An example of such a transient occurring at $\theta_0 = 60.0°$, $d = 5.0°$, and initial conditions $\varphi(0) = 0$, $\dot{\varphi}(0) = 0$ is shown in Figure 11.14.

Not surprisingly, if even a small amount of viscous friction is present in the system, the above-considered asymmetric regimes can be observed only at the initial stage: After a long transient oscillations become symmetric, like those shown in Figure 11.12. Indeed, the asymmetry is caused by the contribution of the second harmonic, which corresponds to natural oscillations with the frequency $\omega_0 = 2\omega$. Mathematically, this second harmonic is the general solution of the homogeneous differential equation. In the presence of viscous friction, these natural oscillations damp out during the transient.

Steady-state non-sticking sub-resonant forced oscillations of the above-considered type exist if dry friction is not strong enough for the given drive amplitude. To find this restriction, it is sufficient to equate the expressions for the lower and upper boundaries of the interval of admissible initial deflections given by expressions (11.27) and (11.29): $\varphi_{0(\text{lower})} = \varphi_{0(\text{upper})}$ (or for extreme elongations $\varphi_{\max} = \theta_0 + d$ and $|\varphi_{\min}| = \frac{5}{3}\theta_0 - d$ at any of the most asymmetric cases). This yields $d_{\max} = \theta_0/3$. At $d = \theta_0/3$ there exists only one (symmetric) limit cycle with the amplitude $\varphi_{\max} = \frac{4}{3}\theta_0$. At greater values of the dead zone width ($d > \theta_0/3$) only steady-state oscillations with sticking are possible.

Similar peculiarities are characteristic of forced oscillations with dry friction, sinusoidally excited at other sub-resonant frequencies of even orders $\omega = \omega_0/(2n)$. In particular, for $\omega = \omega_0/4$ a continuum of non-sticking asymmetric steady-state motions exists for the same values of the system parameters, if the width $d$ of the dead zone does not exceed $\frac{1}{15}\theta_0$, where $\theta_0$ is the drive amplitude. The full angular excursion between extreme elongations equals $\frac{32}{15}\theta_0$. Each of these periodic motions occurs without a transient, if the initial velocity equals zero, and the initial displacement lies in the interval between $\theta_0 + d$ and $\frac{17}{15}\theta_0 - d$.

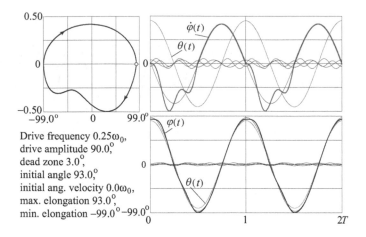

Drive frequency 0.25$\omega_0$,
drive amplitude 90.0°,
dead zone 3.0°,
initial angle 93.0°,
initial ang. velocity 0.0$\omega_0$,
max. elongation 93.0°,
min. elongation −99.0°

Figure 11.15: Phase diagram and time-dependent graphs of $\dot\varphi(t)$ and $\varphi(t)$ with their harmonics for non-sticking asymmetric steady-state oscillations at $\omega = \omega_0/4$ and small width of the dead zone ($d = 3.0°$).

An example of such asymmetric steady-state oscillation is shown in Figure 11.15.

The stability of asymmetric non-sticking solutions at $\omega = \omega_0/(2n)$ was discussed in [64]. The authors claim that in the parameters domain in which these solutions exist they are marginally stable in the third-order approximation.

At sub-resonant drive frequencies of odd orders $\omega = \omega_0/(2n+1)$ non-sticking solutions for an oscillator with dry friction do not exist: At least twice during each cycle of the steady-state motion velocity turns to zero for finite time intervals. An example of steady-state forced oscillation at $\omega = \omega_0/3$ is shown in Figure 11.16. These time-dependent graphs of $\dot\varphi(t)$ and $\varphi(t)$ with their harmonics and the phase orbit are obtained in a computer simulation. For the drive amplitude 80.0° and dead zone width 10.0° oscillations are symmetric with maximum elongation 78.0°. The spectrum contains harmonics only of odd orders: The first harmonic has the amplitude 91.8°, the third 15.6°, the fifth 1.4°. These values are practically the same in cases where the dead zone width lies in the interval 5.0°–25.0°. Such motions consisting of two sliding phases and two sticking phases of a finite duration during each cycle of sinusoidal excitation in a dry friction oscillator are investigated in [71].

## 11.4   Concluding Remarks

In this chapter we concentrated on peculiarities in behavior of a simple mechanical system — a torsion spring oscillator with dry and viscous friction. The intensity of dry friction is characterized by the width $d$ of the dead zone. The amplitude of non-forced oscillations reduces under dry friction during each cycle by the same

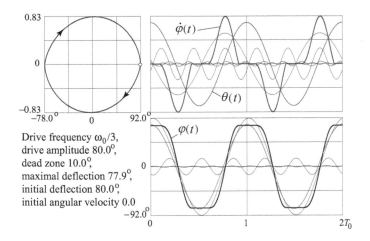

Drive frequency $\omega_0/3$,
drive amplitude $80.0^\circ$,
dead zone $10.0^\circ$,
maximal deflection $77.9^\circ$,
initial deflection $80.0^\circ$,
initial angular velocity $0.0$

Figure 11.16: Phase diagram and time-dependent graphs of $\dot\varphi(t)$ and $\varphi(t)$ with their harmonics for symmetric steady-state oscillations at $\omega = \omega_0/3$ with two sliding and two sticking phases during each cycle ($\theta_0 = 80.0^\circ, d = 10.0^\circ$).

amount $4d$, proportional to the dead zone width, and the oscillator stops dead after a finite time. Under sinusoidal forcing, dry friction cannot restrict the growth of resonant oscillations: At $\omega = \omega_0$ the amplitude of forced oscillations grows indefinitely, increasing in each cycle through $\pi\theta_0 - 4d$, if the drive amplitude $\theta_0$ exceeds the threshold value $4d/\pi$.

In non-resonant cases ($\omega \neq \omega_0$) of harmonic excitation, after a transient of a finite duration, steady-state oscillations of a finite amplitude can establish due to dry friction even in the absence of viscous friction. Generally, such periodic steady-state motion consists of two symmetric non-sticking phases of equal duration $T/2$, if, for the given width $d$ of the dead zone, the drive amplitude $\theta_0$ is large enough to prevent sticking. The steady-state amplitude of these symmetric oscillations for given $\omega$ uniquely depends on $\theta_0$ and $d$.

At sub-resonant frequencies of excitation $\omega = \omega_0/n$ ($n = 2, 4, \ldots$) certain peculiarities of forced oscillations reveal themselves. In particular, in the absence of viscous friction a continuum of different non-sticking steady-state oscillations can exist for the same values of the system parameters $\theta_0$ and $d$. Such oscillations are generally asymmetric: The angular elongation to one side is greater than to the other, though the total angular excursion between the extreme (turnaround) points is the same for a given value of the drive amplitude $\theta_0$ (and is independent of the dead zone width $d$).

The asymmetry of a certain steady-state regime of this continuum depends on the initial conditions. Among each continuum of such solutions coexisting at given values of $\theta_0$ and $d$, there is a single symmetric oscillation. If even a small amount of viscous friction is present in the system, these asymmetric regimes can be observed only at the initial stage: After a long transient the contribution of natural (second) harmonic that causes asymmetry dies out, and eventually the oscillations become symmetric.

# Bibliography

[1] G. L. Baker, J. A. Blackburn, *The Pendulum: A Case Study in Physics*. New York: Oxford University Press, 2005.

[2] Ch. Kittel, W. D. Knight, M. A. Ruderman, *Mechanics*. Berkeley Physics Course, New York: McGraw-Hill Book Company, 1965.

[3] A. B. Pippard, *The physics of vibration* (The simple classical vibrator). New York: Cambridge University Press, 1978.

[4] L. D. Landau and E. M. Lifschitz, *Mechanics*, Moscow: Nauka Publishers, (in Russian), 1976, *Mechanics*, New York: Pergamon, 1988.

[5] K. Johannessen, An approximate solution to the equation of motion for large-angle oscillations of the simple pendulum with initial velocity Eur. J. Phys. **31**, 511–18, 2010.

[6] A. Belendez, J. J. Rodes, A. Hernandez, T. Belendez, Approximation for a large-angle simple pendulum period, Eur. J. Phys. **30**, L25–8, 2009.

[7] C. G. Carvalhaes, P. Suppes, Approximations for the period of the simple pendulum based on the arithmetic-geometric mean, Am. J. Phys. **76**, 1150–4, 2008.

[8] F. M. S. Lima, Simple 'log formulae' for the pendulum motion valid for any amplitude, Eur. J. Phys. **29**, 1091–8, 2009.

[9] F. M. S. Lima, P. Arun, An accurate formula for the period of a simple pendulum oscillating beyond the small angle regime, Am. J. Phys. **74**, 892–5, 2006.

[10] Y. Q. Xin and D. Pei, Comment on Approximation for a large-angle simple pendulum period, Eur. J. Phys. **30**, L79–82, 2009.

[11] E. I. Butikov, The rigid pendulum—an antique but evergreen physical model, Eur. J. Phys. **20**, 429–441, 1999.

[12] G. L. Baker, J. P. Gollub *Chaotic Dynamics. An Introduction*, New York: Cambridge University Press, 1990.

[13] J. W. Miles, Resonance and symmetry breaking for the pendulum, Physica, **D31**, 252–268, 1988.

[14] H. Heng, R. Doerner, B. Hübinger, W. Martienssen, Approaching nonlinear dynamics by studying the motion of a pendulum. I. Observing trajectories in phase space, Int. J. Bifurcation and Chaos, **4**, 751–760, 1994.

[15] R. Doerner, B. Hübinger, H. Heng, W. Martienssen, Approaching nonlinear dynamics by studying the motion of a pendulum. II. Analyzing chaotic motion, Int. J. Bifurcation and Chaos, **4**, 761–771, 1994.

[16] H. W. Broer, I. Hoveijn, M. van Noort, C. Simó, G. Vegter, The Parametrically Forced Pendulum: A Case Study in $1\frac{1}{2}$ Degree of Freedom, J. Dynamics and Differential Equations, **16**, 897–947, 2004.

[17] E. I. Butikov, On the dynamic stabilization of an inverted pendulum, Am. J. Phys. **69**, 755–68, 2001.

[18] R. D. Peters, Resonance response of a moderately driven rigid planar pendulum, Am. J. Phys. **64**, 170–173, 1996.

[19] E. I. Butikov, Subharmonic Resonances of the Parametrically Driven Pendulum, J. Phys. A: Math. and Gen. **35**, 6209–6231, 2002.

[20] E. I. Butikov, Regular and Chaotic Motions of the Parametrically Forced Pendulum: Theory and Simulations, *Computational Science – ICCS 2002*, Springer Verlag, LNCS 2331, 1154–1169, 2002.

[21] E. I. Butikov, An improved criterion for Kapitza's pendulum stability, J. Phys. A: Math. and Theor. **44**, 295202 (16 pp), 2011.

[22] S. M. Curry, How children swing, Am. J. Phys. **44**, 924–926, 1976.

[23] P. L. Tea Jr., H. Falk, Pumping on a swing, Am. J. Phys. **36**, 1165–1166, 1968.

[24] W. Case, The pumping of a swing from the standing position, Am. J. Phys. **64**, 215–220, 1996.

[25] M. A. Pinsky, A. A. Zevin, Oscillations of a pendulum with a periodically varying length and a model of swing, Int. J. Non-Linear Mech. **34**, 105–109, 1999.

[26] D. Stilling, W. Szyskowski, Controlling angular oscillations through mass reconfiguration: a variable length pendulum case, Int. J. Non-Linear Mech. **37**, 89–99, 2002.

[27] E. I. Butikov, *Pendulum with a square-wave modulated length* (simulation program), http://butikov.faculty.ifmo.ru/Applets/PendParSquare.html, 2013.

[28] E. I. Butikov, Parametric excitation of a linear oscillator, Eur. J. Phys. **25**, 535–554, 2004.

[29] E. I. Butikov, Parametric resonance in a linear oscillator at square-wave modulation, Eur. J. Phys. **26**, 157–174, 2005.

[30] J. B. McLaughlin, Period-doubling bifurcations and chaotic motion for a parametrically forced pendulum, J. Stat. Physics **24**, 375–388, 1981.

[31] B. P. Koch, R. W. Leven, B. Pompe, C. Wilke, Experimental evidence for chaotic behavior of a parametrically forced pendulum, Phys. Lett. A **96**, 219–224, 1983.

[32] R. W. Leven, B. Pompe, C. Wilke, B. P. Koch, Experiments on periodic and chaotic motions of a parametrically forced pendulum, Physica D **16**, 371–384, 1985.

[33] W. van de Water, M. Hoppenbrouwers, Unstable periodic orbits in the parametrically excited pendulum, Phys. Rev. A **44**, 6388–6398, 1991.

[34] J. Starrett, R. Tagg, Control of a chaotic parametrically driven pendulum, Phys. Rev. Lett. **74**, 1974–1977, 1995.

[35] S.-Y. Kim, B. Hu, Bifurcations and transitions to chaos in an inverted pendulum, Phys. Rev. E **58**, 3028–3035, 1998.

[36] A. Stephenson, On an induced stability, Phil. Mag. **15**, 233–236, 1908.

[37] P. L. Kapitza, Dynamic stability of the pendulum with vibrating suspension point, Soviet Physics – JETP **21**, 588–597, 1951 (in Russian), see also Collected papers of P. L. Kapitza, edited by D. Ter Haar, London: Pergamon, v. 2, pp. 714–726, 1965.

[38] P. L. Kapitza, Pendulum with an oscillating pivot *Sov. Phys. Uspekhi*, **44**, pp. 7–20, 1951.

[39] L. D. Landau and E. M. Lifschitz, Mechanics, Moscow: Nauka Publishers, 1988 (in Russian), *Mechanics*, New York: Pergamon, 1976, pp. 93–95.

[40] F. M. Phelps, III, J. H. Hunter, Jr. An analytical solution of the inverted pendulum, Am. J. Phys. **33**, 285–295, 1965, **34**, 533–535, 1966.

[41] D. J. Ness, Small oscillations of a stabilized, inverted pendulum, Am. J. Phys. **35**, 964–967, 1967.

[42] H. P. Kalmus, The inverted pendulum, Am. J. Phys. **38**, 874–878, 1970.

[43] M. M. Michaelis, Stroboscopic study of the inverted pendulum, Am. J. Phys. **53**, 1079–1083, 1985.

[44] J. A. Blackburn, H. J. T. Smith, N. Groenbech-Jensen, Stability and Hopf bifurcations in an inverted pendulum, Am. J. Phys. **60**, 903–908, 1992.

[45] H. J. T. Smith, J. A. Blackburn, Experimental study of an inverted pendulum, Am. J. Phys. **60**, 909–911, 1992.

[46] M. J. Moloney, Inverted pendulum motion and the principle of equivalence, Am. J. Phys. **64**, 1431, 1996.

[47] W. T. Grandy, Jr., M. Schöck, Simulations of nolinear pivot-driven pendula, Am. J. Phys. **65**, 376–381, 1997.

[48] J. G. Fenn, D. A. Bayne, B. D. Sinclair, Experimental investigation of the 'effective potential' of an inverted pendulum, Am. J. Phys. **66**, 981–984, 1998.

[49] E. I. Butikov, On the dynamic stabilization of an inverted pendulum, Am. J. Phys. **69**, 755–768, 2001.

[50] P. S. Landa, Nonlinear Oscillations and Waves in Dynamical Systems. Dordrecht: Kluwer Academic Publishers, 1996.

[51] D. J. Acheson, Multiple-nodding oscillations of a driven inverted pendulum, Proc. Roy. Soc. London **A 448**, 89–95, 1995.

[52] E. I. Butikov, Subharmonic Resonances of the Parametrically Driven Pendulum, J. Phys. A: Mathematical and General, **35**, 6209-6231, 2002.

[53] B. Feeny, A. Guran, N. Hinrichs, K. Popp, A historical review on dry friction and stickslip phenomena, ASME Applied Mechanics Reviews, **51**, 321–341, 1998.

[54] E. I. Butikov, *Physics of Oscillations*, User's Manual. New York: American Institute of Physics, Physics Academic Software, 1996.

[55] L. M. Burko, A Piecewise-Conserved Constant of Motion for a Dissipative System, Eur. J. Phys. **20**, 281–288, 1999.

[56] A. Marchewka, D. S. Abbott, R. J. Beichner, Oscillator damped by a constantmagnitude friction force. Am. J. Phys. **72**, 477–483, 2004.

[57] J. P. Den Hartog, Forced vibrations with combined Coulomb and viscous damping, Transactions of the American Society of Mechanical Engineers, **53**, 107–115, 1930.

[58] E. I. Butikov, *Torsion Pendulum with Dry and Viscous Friction*. Web page with an embedded simulation program (Java applet), see on the web at http://butikov.faculty.ifmo.ru/Applets/DryViscOsc.html, 2013.

[59] S. W. Shaw, On the dynamic response of a system with dry friction, J. Sound Vib. **108**, 305–325, 1986.

[60] H.-K. Hong, C.-S. Liu, Non-sticking oscillation formulae for Coulomb friction under harmonic loading, J. Sound Vib. **244**, 883–898, 2001.

[61] J. W. Liang, B. F. Feeny, Identifying Coulomb and Viscous Friction in Forced Dual-Damped Oscillators, Journal of Vibration and Acoustics, **126**, 118–125, 2004.

[62] A. C. J. Luo, B. C. Gegg, Stick and non-stick periodic motions in periodically forced oscillators with dry friction, J. Sound Vib., **291**, 132–168, 2006.

[63] G. Csernák, G. Stépán, On the periodic response of a harmonically excited dry friction oscillator. J. Sound Vib., **295**, 649–658, 2006.

[64] G. Csernák, G. Stépán, S. W. Shaw, Sub-harmonic resonant solutions of a harmonically excited dry friction oscillator, Nonlinear Dynamics, **53**, 93–109, 2007.

[65] S. Chatterjee, Resonant locking in viscous and dry friction damper kinematically driving mechanical oscillators, J. Sound Vib., **332**, 3499–3516, 2013.

[66] E. Pratt, A. Léger, X. Zhang, Study of a transition in the qualitative behavior of a simple oscillator with Coulomb friction, Nonlinear Dynamics, **74**, 517–531, 2013.

[67] A. C. J. Luo, *Regularity and Complexity in Dynamical Systems*, Springer, 2011.

[68] F. J. Elmer, Nonlinear dynamics of dry friction, J. Phys. A: Math. Gen. **30**, 6057–6063, 1997.

[69] J. Awrejcewicz, P. Olejnik, Analysis of dynamic systems with various friction laws, Applied Mechanics Reviews, **58**, 389–411, 2005.

[70] A. F. Filippov, *Differential Equations with Discontinuous Righthand Sides*, Mathematics and its Applications (Soviet Series), v. **18**, Dordrecht: Kluwer, 1988.

[71] S. P. Yang, S. Q. Guo, Two-stop-two-slip motions of a dry friction oscillator, Sci. China Tech. Sci. **53**, 623–632, 2010.

# Index